T0315160

Contemporary Ergonomics and Human Factors 2014

Contemporary Ergonomics and Human Factors 2014

Editors

Sarah Sharples

University of Nottingham

Steven T. Shorrock

Eurocontrol and University of New South Wales

CRC Press

Taylor & Francis Group

Boca Raton London New York

CRC Press is an imprint of the
Taylor & Francis Group, an **informa** business

A TAYLOR & FRANCIS BOOK

Institute of
Ergonomics &
Human Factors

CRC Press
Taylor & Francis Group
6000 Broken Sound Parkway NW, Suite 300
Boca Raton, FL 33487-2742

First issued in hardback 2017

© 2014 Taylor & Francis Group, LLC
CRC Press is an imprint of Taylor & Francis Group, an Informa business

A framework for human assessment in the defence system engineering lifecycle
J. Astwood, K. Tatlock, K. Strickland & W. Tutton
© Crown copyright 2014/DSTL

Development of a Human Factors assessment framework for land vehicles
R. Saunders Jones, K. Strickland & W. Tutton
© Crown copyright 2014/DSTL

Typeset by MPS Ltd, a Macmillan Company, Chennai, India

No claim to original U.S. Government works

ISBN-13: 978-1-138-02635-3 (pbk)
ISBN-13: 978-1-138-43088-4 (hbk)

Visit the Taylor & Francis Web site at
http://www.taylorandfrancis.com

and the CRC Press Web site at
http://www.crcpress.com

Contents

Preface   XI

Donald Broadbent Lecture
Human Capability   3
K. Carr

Institute Lecture
Ergonomics, resilience and reliability – to ERR is human ...   7
J. Berman

Keynote Lecture
Making better things: The role of human factors in user experience design
(or vice versa?)   11
D.J. Gilmore

Plenary Lectures
Socio-technical thinking and practice – Being braver!   15
C. Clegg

The global economic recession: Impact on workplace safety and health   16
J. Devereux

Culture, control and leadership
Insider attacks: The human factors dimension – Learning conflict
resolution lessons from industry to counter insider attacks in Afghanistan   27
R. Meeks, P. Newman & C. Skelton

Visual meaning for leadership – A case study   35
P. Nock & J. Burton

Hazards, risk and comfort
Perception of self-motion headings when subjected to near-threshold
visual and inertial motion   41
B. Bao & R.H.Y. So

Factors contributing to the perceived slipperiness rating   49
W.-R. Chang, M.F. Lesch, C.-C. Chang & S. Matz

The study on effectiveness of bamboo charcoal for improving work
efficiency in learning environment   53
H. Nishimura, Z. Wang, A. Endo, A. Goto & N. Kuwahara

Predictors of injury amongst white-water raft guides working in the UK   59
I. Wilson, H. McDermott & F. Munir

HF in high value manufacturing systems
Using SHERPA to inform the design of intelligent automation
for TIG welding   69
W. Baker

Human Factors analysis in risk assessment: A survey of methods and
tools used in industry   77
N. Balfe & M.C. Leva

Investigating the impact of product orientation on musculoskeletal risk
in aerospace manufacturing   85
S.R. Fletcher & T.L. Johnson

A computer software method for ergonomic analysis utilising non-optical
motion capture   93
T.L. Johnson & S.R. Fletcher

Relationships between degree of skill, dimension stability, and mechanical
properties of composite structure in hand lay-up fabrication method   101
T. Kikuchi, Y. Tani, Y. Takai, A. Goto & H. Hamada

Ergonomics issues arising from the 'Next Manufacturing Revolution'   110
M.A. Sinclair & C.E. Siemieniuch

Users' understanding of industrial robot gesture motions and effects
on trust   116
G. Tang, G. Charalambous, P. Webb & S.R. Fletcher

An evaluation of the ergonomic physical demands in assembly work
stations: Case study in SCANIA production, Angers   124
M. Zare, M. Croq, A. Malinge-Oudenot & Y. Roquelaure

HF in product design
Referability: Making things easy to refer to   135
G. Andresen & Ø. Veland

Analysis of human factors issues relating to the design and function
of bicycles   143
D. Covill, E. Elton, M. Mile & R. Morris

Human Factors and Product Design: Uncomfortable allies   151
E. Elton

Playing in the park: Observation and co-design methods appropriate
to creating location based games for children   159
C. Grundy, L. Pemberton & R. Morris

Beyond box ticking – The role of Human Factors in design   167
D.P. Jenkins

Human Factors contributions to consumer product safety 173
B.B. Novak

Lifelong health and wellbeing
Evaluating the universal navigator with consumers with reduced mobility,
dexterity & visual acuity 183
P.N. Day, J.P. Johnson, M. Carlisle & G. Ferguson

Healthy ageing in the construction industry; background and
preliminary findings 191
S. Eaves, D.E. Gyi & A.G.F. Gibb

Efficacy evaluation of recreational picture colouring in elderly nursing
home patients 195
S. Kawabata, M. Nasu, A. Yamamoto & N. Kuwahara

User-centred design of patient information for hospital admissions and
patient experience 203
A.R. Lang, T.C. Kwok, E. Fioratou, I. LeJeune & S. Sharples

Retained guidewires in central venous catheterisation: An analysis of
omission errors 211
Y.-C. Teng, J. Ward, T. Horberry, V. Patil & J. Clarkson

NHS managers' utilisation of information resources to support
decision-making in shift work 215
E. Mowbray, P. Waterson & R. McCaig

As I am now, so shall you become 219
A. Woodcock

Musculoskeletal disorders
Ergokita: Analysis of the musculoskeletal strain of nursery school teachers 229
E.-M. Burford, R. Ellegast, B. Weber, M. Brehmen, D. Groneberg,
A. Sinn-Behrendt & R. Bruder

Physiological workload and perceived exertion of female labourers in
harvesting activities 236
R. Salunke, S. Sawkar, R. Naik, P.R. Sumangala & K.V. Ashalatha

Predicting isometric rotator cuff muscle activities from arm postures
and net shoulder moments 241
X. Xu, R.W. McGorry & J.-h. Lin

Smart environments & sustainability
How do malfunctions in 'smart environments' affect user performance? 251
K.F.K. Narul & C. Baber

Creating a lasting commitment to sustainable living 259
H. Taylor, W.I. Hamilton & C. Vail

Systems: Preparing for the future

Introduction to Human Factors Integration and a brief historical overview   269
R.S. Bridger & P. Pisula

HFI training for systems engineering professionals   277
E.-M. Hubbard, L. Ciocoiu & M. Henshaw

Managing the lifecycle of your robot   283
M.A. Sinclair & C.E. Siemieniuch

A framework for human assessment in the defence system engineering
lifecycle   290
J. Astwood, K. Tatlock, K. Strickland & W. Tutton

Transport: Journey experience

Will autonomous vehicles make us sick?   301
C. Diels

Car and lift sharing barriers   308
A. Woodcock, J. Osmond, J. Begley & K. Frankova

Measuring quality across the whole journey   316
*A. Woodcock, N. Berkeley, O. Cats, Y. Susilo, G.R. Hrin, O. O'Reilly,
I. Markucevičiūtė & T. Pimentel*

Transport: Operations

Designing visual analytics for collaborative activity: A case study of
rail incident handling   327
T. Male & C. Baber

The effect of driver advisory systems on train driver workload and
performance   335
D.R. Large, D. Golightly & E.L. Taylor

ERTMS train driving-incab vs. outside: An explorative eye-tracking
field study   343
A. Naghiyev, S. Sharples, M. Carey, A. Coplestone & B. Ryan

Vigilance evidence and the railway lookout   351
L. Pickup, E. Lowe & S. Smith

Understanding human decision making during critical incidents in
dynamic positioning   359
L.J. Sorensen, K.I. Øvergård & T.J.S. Martinsen

Operators in tunnel control – Analyzing decisions, designing decision
support   367
S. Spundflasch & H. Krömker

User systems

The analysis of incidents related to automation failures and human factors
in flight operations 377
W.-C. Li, D. Durando & L.-Y. Ting

Using N squared analysis to develop the functional architecture design for
the High Speed 2 project 385
P. Nock, E. Walters, N. Best & W. Fung

Human Factors aspects of decommissioning: The transition from operation
to demolition 390
C. Pollard

East: A method for investigating social, information and task networks 395
N.A. Stanton

User systems architectures, functions and job and task synthesis 403
M. Tainsh

Vehicle design

Better vehicle design for all 413
S. Karali, D. Gyi & N. Mansfield

The motorcycle: A human operator influenced workstation 417
*M.I.N. Ma'arof, H. Rashid, A.R. Omar, S.C. Abdullah, I.N. Ahmad &
S.A. Karim*

Motorcyclist muscle fatigue index: An effort to help reduce
motorcycle accidents 425
*H. Rashid, M.I.N. Ma'arof, A.R. Omar, S.C. Abdullah, I.N. Ahmad &
S.A. Karim*

Workload & vigilance

Speaking under workload in first and second languages: Implications
for training and design 435
C. Baber & H. Liu

Alarm vigilance in the presence of 80 dBa pink noise with negative
signal-to-noise ratios 443
B. Karunarathne, R.H.Y. So & A.C.S. Kam

Continuous detection of workload overload: An fNIRS approach 450
H.A. Maior, M. Pike, M.L. Wilson & S. Sharples

Poster

Proposed decision model for contemporary memorials 461
P. Davies, C. Williams & P.W. Jordan

Relationship between visual aesthetics evaluation and selling price to
Kimono of Yuzen-Zome   469
T. Furukawa, Y. Takai, A. Goto, N. Kuwahara & N. Kida

Level of comfort of customers affected by expert manicurist and
non-expert manicurist   477
S. Isobe, A. Endo, T. Ota & N. Kuwahara

Ergonomics of parking brake application – Factors to fail?   481
V.G. Noble, R.J. Frampton & J.H. Richardson

Development of a Human Factors assessment framework for land vehicles   489
R. Saunders Jones, K. Strickland & W. Tutton

Author index   497

Preface

This book contains the proceedings of the International Conference on Ergonomics and Human Factors, held in April 2014 at the Grand Harbour Hotel, Southampton. The conference is a major international event for ergonomists and human factors professionals and attracts contributions and delegates from around the world.

Papers are subject to a peer review process before consideration by the Programme Committee. The Programme Committee selects papers for publication in proceedings as full or short papers, as well as identifying the groupings of papers to form dedicated sessions at the conference. In addition to the open call for papers, many authors respond to calls for symposia – these papers are still subject to peer review but are included within dedicated specialist sessions. This conference included symposia on *Human Capability, Human Factors in Product Design, Human Factors in High Value Manufacturing Systems* and *Systems: Preparing for the Future.*

This is the first year that we have acted as editors of *Contemporary Ergonomics & Human Factors.* The diversity of topics within the papers is apparent and demonstrates not only the breadth of areas to which human factors is applicable but also the range of skills now exhibited by practising ergonomists. The diversity is reflected in the target domains of the papers, which include transport, defence, manufacturing, consumer products, health and wellbeing. The activities considered within each domain are similarly diverse, reflecting the roles of front-line operators to managers and leaders. The methods used and reported in the papers cover all aspects of ergonomics – physical, cognitive, social, organisational and societal.

Papers on *human factors in high value manufacturing systems* include some important themes such as lean ergonomics, human factors analysis in risk assessment, and the 'next manufacturing revolution'. More specific applications include assembly work stations, robots and intelligent automation, aerospace manufacturing and hand fabrication. The issues explored in the papers are diverse, and include posture and musculoskeletal risk, skill, trust and safety.

Lifelong health and wellbeing has emerged as an increasingly important concern to society and therefore to ergonomics and human factors. This year, specific hospital-related papers include the design of patient information, information to support managers' decision-making regarding shift design, and retained guidewires in central venous catheterisation. The conference also included a workshop on human factors and healthcare. Other papers consider device accessibility and ageing with respect to recreation, work and life generally. Turning to the future, a smart kitchen environment was simulated to consider how malfunctions in 'smart environments' affect user performance. More broadly, the need for sustainability was explored, considering the need to create a lasting commitment for sustainable living.

Product design was another theme addressed by authors. Papers explore a number of general issues such as the role of human factors and the relationship between human factors and product design, and human factors contributions to consumer product safety. Specific issues include the concept of referability (or making things easy to refer to), the design and function of bicycles, and design of location based games for children. During the conference, there were also presentations for a student design exhibition. The role of human factors in user experience design was also the theme of a keynote presentation by David Gilmore this year.

A major theme on *systems* emerged. The papers will be of interest to many practitioners in safety-related industries, and consider several frameworks, approaches and methodologies, including human factors integration, user systems architecture, MODAF and ISO standards, and methods for investigating social, information and task. Specific issues addressed of concern to industry include decommissioning and the robot lifecycle. During the conference, there was also a workshop on human factors for systems of systems.

Transport has been a consistent theme in *Contemporary Ergonomics & Human Factors* over the years, and this year is no exception. Most sectors are represented, including road, rail, aerospace and maritime. Automotive ergonomics papers include many forms of transport, including car, motorbikes and pushbikes. Emerging themes include the journey experience (e.g. autonomous vehicles, lift and car sharing, measurement), vehicle design, and motorcycle ergonomics. Rail remains a strong sector for ergonomics and human factors. Papers address safety systems for train driving such as driver advisory systems and the European Rail Traffic Management System. Other activities addressed include rail incident handling and user interfaces for decision support. The UK's 'High Speed 2 Project' also featured in papers on functional architecture design and the use of pictures and visual metaphors to explore complex problems. Key human factors concepts addressed include workload, drivers' allocation of visual attention and vigilance. In the maritime domain, decision making during critical incidents in dynamic positioning is considered, while in aerospace, papers consider aerospace manufacturing, as well as automation failures and human factors in flight operations.

Defence-related papers included learning conflict resolution lessons from industry to counter insider attacks in Afghanistan, and human assessment in the defence system engineering lifecycle. *Manufacturing* also features as a domain in many papers, covering traditional musculoskeletal considerations as well as interactions with robotics and intelligent automation.

Across all domains, workload, vigilance and human error remain key concepts in ergonomics and human factors. Papers this year include speaking under workload in first and second languages, the detection of workload overload, alarms, vigilance and noise. There was also a workshop on the concept of human error in human factors and society.

Collectively, the papers cover a range of key ergonomics criteria, including safety, health, wellbeing, comfort, productivity, efficiency, and sustainability, demonstrating the continued wide applicability of ergonomics and human factors and the range of domains for which we must ensure there are appropriate methods and tools.

Editors

Sarah Sharples & Steven T. Shorrock, January 2014

Invited and Plenary Lectures

Donald Broadbent Lecture: "Human Capability", Karen Carr

Keynote Lecture: "Socio-technical thinking and practice – Being braver!", Chris Clegg

Plenary Lecture: "The Global Economic Recession: Impact On Workplace Safety And Health", Jason Devereux

Institute Lecture: "Ergonomics, Resilience and Reliability – to ERR is human ..." Jon Berman, Greenstreet Berman Ltd

Keynote Lecture: "Making better things: The role of human factors in user experience design", David Gilmore

Workshops

Several workshops were held, reflecting the need to maximise interactivity and engagement at conferences. This year, there were workshops and debates on the following:

* Human factors for systems of systems (Huseyin Dogan)
* Human capability: science, engineering, art and imagination (Glenn Hunter & Emma Sparks)
* Quick tricks in psychometrics (Bob Bridger)
* Human factors and healthcare (Patrick Waterson)
* Behaviour change (Claire Williams)
* Is 'human error' the handicap of human factors? (Steven Shorrock)
* Continual professional development (Adrian Wheatley)
* Hazards, risk & comfort (Ken Parsons)
* Design debate (Eddy Elton & Laurence Clift)

Programme Committee

Professor Sarah Sharples (Chair), University of Nottingham
Dr Steven T. Shorrock, Eurocontrol and University of New South Wales
Dr Murray Sinclair, Loughborough University
Dr Margaret Hanson, WorksOut

Dr Patrick Waterson, Loughborough University
Professor Richard So, Hong Kong University of Science & Technology
Dr Wen Ruey Chang, Liberty Mutual
Dr Sue Hignett, Loughborough University
Dr Carole Deighton, DSTL
Dr David Golightly, University of Nottingham
Dr Mike Tainsh, Krome
Dr Lisa Kelly, Thales
Dr Will Baker, Cranfield University
Dr Eddy Elton, Brighton University
Dr Sarah Fletcher, Cranfield University

IEHF Organising Team

James Walton, Tina Worthy

DONALD BROADBENT
LECTURE

HUMAN CAPABILITY

Karen Carr

Cranfield University

This lecture will explore ways of thinking about ourselves and the systems we create and use, and what they mean for Ergonomics and Human Factors.

During the last century, interesting developments in understanding the world have arisen from the embracing of uncertainty. Rather than avoiding uncertainty in a deterministic fashion, these world views are non-deterministic, non-linear and holistic. Such ways of thinking and understanding help us to deal with complexity. How does the discipline of Ergonomics and Human Factors work in such worlds?

The late Professor John Wilson recognised the implications that a 'Systems' world view has for the Ergonomics and Human Factors development teams. In order to develop a system that has non-deterministic or emergent properties, the team developing it has to have certain properties such as flexibility and interconnectedness in order to recognise emerging interdependencies. Yet we know that many organisations that are tasked with developing complex systems are not structured in a manner that is amenable to working in this way.

The MOD has the task of developing many complex systems – or at least managing the development of them. They have taken a 'Capability' approach in order to achieve the agility necessary for dealing with complex needs and uncertain contexts. Does this mean the MOD organisation allows the sort of flexibility and inter-connectedness needed to take a systems approach? This issue will be explored, and some potentially useful ways of thinking about people and Defence Capability presented. This has implications for MOD's organisational and process infrastructure, in order for military capability to gain maximum value from its people and the attributes they can bring to the whole Defence enterprise.

INSTITUTE LECTURE

ERGONOMICS, RESILIENCE AND RELIABILITY – TO ERR IS HUMAN …

Jonathan Berman

Greenstreet Berman Ltd

Ergonomics as a discipline has a long and distinguished history. Its origins are well-known, coming from a recognised need, initially in a military context, to address the apparent limitations in system performance represented by human capabilities – how to design systems to mitigate or be tolerant of those limitations. Over time, this focus evolved into the discipline in its present form, focusing not only on the person as part of a complex system, but also on all aspects of our interactions with the work and leisure environment, considering both system performance and also health and wellbeing, and focusing on the positive contributions as well as human limitations.

As the discipline has evolved, so has its focus. In addition to continued attention to the performance of the individual, whether in respect of physical capabilities, environmental stressors, or cognitive demands, there has been increasing attention to system performance and organisational capabilities, and a reinforcement of the position that shortfalls in performance are rarely if ever due solely to a failure by an individual.

The roll call of major accidents and disasters – or 'learning opportunities' as some might describe them – is both depressing and extensive. From Flixborough to Deepwater Horizon via Buncefield, or from Chernobyl to Fukushima via Kings Cross, these are events that need no additional description – their names are sufficient to capture both the magnitude of the catastrophe and the extent of the organisational complexity that led to their occurrence. But if we consider these accidents there are repetitive themes in the root causes: failure to learn from past events or even to recognise them, unrecognised complexity in the system, conflicting goals, unwarranted assumptions, complacency… Has Ergonomics been fully engaged in addressing these aspects of system performance? Has Ergonomics been successful in improving reliability?

This paper suggests that it has, that it continues to do so, and that it is well-placed to do so. Ergonomics provides a scientific, evidence-based approach to the synthesis of knowledge from across related domains in support of its application to the work environment. But perhaps it hasn't been sufficiently visible in its role of supporting system reliability. The paper discusses the role and contribution of an ergonomics focus to such topics as organisational drift, resilience, and sustained high performance. It looks at the emerging theme of learning from success – gaining greater clarity as to how high-performing systems can remain so.

These topics might benefit from a re-orientation of the external perception of what Ergonomics offers. As a discipline, Ergonomics continues to draw on such diverse knowledge domains as Cognitive and Organisational Psychology, and Human Physiology. However, the Ergonomics agenda extends beyond these areas. By adopting and promoting a systems-based perspective on human performance it enables complex systems to be enhanced and their reliability increased. In order to prevent another disaster it is not sufficient to focus only on, for example, organisational psychology. But nor is it sufficient to focus on engineering or physiology. Whilst each is important, high reliability systems need to consider all of these strands in a coherent manner. This is not a play to promote Ergonomics – rather it's a plea for avoidance of narrow or parochial approaches to solutions to the challenges of complex high-reliability systems. Addressing Human Performance will not resolve a poor design. Focusing on Safety Culture will not on its own resolve inadequate competence management. Focusing on behaviour change will not on its own resolve excessive workload demands.

We rely on complex systems and high-hazard processes, and will continue to do so. This paper discusses the role of Ergonomics in providing reassurance that reliance on such systems is not misplaced, whilst also maintaining a robust intolerance of poor performance. If resilience is about the design, management and operation of the system, then it is firmly within the domain of Ergonomics. If Ergonomics is fully to support resilience and high reliability, then it must continue to engage at all levels within the socio-technical system – and demonstrably offer benefits at each level.

KEYNOTE LECTURE

MAKING BETTER THINGS: THE ROLE OF HUMAN FACTORS IN USER EXPERIENCE DESIGN (OR VICE VERSA?)

David J. Gilmore

Senior UX Scientist, Connected Experience Labs, GE Global Research, San Ramon, CA, USA

Human Factors and User Experience design have developed relatively independently and from different traditions and yet they both share the all important goal of making better things for people to use. Human Factors pre-dates computer systems and grew from industrial engineering and psychology, with a focus on large socio-technical systems, whereas user experience design has grown from interaction design with an emphasis on internet experiences and e-commerce. At GE (and elsewhere) these worlds are colliding, as the consumeriation of IT brings mobile devices and expectations of consumer-level simplicity to industrial work environments. Which leads to a key question – what is the relationship between human factors and user experience design? How do they work together? What do they each contribute to the other? In this talk, I intend to explore these questions through looking at a variety of projects I've been involved with, or close to, over the past 30 years.

PLENARY LECTURES

SOCIO-TECHNICAL THINKING AND PRACTICE – BEING BRAVER!

Chris Clegg

Professor of Organisational Psychology
Director, Socio-Technical Centre
Leeds University Business School
www.SocioTechnicalCentre.org

The objectives of this paper are to:

- Describe a socio-technical systems framework and its application in a number of different domains
- Make the case for the use of systems thinking to address some of the 'grand challenges' facing society
- Argue for the use of such ideas predictively (rather than simply after the event)
- Outline the need for the development of methods and tools to support such work, and, in short
- Argue the case for being 'braver'

In the talk I will describe some work in progress on which colleagues and I from the Socio-Technical Centre are working. These projects include work with Arup, Rolls-Royce, Yorkshire Water, NHS, Go Science (BIS), DCLG and the Cabinet Office.

I will include analyses of a number of disasters including those at the Hillsborough Football Stadium, the King's Cross Underground Fire, the Bradford City Stadium Fire, and the Deepwater Horizon Oil Spill.

THE GLOBAL ECONOMIC RECESSION: IMPACT ON WORKPLACE SAFETY AND HEALTH

Jason Devereux

Lloyd's Register Consulting – Energy Limited
Business Psychology Unit, University College London

Multinational enterprises (MNEs) experienced the worst financial crisis and economic recession in recent history between 2007–2009. To identify the potential effects from corporate restructuring on workforce health and safety and system safety, a scientific literature review was conducted, the annual reports of 45 MNEs were analysed, and 13 MNEs were investigated using interviews and thematic/documentation analysis. Distraction from health and safety due to focus on growth, increase in psychosocial risks among layoff survivors such as work intensification without sufficient resources, budget cuts, and losing skills and experience needed to maintain system integrity were emergent themes. This research formed the basis for recommendations by the ILO.

Introduction

The General Assembly of the United Nations endorsed in July 2009 the statement agreed among experts that the world has been confronted with the worst financial and economic crisis since the Great Depression (United Nations General Assembly, 2009). The International Labour Organization (ILO), a United Nations Agency, is concerned that the financial and economic crisis has led to restructuring that could have compromised workplace safety and health within multinational enterprises (MNEs). The ILO has provided ten key recommendations based on a research study commissioned in 2010 (ILO, 2013).

It is the aim of this paper to provide the findings of the scientific research study that formed the basis for ILO's recommendations.

Methodology

A review and synthesis of the scientific and grey literature was conducted on the "Great Recession" (2007–2009). Corporate annual reports (2008–2010) for 45 MNEs, headquartered across the USA, UK, France and Germany, were also reviewed to determine the corporate responses to the crisis (refer to ILO, 2013).

Semi-structured phone interviews were then conducted by Devereux between June-Sept 2010, by using a thematic template (see ILO, 2013). In total, representatives from thirteen MNEs, providing within the annual report(s) points of interest in relation to the study objectives, agreed to be interviewed. The MNEs were spread across all nine major sectors (energy, financial, materials, industrials, consumer discretionary, IT, utilities, telecommunications, health care and consumer staples). The thirteen MNEs were headquartered in France (N = 3), Germany (N = 2), U.K. (N = 5) and the U.S.A. (N = 3). The average employee size across this group of MNEs was 108,000 employees.

The data gathered (using voice recording and/or scribing) from each interview (1.5–2.0 hours in length with a safety representative) was compared with the documented data from various sources including the company website, annual reports and presentations, published literature or documentation provided by the company upon request. This data triangulation was performed in order to check consistency and to offer opportunities for deeper insight. The following questions were asked in each interview:

- Impact of the recession on the MNE and the workforce.
- Effects on occupational safety and health management systems.
- Potential threats to safety and health from the recession.

After 13 interviews, no further data codes were generated indicating saturation (data shown in ILO, 2013). The number of interviews used for generating themes falls within the recommended research guidelines (King, 2004). A thematic template analysis was used to construct a long list of codes and a smaller number of themes under each question in accordance with the method described by King (2004). The final thematic template was verified for credibility by an independent researcher using the coded data.

Results

Market forces

MNE market funds across major economies declined over 50% between 2007 and 2009 from their peak to their trough reflecting significant declines in market capitalization and increasing pressure to restructure assets and the labour force. Revenues and profit margins, including MNEs in the normally resilient energy sector, were decimated in 2009 compared to the two previous years (about a 30% drop in revenues for BP, Total and Exxon-Mobile, as an example).

The basic material and industrial sectors were the two most affected by the recession. These sectors also include high hazard industries like forestry, mining, transportation and construction. Some sectors like healthcare had already begun to experience economic declines before the recession began.

For the 13 MNEs that were interviewed, the average revenue decline was 8.6%. The average profit decline was 28%. The average decline in employee size was 4.9%. Revenue and employee size had declined in 9 of the 13 MNEs (69%).

Organizational restructuring and impact on Layoff Survivors

The various types of restructuring mentioned during the analysis of 45 MNE annual reports were grouped into ten types of activity:

- Unpaid leave (n = 1).
- Applying bonus freezes (n = 1).
- Changing job roles (n = 1).
- Outsourcing and subcontracting essential work (n = 2).
- Controlling unnecessary expenses, freezing pay or reducing pay (n = 2).
- Re-distributing some of the workforce to facilities that are more productive (n = 2).
- Closing factories (n = 3).
- Decreasing production (n = 3).
- Reducing working hours (n = 4).
- Downsizing the workforce using early retirement and other schemes (n = 11).

Workforce downsizing (or "Crisis-sizing") was the most commonly reported restructuring activity using early retirement and other schemes. The scientific evidence investigating layoff survivors shows that restructuring and employee downsizing are mostly associated with increased risk for sickness absence and negative health outcomes including cardiovascular disease, mental ill-health and work-related musculoskeletal disorders (Westgaard and Winkel, 2011).

Effects of restructuring on exposure to psychosocial workplace hazards

A number of psychosocial workplace hazards could be affecting layoff survivors. These hazards are considered in light of the restructuring activities previously listed. They include:

Workload intensification – important both during a recession and recovery. An imbalance from increased demands with fewer organizational or individual resources can significantly affect psychological distress, burnout, motivation and workforce engagement (Crawford *et al.*, 2010; Devereux *et al.*, 2011). Workload is also recognized as a cause of accidents (Amati and Scaife, 2006).

Role stress – changing job roles during restructuring can create role stress, defined as a combination of role conflict, job ambiguity and increasing workload. Role ambiguity occurs when one is unclear about role requirements, performance standards and responsibilities. Role conflict occurs when two or more requirements of one's role are conflicting perhaps from incongruent instructions from managers or conflict with personal values or resources (Rydstedt *et al.*, 2011). Role ambiguity is recognized as a contributor to accident causation (Amati and Scaife, 2006).

Distrust of management, lack of respect, feelings of betrayal – associated with unpaid leave, bonus freezes, reduced working hours and outsourcing/contractual work. This can result in the "Dark side of work behaviour" including lying and deceit, insubordination, sabotage affecting production, drug/alcohol abuse, bullying and even assault (Furnham and Taylor, 2004).

Uncertainty regarding future restructuring and job loss even in the absence of actual material change – Westgaard and Winkel (2011) state that repeated downsizing over time further increases the risk of worsening health and work performance.

Little reward for the effort put into work – associated with pay freezes or reducing pay or incongruent status. There is consistent evidence showing that combinations of high demands and low decision latitude or combinations of high work effort and low reward are risk factors for the onset of common mental disorders (Stansfeld and Candy, 2006).

Reduced job commitment and lower job involvement – perhaps associated with forced employee redistribution to more productive facilities. Job involvement and communication are an important part of a participative management approach for improving health and safety (Devereux and Manson, 2008).

Feeling undervalued and lower morale – most likely from a combination of the issues listed above. Layoff survivors not at risk of unemployment have an increased likelihood of voluntary turnover after employee downsizing (Datta *et al.*, 2010). Employee downsizing results in deterioration in the social work environment and this negatively affects morale and creativity, which are needed for solving problems and for developing new growth initiatives.

Recessionary Effects on Safety and Health Management Systems
According to the data analysis of 45 MNE annual reports, specific changes were made to management systems including the establishment of health committees, psychosocial risk assessment and management, stress or fatigue prevention programmes, ergonomics prevention or continuous improvement programmes (e.g. EDF Energy, La Farge, Renault, Peugeot, Michelin, BNP Paribas, Exxon Mobil, GlaxoSmithKline and BT).

Based on the 13 interviews with selected MNEs, eight higher order themes emerged about the impact on safety and health management systems from the economic recession. These themes are shown below and a brief overview is provided.

1. Budget cuts at headquarters and field offices.
2. Policy changes.
3. Worker participation.
4. Management responsibility and accountability.
5. Staff levels and training.
6. Planning, development and implementation procedures (risk reduction).
7. Investigation processes (continuous improvement processes).
8. Audit procedures (performance monitoring and review).

In regard to these themes, the focus on organisational restructuring to meet budget cut objectives could have a negative impact on the development of a strong corporate culture on safety affecting prevention policy. Both managers and employees could become "distracted and distressed" by events during restructuring from the recession, undermining participative problem-solving strategies. For example, employees anxious about losing their jobs may be less likely to report a near miss (an unplanned event that did not result in harm or injury but had the potential to do so); hence, such occurrences may not be investigated and prevented in the future.

It was stated that the economic recession has negatively affected health and safety resources. Budgets, the number of professionals and staff training have been reduced due to the recession. However, the biggest drain on resources was stated to be the increase in workload from individual employee requests and corporate initiatives on psychosocial risk management.

Individual employee requests arose from concerns about mental stress and health and safety implications brought about by redundancy action. The response time had increased as a result of dealing with a higher number of these requests.

A number of MNEs have introduced or improved workplace health or stress management policies in the workplace; For example, MNEs were forced into prevention by the French Government due to the increasing number of employee suicides committed during the recession. Interventions have led to extra workload in planning on how to assess psychosocial risks, performing audits and attending safety and health committees. This workload combined with the need to manage existing or emerging risks (from noise, vibration, heat, the human-machine interface, chemicals etc.) creates a significant challenge for health and safety managers.

According to the interview data, there is a possibility that the risk management process may be compromised if adequate resources are not made available or if there is too strong a reliance on outsourcing of health and safety issues. This is supported by data from the European risk observatory in 2010 showing that the main difficulties in managing health and safety among larger enterprises was a lack of resources such as time, staff and finance and the lack of awareness and safety culture within the organization (EU-OSHA, 2010).

According to the scientific literature, most accidents occur during corrective maintenance activities (Milczarek and Kosk-Bienko, 2010). Maintenance could be compromised due to early retirement, redundancy or the insidious turnover that can occur among layoff survivors. There is a risk that the organizational learning or "Tribal knowledge" with respect to a company's operations and maintenance history may be lost. Furthermore, maintenance tasks may be outsourced and this can lead to increased safety and health risks among workers and can also affect system integrity. Johnstone *et al.*, (2005) cite examples where several explosions at petrochemical plants in the late 1980s were clearly linked to the use of independent contractors on maintenance activities. A similar situation arose in the Macondo well blowout in April 2010; neglected maintenance was cited as one of the system failures (Deep Water Horizon Study Group, 2011).

Comparison with previous economic recessions

What can be drawn from the data from previous recessions to confirm these findings? The financial data from 4,700 U.S. companies during three milder recessionary periods 1980–1982, 1990–1991 and 2000–2002 were analysed by researchers at Harvard Business School and Kellogg School of Management (Gulati *et al.*, 2010). As one might expect, reduced operating costs, discretionary expenditures, rationalization of business portfolios, employee downsizing and preservation in cash flow were also identified as recessionary corporate strategies. In effect, doing more work with fewer resources was the general conclusion of the researchers. However, the surprising result was that corporations that cut costs faster and deeper than rivals do not necessarily flourish. In fact, they have the lowest probability (21%) of pulling ahead of the competition when times get better according to Gulati *et al.*, (2010).

Conclusion

A number of potential threats exist for workforce health and safety and system safety during the economic recovery due to the "Great Recession" including:

- Focus on productivity and growth leading to a distraction on health and safety, including emerging risks (including the human-machine interface from the complexity of new technologies).
- Increase in psychosocial risks; especially work intensification with insufficient resources.
- Loss of vital skills and experiential knowledge needed to maintain system and asset integrity due to employee downsizing.
- Future budget cuts affecting health and safety due to ongoing restructuring activity.

The economy is beginning to show signs of recovery and when it gathers greater momentum, it is likely that the priority and focus among senior management will be on production and work intensification, and not health and safety at work. For enterprises that have reduced the size, experience and skill levels of their workforce; this could create a window of opportunity for organizational or operational hazards resulting in serious accidents and problems affecting health and sustainable well-being. This is a particular concern in high hazard industries; some of which have been affected the most by the recession. To quote one MNE safety representative from a high hazard industry: –

"I have seen it too many times when a company loses its focus on occupational safety and health, risks creep in and more accidents happen."

Acknowledgements

Special thanks go to Professor Mark Cropley, Department of Psychology, University of Surrey, UK, for being the independent thematic analysis reviewer, and also to

Professor Adrian Furnham, Business Psychology Unit, University College London, UK, for his valuable comments on the study design and draft of the 150 page research report for ILO. This research was funded by Safework within the International Labour Organization, Geneva. A special note of gratitude goes to all the companies that agreed to provide their commitment, documentation and time for interviews during a difficult time for all.

References

Amati, C. and Scaife, R. 2006. *Investigation of the links between psychological ill-health, stress and safety*. HSE Research Report 488 (Health and Safety Executive Book: Sudbury, Suffolk).

Crawford, E., Lepine, J., Rich, B. 2010. "Linking job demands and resources to employee engagement and burnout: A theoretical extension and meta-analytic test." *Journal of Applied Psychology* 95(5): 834–848.

Datta, D., Guthrie, J, Basuil, D., Pandey, A. 2010. "Causes and effects of employee downsizing: A review and synthesis." *Journal of Management*: 36(1): 281–348.

Deepwater Horizon Study Group 2011. Final report on the investigation of the Macondo Well Blowout. http://ccrm.berkeley.edu/pdfs_papers/bea_pdfs/DHSG FinalReport-March2011-tag.pdf

Devereux, J., Rydstedt,L., and Cropley, M. 2011. "Psychosocial work characteristics, need for recovery and musculoskeletal problems predict psychological distress in a sample of British workers." *Ergonomics* 54(9): 840–848.

Devereux, J. and Manson, R. 2008. "Using participatory change to reduce lost time injury and illness." In P. Bust (ed). *Contemporary Ergonomics 2008*, (Taylor & Francis: London), 79–83.

EU-OSHA, 2010. *European Survey of enterprises on new and emerging risks:Managing safety and health at work. European Agency for Safety and Health at Work: Bilbao.* (Publications Office of the European Union, Luxembourg).

Furnham, A. Taylor, J. 2004. *The dark side of behaviour at work: Understanding and avoiding employees leaving, thieving and deceiving.* (Palgrave Macmillan: Hampshire).

Gulati, R., Nohria, N., Wohlgezogen, F. 2010. "Roaring out of recession." *Harvard Business Review*, March issue: 1–8.

International Labour Organization. 2013. *Protecting workplace safety and health in difficult economic times – the effect of the financial crisis and economic recession on occupational safety and health.* (International Labour Office, Geneva, Switzerland). ISBN 978-92-2-127054-6.

Jonstone, R. Mayhew, C., Quilan, M. 2010. "Outsourcing risk? The regulation of occupational health and safety where subcontractors are employed." *Comparative Labor Law and Policy Journal* 22: 351–393.

King, N. 2004. Using templates in the thematic analysis of text. In C. Cassell & G. Symon (eds.) *Essential Guide to Qualitative Methods in Organizational Research*, (Sage Publications: London), Ch. 21, 256–270.

Milczarek, M., Kosk-Bienko, J. 2010. *Maintenance and occupational safety and health: A statistical picture.* European Risk Observatory Report. (Office for Official Publications of the European Communities, Luxembourg).

Rydstedt, L., Cropley, M., Devereux, J. 2011. "Long-term impact of role stress and cognitive rumination upon morning and evening cortisol secretion." *Ergonomics* 54(5): 430–435.

Stansfeld, S., Candy, B. 2006. "Psychosocial work environment and mental health-a meta analytic review." *Scandinavian Journal of Work, Environment and Health* 32(6): 443–62.

United Nations General Assembly. 2009. Outcome of the conference on the world financial and economic crisis and its impact on development. A/RES/63/303.

Westgaard, R. Winkel, J. 2011. "Occupational musculoskeletal and mental health: Significance of rationalization and opportunities to create sustainable production systems – A systematic review." *Applied Ergonomics* 42: 261–296.

CULTURE, CONTROL AND LEADERSHIP

INSIDER ATTACKS: THE HUMAN FACTORS DIMENSION – LEARNING CONFLICT RESOLUTION LESSONS FROM INDUSTRY TO COUNTER INSIDER ATTACKS IN AFGHANISTAN

Ryan Meeks[1], Pam Newman[1] & Colin Skelton[2]

[1]*Frazer-Nash Consultancy Ltd*
[2]*Defence Science and Technology Laboratory (DSTL)*

Insider "'Green-on-Blue' Attacks are a growing problem for deployed personnel in Afghanistan, accounting for 15% of Coalition deaths in 2012" (SIGAR, 2013). They are driven by a complex mix of socio-cultural, behavioural and organisational factors that make them very much a 'systems problem', necessitating the need to counter them with a thorough systems approach. This work explores how organisations within healthcare, prisons, peacekeeping agencies and others train conflict resolution skills and techniques to identify lessons that could be exploited in the context of countering Insider Attacks. The output from this work demonstrates the benefit of amalgamating knowledge from different contexts to derive 'best practice' in order to solve a similar challenge in a novel context. It also highlights the potential for military applications to re-inform industry; in this case strengthening conflict resolution training best practice in a range of applications.

Introduction

Insider attacks (IA), also called "Green on Blue" attacks, occur when a person or persons in a position of trust with the International Security Assistance Force (ISAF) and/or the Afghan National Security Forces (ANSF) initiates an act of violence against the ISAF/ANSF. Such Insider Attacks are a major problem for deployed personnel in Afghanistan, accounting for a significant number of casualties. This paper outlines research conducted on behalf of the UK Ministry of Defence (MoD) to tackle this problem.

Insider Attacks may have complex socio-cultural, behavioural and organisational factors that make them very much a systems problem, necessitating the need to address the issue with a systems view. This work involved identifying the competences, techniques, methods and strategies (TMS) that front-line personnel in Afghanistan, responsible for mentoring their Afghan counterparts, should be equipped with through suitable training in order to prepare them for dealing with the Insider Threat. The competences and TMS aim to address a number of underlying system factors implicit in insider attacks. This work explored how conflict resolution, conflict management and violence prevention are trained across a range of different organisations and industries, including healthcare, prisons, aid and peacekeeping agencies, security services and others, to identify any lessons and/or methods/tactics that could be exploited in the Defence context.

One of the main benefits of this approach is that the military can learn from 'best practice' across a number of different organisations and industries. The underlying generic system factors inherent in conflict renders it generic in a number of different contexts, and so key lessons in terms of mitigation and training can be applied in relation to the counter IA challenge. Subsequent findings concerning the military application of the recommended TMS will re-inform industry and help to establish a well tested 'best practice' approach to conflict resolution training.

Background

This work was primarily concerned with the reactive, critical, process-based skills required to deal with fast-emerging escalating Insider Attack situations, as well as some more proactive mitigations. Personnel must be able to react quickly, make fast informed decisions, rapidly problem-solve and communicate effectively with the threat in order to dissuade, defuse and control the situation. Training recommendations contain detail on the specific TMS for training people in six key training areas; verbal, non-verbal, conflict resolution models, risk, self-control and observation techniques.

Initially, this improvement in training and awareness of the required competences will allow UK Forces to mitigate against Insider Attacks up until the end of 2014, when security responsibility transfers over to the ANSF. UK Forces will be particularly vulnerable during this period and so there is an immediate need to rapidly prepare personnel with the desired skills and competences.

The specific TMS proposed as part of this project will facilitate rapid exploitation of their use in current and future training programmes, across various courses in the military training community. There is a focus on simple techniques with a low trainability demand, to enable their effective use on basic training courses.

Due to the generic nature of the skills and competences utilised for dealing with Insider Attacks, the developed training programmes and tools resulting from this work will be able to be transferred to the ANSF once they take ownership of security responsibility at the end of 2014. In addition to the threat of Taliban insurgence, the delicate and potentially unstable relationship between people of different Afghan tribal, religious, regional and cultural groups within the ANSF could lead to Insider Threats evolving. As part of UK Forces' efforts to transfer skill, knowledge and competence to develop ANSF independence, counter-Insider Attack training programmes and tools could form a vital component.

Methodology

The novelty in the approach of this work is to bring together TMS from various industries, to determine what generic transferrable skills could be used in the military context. Generic aspects of Insider Attack situations such as dealing with conflict, dissuading people from taking certain actions, defusing tension and intent and reacting in the correct manner are also present across a number of different types of organisations. In roles and environments such as policing, prisons, hospital A&E, retail customer service, nightclub security, peacekeeping organisations, banks and rail, people possess relevant skills sets. This work identified how these skills are trained to determine whether they could be successfully translated into the Defence environment. The work was undertaken in three phases.

Context analysis – Defining the problem space

An understanding of the context and nature of IA incidents makes it easier to determine what TMS and training methods are best suited for exploitation in the military context. As this is such a unique context and novel application, it was essential for the project to define and understand the 'problem space' and assess levels of transferability with the wider organisational/industrial contexts, before attempting to outline and recommend potential training solutions.

Information was collated from unclassified open source web resources, including armed forces websites and expert blogs. The information collected has been based on an analysis of 93 insider attack incidents between July 2007 and September 2013. The following are the key points related to the Insider Attack and Conflict context:

- The majority of Insider Attacks occur on joint/combined bases;
- Personal arguments and disputes precede many Insider Attacks (in at least 40 recorded incidents);
- Trainers and mentors may become increasingly vulnerable and exposed to the risk of Insider Attacks during the Security Transition Period in 2014, so they will require Conflict Resolution training;
- Threats to Islam and personal frictions are influencers to Insider Attacks;
- Personal frictions are a key influencer to Insider Attacks;
- Stress plays a major contributing, but little understood, role in Insider Attacks;
- Inappropriate behaviour has been noted as triggering a number of Insider Attack incidents;
- Insider Attacks generally occur at smaller, more remote locations such as on guard and sentry duties;
- The role of Interpreters should be carefully considered, in terms of their opportunity to escalate/de-escalate situations and their role in communication;
- Cultural aspects remain a key factor;
- Establishing positive personal relationships is a key mitigation against Insider Attacks.

Competency framework – Defining competences to deal with conflict

The Competency Framework outlines the knowledge, skills, attributes and attitudes (KSAs) required for training to counter IAs, based on a cross-industry review of conflict resolution training and using the context-specific insight gathered from the Context Analysis. Information provided in the Competency Framework has been collated taking into account cross-organisational practices, 'best practice' and proven methods for managing conflict. Consideration of the operational environment and specific military context should allow for the development of specific solutions to reduce the Insider Threat. The Competency Framework is presented alongside the Conflict Intensity Continuum (Robbins, 2005), to show how KSAs can be applied at different levels of conflict escalation.

Key knowledge competency areas

The following are the key knowledge areas, identified from the pan-organisation review and engagement with Subject Matter Experts, that military personnel require to support them in dealing with conflict related to the IA context:

- Human psychological and physiological responses, such as the key observable behavioural and physiological indicators (e.g. disproportionate emotional reactions, red face, heavy sweating, agitation, disorientation, hesitation, nervousness, fidgeting, etc.);

Figure 1.   Robbins' Conflict Intensity Continuum (2005).

- Human behaviour and behavioural triggers, such as phrases to avoid, individual behavioural red flags, social and behavioural faux pas and common conflict escalation pitfalls, appropriate language and phrases and an understanding of barriers and enablers to effective communication, including trigger phrases, irony and sarcasm;
- Self-awareness of prejudices, perceptions, biases, cognitive distortions and propensity for error, to help people regulate their behaviour and maintain control;
- Cultural awareness and understanding, including cultural grievances, cultural events and landmarks, common myths and stereotypes and its influence on behaviour and attitudes;
- Knowledge of ANSF colleagues, organisational structure and historical context, to aid in building rapport, trust and relationships and in helping to integrate forces and facilitate effective team working;
- Individual roles and responsibilities, including policy, directives, legislations and procedures;
- Behavioural and situational baselines and knowledge of normal community patterns of life.

Key skills competency areas

The following skills were identified as being of particular use in conflict situations:

- Empathy ('be able to think like your adversary');
- Self-awareness and self-control (emotional, cognitive and physical aspects);
- Listening skills, to demonstrate attention, care and concern;
- Co-operation, interpersonal and team working skills;
- Language skills, such as local Pashtu and Dari;
- Non-verbal 'tactical' communication, including body language, posture, presence, stance and positioning;
- Conflict resolution style and delivery, including flexibly applying different styles (competitive, collaborative, compromising, accommodating, avoiding);
- Flexibility and cultural adaptability, enabling people to fully embrace, understand and immerse themselves in local culture and custom;
- Rapport and trust building, to facilitate closer working and personal relationships with ANSF colleagues to both increase the chance of identifying emerging threats at the earliest opportunity and to reduce the perception of oneself as a 'target';
- Negotiation and mediation, as key conflict resolution skills;
- Observational skills, including behavioural identification, recognition and analysis, situational awareness, attention, search and identification, alertness, vigilance and anticipation.

Key attribute/attitudes competency areas

- Professionalism, primarily as a method of establishing mutual respect;
- Self-confidence, to help personnel have trust in their instinct and intuition and to empower them with the confidence to tackle potentially dangerous situations;

0 - No conflict	1 - Minor disagreements or misunderstandings	2 - Overt questioning or challenging of others	3 - Assertive verbal attacks	4 - Threats and ultimatums	5 - Aggressive physical attacks	6 - Overt efforts to destroy the other party
			Safer assessment, approach and positioning skills to reduce potential for assault (NHS)			
	Initial response skills (Security)					
				Rapid response skills (ISAF AF)		
					Advanced weapons proficiency - enhanced speed of response, controlled weapons handling and accurate shooting (ISAF, ISAF AF)	
	Effective conflict resolution style and delivery - competitive, collaborative, compromising, accommodating, avoiding (Mind Tools, VA)					
Flexibility in approach and style (Mind Tools, VA)						
Rapport and trust building, based on mutual understanding, to develop close relationships (Peace-keeping, TR 1, TR 3, TR 5, TR 7, ISAF, ISAF AF, DLW)						
Elicitation (TR 6)						
Cultural adaptability and sensitivity (ISAF, DLW)						
Negotiation (Education)						
	Mediation (Education)					
Decision making (Education, TR 2)						
Rational thinking (Service Alternatives)						
Intelligence fusion (ISAF)						
Observation (Dstl, TR 2, TR 4, TR 5, TR 7, TR 8, ISAF, ISAF AF, ISAF AH)						
Behavioural identification, recognition and analysis (TR 1, TR 5, TR 6, TR 7, TR 8, ISAF)						
Situational awareness (ISAF)						
Personality identification and categorisation (VJT)						

Figure 2.   Competency Framework Table.

- Dignity, respect and courtesy; the 'cornerstones' of Afghan culture and society;
- Aptitude and motivation, to recognise the qualities of ANSF colleagues and to continually pursue opportunities to improve the relationship in times of conflict.

Figure 2 below shows how the identified competences were mapped against Robbins' Conflict Intensity Continuum levels in a table format;

Training recommendations – Techniques, methods and strategies (TMS) for conflict resolution training

Training recommendations are focussed on a key set of TMS that could be rapidly integrated into current and future training programmes to help counter the Insider Threat. TMS were identified from across various organisations, compared against Robbins' conflict escalation levels and the Competences and selected for their relevance, suitability and practicality. Training courses were reviewed from domains such as the Prison Service, policing, healthcare and peacekeeping organisations, as well as from non-UK military, to determine a core set of suitable TMS that would equip personnel with a better ability to handle conflict situations in the counter Insider Threat context. Table 1 outlines the core training areas and TMS identified:

Conclusion

The various industries of interest for this work have not been previously explored to inform Defence training in this context, and so there may be the opportunity

Table 1.   Key Training Areas and TMS for Conflict Resolution.

Training Area	Technique, Method or Strategy	Specific and/or Underpinning Relevance
Verbal Techniques	Active Listening (echoing, labelling and paraphrasing) Language Style (tone of voice, speed, pitch, volume, modulation, assertiveness)	Specific
Non-Verbal Techniques	Body Language (non-confrontational approach, personal space, gestures and posture) Assertiveness and self-confidence	Specific
Conflict Resolution Models and Concepts	The Betari Box Concept Step-by-step assessment and coping models such as COPING, LEAPS and CAPE	Underpinning
Observation Techniques	Human Behavioural Analysis (identification of warnings, cues and indicators at a personal level) Background atmospherics (contextual cues and indicators from the environment and Pattern of Life)	Specific and Underpinning
Risk Techniques	Risk assessment measures including risk avoidance, reduction, retention and transfer techniques and strategies	Underpinning
Self-Control Techniques	Techniques to improve the recognition of the indicators and effects of stress, fear and anxiety on oneself and others	Specific

for a wealth of useful new knowledge from the wider organisational and industrial context to be exploited for military purposes. This project is an example of how the military training community can invest in innovative research to make use of wider best practice methods and learn from the successes of others.

The main outputs from this work are the competences and training recommendations that can be built into training policy, requirements and programme content, to inform decision makers and allow the MoD training community to prepare UK Forces against the continued risk of Insider Attacks in Afghanistan. The competences will also become increasingly important in future deployments beyond Afghanistan such as Syria or Mali, where forces may need to interact closely with local Forces and populations, increasing the risk of Insider Attacks. The outputs of this work could also be utilised effectively in environments outside of Defence.

There are future challenges concerning the implementation of training including aspects such as universal adoption, subsequent validation and determining levels of proficiency. However, the coherence of the approaches across the various application contexts provides reassurance that the recommended TMS have a wide exploitation potential both within and outside of the Defence context.

References

Special Inspector General for Afghanistan Reconstruction (SIGAR). *Quarterly Report to Congress*, April 30th 2013. www.sigar.mil/quarterlyreports [accessed 6/12/2013]

Robbins, S. (2005). *Organisational Behaviour.* (Prentice Hall, New Jersey, USA)

VISUAL MEANING FOR LEADERSHIP – A CASE STUDY

Peter Nock[1] & Julian Burton[2]

[1]*Operations Concepts Manager, HS2 Ltd*
[2]*Director, DELTA 7 Change Ltd*

High Speed 2 is the largest infrastructure project in Europe. Large numbers of people from a wide range of disciplines and organisations are involved in its design. Its long timescale and complexity means that many more people will come to work on it in the future. How do you ensure that there is a common goal, a common aim that can be accessed readily by all? How do you change natural behaviours which tend to drive people into discipline silos and create fully joined up thinking? HS2 worked with DELTA 7 to use pictures and visual metaphors to create meaning and to offer teams a different way of exploring complex problems. We attempted to create understanding of the bigger picture in order to generate a different kind of meaning, knowing and understanding.

Introduction

Spring 2013 saw the establishment of an Operability Team at the High Speed 2 project. At that time HS2 described itself as the biggest infrastructure project in Europe and was dominated by Engineering disciplines. Existing documentation and plans focussed on the building of the infrastructure. There was little focus on the operation of a railway and even less on the passenger. A method of connecting people to the purpose of the project was required, a way of getting people to see and understand the big picture. The solution? Create a big picture!

Visual meaning as a tool

Leadership researchers such as Kotter and Schein suggest that the key role of a leader is to step back, make sense of the bigger picture and create shared meaning for their people. They do this best by telling a compelling story of change of their organization – its identity, purpose, vision and direction.

However, leadership teams are themselves driven by complex responsibilities and competing demands. Having to make difficult decisions about intractable problems causes stress and overload. And we all know that without leadership clarity, communication to the rest of the organization quickly becomes a series of impenetrable documents and Powerpoint presentations.

Using visual metaphors to represent complex strategic issues can bring more areas of the brain into play and help to create a shift in the way that the problem is seen, talked about and made sense of. We will look at how pictures differ from words, then examine two ways that pictures can act as a leadership tool: first, as a sense-making tool for the leaders themselves, and secondly as a way of engaging the rest of the organization with change.

Pictures and words

The dominant representational system in organizations today is verbal. Leadership teams generally make sense together and communicate strategic change using words. Words are a powerful communication tool, yet they do have their limitations. First, they are processed by the left, more rational side of the brain. Secondly, they are abstract: it is only by arbitrary convention that words and what they represent are linked. Finally, they are linear in that they can only describe one thing at a time. As the world becomes more complex they can become inadequate for representing emotionally charged, complex, or multi-dimensional problems. Imagine trying to explain something as complicated as the London underground system in words!

The problem is not just in language itself, but in the *type* of language that is typically found in organizations. In many change projects a consistent problem is that the words used to discuss and communicate change can be highly abstract and distant from experience. Real and complex challenges can get smothered in abstractions and generalisations. The result is that a new change strategy or a critical directive that is obvious to the board can be very disconnected from employee's lived experience. And if the strategy is communicated one-way, in can be mis-communicated, leading to a lack of clarity, confusion and potentially little or no action. Difficult but important subjects are avoided, or made safe and unthreatening, so the truth gets hidden away and lost. This way of communicating is so ingrained that people are barely aware that they do it.

Communication through verbal language has its limitations, so what are the alternatives? There is a growing body of research that advocates the use of visualisation, metaphor and narrative as group 'meaning-making' tools that can help senior teams make sense of and articulate strategy in a way that employees can understand. Pictures have been used for thousands of years to tell stories, create meaning and represent what is most important for people in communities. The visual medium is able to show complex inter-relationships in an instant, which means that the observer can have an overview of a whole. Metaphors are memorable, and allow us to re-present complex and tacit information in visceral, meaningful ways.

Picture creation as a tool for sense making

The western corporate world's reliance on linguistic and rational intelligences is perhaps nowhere more evident than in the boardroom, where it is clear that conventional thinking about change is no longer enough. Prof Keith Grint (Grint, 2010) proposed the idea of 'wicked problems', where there are no linear causes or linear

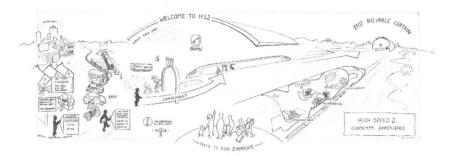

Figure 1.   The Big Picture.

solutions, and it is these kinds of challenges that business leaders are increasingly facing. Rational processes are simply not robust enough to address wicked problems. A different, less rational level of thinking is needed – a different way of knowing. It is through this 'different way of knowing' that visual metaphors can help leadership teams discover, bringing more creativity to the boardroom in the process.

Visual art, used properly, is a particularly powerful way of representing important themes symbolically, non-verbally and aesthetically. Visual metaphors and stories present knowledge in a more concrete form than words alone, stimulating more spontaneous and grounded conversations, the sort in which new meaning or knowing can arise naturally within the group.

Creating the big picture

Two skilled change consultant and artist from DELTA 7 interviewed key people in HS2, those people who had the real vision of what the project will deliver for passengers. In depth questioning enabled them to isolate the really key elements and then their artistic skills enabled them to render these ideas as a visual representation of a complex, multi-faceted story (see figure 1). It really is a "big" picture. The version regularly used by HS2 is actually some 8 ft × 3 ft!

Using the big picture

The leaders of this work now had a visual representation of their thinking. The process had allowed them to crystallise their own thoughts and articulate the key issues. Now the picture could be used to communicate those ideas across the organisation. Workshops were held where the picture was presented in a very interactive way. Large versions allowed staff to add their own ideas and comments. The physical nature of the delivery of the pictures encouraged active participation and drove spontaneous conversations.

Figure 2.   HS2 O&M Big Picture.

Sharing the vision across the organisation was the original intent. However, it soon became clear that this was also a way of engaging with stakeholders and potential passengers. Some 12 to 18 months after its creation, the big picture is still used in corporate communications and presentations to explain what the project is attempting to create and to engage people in the story of HS2.

A success story

The continued use of the Big Picture and reactions to it indicate success. Its use as an external communication tool exceeded original expectations. But perhaps the most telling measure is the fact that the organisation has used the techniques again.

The operations team wanted to drive behaviour change by getting design and engineering teams to understand that HS2 is a single system. Traditional siloed approaches to design hamper the realisation of such a systems based approach. Another big picture was created to communicate this idea and to engage the design community (see figure 2).

In addition, 2 other teams have used the same techniques to communicate the complete Health and Safety picture and Building Information Management (BIM) techniques. Two further teams are considering the techniques to engage on The Design Process and the Skills and Opportunities agenda.

Reference

Grint, K. (2010) Wicked problems and clumsy solutions: the role of leadership. *The new public leadership challenge* 11, 169–186.

HAZARDS, RISK AND COMFORT

PERCEPTION OF SELF-MOTION HEADINGS WHEN SUBJECTED TO NEAR-THRESHOLD VISUAL AND INERTIAL MOTION

Beisheng Bao & Richard H.Y. So

The Department of Industrial Engineering and Logistics Management,
The Hong Kong University of Science and Technology, Hong Kong, PRC

Understanding human limitations in discriminating self-motion heading directions can contribute to the design of better steering interfaces. This study investigated whether a previously reported vestibular over-weighting phenomenon still exist when participants were exposed to weak (near perception threshold) visual motion and inertial motion. A previous theory predicted that such phenomena should not exist but our preliminary results indicate they do. This discrepancy can be explained by a known suppression of cortical visual response during inertial motion. Potential applications include assessing and predicting the heading discrimination ability of a driver, who remotely controls a vehicle from another moving platform.

Introduction

Background

Drivers are regularly engaged in self-motion heading discrimination tasks. When driving a remote-control vehicle through a closed-circuit TV, while physically travelling in another moving vehicle, drivers may be exposed to conflicting and non-conflicting visual and inertial heading cues. Modelling of how humans integrate visual and vestibular heading information to estimate their self-motion heading has been studied extensively (e.g., Fetsch et al. 2009; Butler et al. 2010; de Winkel et al. 2010). Heading discrimination thresholds (HDT) has been used as a quantifying measure of the strength of a heading cue. HDT has been defined as the smallest rightward deviations (degrees) from the straight ahead direction by a heading cue that can be correctly detected with an accuracy of 84%. A strong cue will have a small threshold while a weak cue will have a large threshold (i.e., larger deviation is needed to achieve 84% correct detection rate). The choice of 84% accuracy level was originated in Gu et al. (2008) when he proposed a Bayesian model to predict how humans integrate visual and vestibular heading information. This 84% for defining the HDT was adopted in Fetsch et al. 2009 and also in this study. Participants were repeatedly exposed to the heading stimuli of the same strength but with different heading directions ranging from $-16°$ (left) to $+16°$ (right). For each cue, participants were asked to judge whether the cue is to the left or to the right.

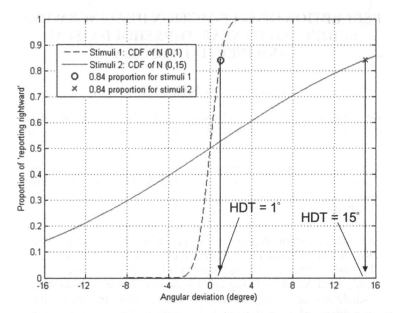

Figure 1. HDTs extraction from plots of percentages of rightward as functions of heading directions. Two hypothetical curves were plotted with HDTs of 1° and 15°. Strong heading cues were used in stimuli 1 and weak cues were used in stimuli 2.

The percentages of a perceived rightward direction were plotted against the heading directions (Figure 1). When 50% level is aligned to 0 degree, the degree at 84% level would be the HDT. In this study, the same method was used to determine HDTs.

The self-motion heading is a weighted summation of perceived visual heading and inertial heading (Butler et al. 2010). The optimal estimation of self-motion headings could be achieved by a Bayesian inference of perceived heading information. Humans have been shown to optimally estimate self-motion headings (e.g., Fetsch et al. 2009) and Bayesian inference has been used to model the human visual-vestibular integration process in the heading discrimination. The weighting of each modal (visual or inertial) relies on the relative reliability of each heading cue. In this experiment, uni-modal HDTs were used to estimate the relative reliability of the corresponding heading cues and to further calculate the corresponding weightings. This method was proposed by Gu et al. (2008).

Problem statement and gap

The predictions of the Bayesian model have been partially verified (Fetsch et al. 2009; Butler et al. 2010) and a prediction bias had been found when the model was used to predict bimodal HDTs (Fetsch et al. 2009; de Winkel et al. 2010). According to the model, the presence of both visual and vestibular cues (i.e., bimodal) should give a HDT lower than the HDT associated with visual cues alone. In order words, bimodal cues should improve heading perception. However, past studies indicated the opposite. When visual HDT is much smaller than the vestibular HDT, the bimodal HDT was higher than the visual HDT but lower than the vestibular HDT. In order words, the bimodal HDT "biased" towards the vestibular HDT. This phenomenon has been referred to as 'vestibular over-weighting' because the bias could be explained by an increase in the relative weightings associated with the vestibular cues in the model. In Gu's model, the contributions of the visual heading perception and inertial (vestibular) heading perception were weighted with W_{vis} and W_{ves} in which the sum of W_{vis} and W_{ves} are always unity (Gu et al. 2008).

Following the above logic concerning the 'vestibular over-weighting', we have come up with an interesting research question: if a participant is exposed to both visual and vestibular heading cues but the vestibular cue was very weak (near perception threshold), according to the current theory, the effects of 'vestibular over-weighting' should be small. A review of literature found no studies to address the above research question. Hence, a research gap exists.

Methodology

Objectives and hypotheses

In order to fill in the research gap, we designed a heading discrimination task with three near-threshold inertial motion (9, 14, 19 mg) and visual motion with 8% of visual coherence (i.e., 8% of dots on the screen will move according to the pre-determined heading and the rest of the dots will move randomly).

We chose 8% for two reasons: (i) 8% visual coherence had been shown as a valid and effective visual heading cue (Fetsch et al. 2009); and (ii) we would like to lower the visual HDTs in this study so that it would be higher than the vestibular HDT of the 9 mg inertial motion but lower than the vestibular HDT of the 19 mg stimulus. The latter point, if achieved, would allow us to test our research question in situations where the visual cue is stronger than the vestibular cue and vice versa.

The three peak acceleration levels used in the inertial motion cues (9, 14 and 19 mg) were chosen because they were just higher than the average lateral inertial heading perception threshold of 7.7 mg (at 0.2 Hz, Benson et al. 1986). Hence, they were near threshold inertial cues. Both the visual and vestibular cues had the same headings directions. We tested the following hypotheses. The testing of these hypotheses will eventually lead to improvement of modeling bimodal heading perception, which in turn will help to design better navigation interfaces.

- H1: the visual HDT is significantly lower than the vestibular HDT of the 9 mg motion cue but significantly higher than the vestibular HDT of the 19 mg motion cue. If H1 is proven, this suggests that our study can test H2 in situations where visual cues are stronger or weaker than the vestibular cues. If not proven, our study would have a reduced scope and can test H2 in only one particular situation.

- H2: the bimodal (i.e., combined visual and vestibular cues) HDT will be significantly larger than the visual HDT (i.e., the 'vestibular over-weighting' bias phenomenon will be present in this study).

Subjects

University students were recruited for a one-hour training session which contains all conditions of the formal experiment. If the subject fully understood the experiment procedure and was capable to follow the instructions, he/she was allowed to take part in the formal experiment. So far, six subjects (five male and one female) with a mean age of 24 participated the study and their data are presented in this paper. Each participant was compensated with HK$50/hour for their time. All subjects had a normal or corrected-to-normal vision. They were not receiving any medical treatment and were naïve to the experiment. The experiment was approved by the Human Subject and Research Ethics Committee at the Hong Kong University of Science and Technology.

Apparatus and stimuli generation

The experiment was conducted inside a dual-axis motion simulator (Kwok et al. 2009). Subjects sat securely in a chair with a four-point-harness. A head-restraint and a chin rest were used to limit head movement. Subjects were exposed to 75 dBA white noise through a pair of canal phones to reduce the chances of hearing auditory cues related to the movement of the motion simulator. Two accelerometers were mounted on the chair and the forehead of the subject to monitor the acceleration of motion simulator and the head. The visual heading cues were displayed on a 46 inches LCD TV. With a viewing distance of 70 cm, the FOV (field of view) was 80° (horizontal) × 45° (vertical). The display had a resolution of 1920 × 1080 pixels and a refresh rate of 60 frames per second. The visual heading cues were synchronized with the inertial heading cues to within 100 ms for the bimodal condition. A black curtain was used to enclose the TV and the subject to reduce the influence of environment. The experiment was conducted with light off. All control and display programs used for generating visual motions were coded with OpenGL and C++.

The 5s inertial heading cue was a uni-directional linear motion whose acceleration followed a 0.2 Hz sinusoidal function. The displacement, velocity and acceleration are illustrated in Figure 2. A 5s homing back motion was performed after every inertial stimulus. It shared the same motion profile as the stimulus but in reversed direction. Subjects were not asked about the heading directions during the homing motion. Three levels of acceleration magnitude (9 mg, 14 mg, and 19 mg) were used there with nine directions ($\pm16°$, $\pm9.2°$, $\pm3°$, $\pm1.7°$ and $0°$) per magnitude. The $0°$ was set to be the facing direction of the subject and positive angle refers to heading motions which were heading to the right of the subject. In each repetition, the nine heading cues with different directions were presented in a random sequence. In the bimodal condition, the directions of the visual motion cue and inertial motion cue were kept the same. In other words, no conflicting cues.

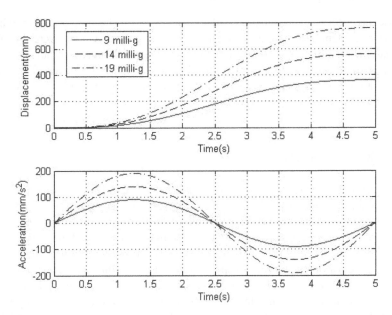

Figure 2.   Acceleration and displacement plots of the inertial heading cues.

The visual stimuli, also last for 5 seconds, was a zooming star-field made up of 900 balls presented in perspective. The FOV was of the visual heading cues was 80° (horizontal) × 45° (vertical). The trajectory of the virtual view-point was the same as the simulated inertial motion trajectory (Figure 1). The ball would appear at the far end and disappear at the cutting plane to avoid extremely large balls appearing near the edge of the display. The coherence of the visual motion was 8%, which means there are 92% balls in the space moved randomly. Reasons for choosing 8% were explained in the above text.

Experiment design and procedure

A two factor (visual heading: on or off; inertial heading: 9, 14, 19 mg) full factorial within-subject design with one exempted condition (no visual and no inertial heading cue) was used. In total, there were seven conditions. In each condition, subjects were exposed to 270 trials presented in random order (30 repeats for 9 different heading directions). There were 5s short rests between consecutive trials and a long rest (1–5 minutes) between every 90 trials. Each condition lasted for an hour and each subject was only exposed to one condition (about 1 hour) per day to avoid tiredness. During the one hour, subjects sat in front of the LCD and were instructed to keep their eyes on a red fixation cross at all-time except resting. During exposure to each of the 270 5s heading stimuli, subjects were required to report whether the perceived heading direction was to the left or to the right via a keyboard. For example, in a visual-on and inertial 19 mg condition, the subject will be exposed to both visual and 19 mg inertial heading stimuli with the same direction. This presentation will be repeated with 9 heading directions whose order of presentation was random. During a 5s stimulus, subjects were forced to make a choice between 'left' and 'right'. A 5s rest follows the stimulus and the subject could take a rest and prepare for the next stimulus. After every 90 consecutive stimuli, there was a 1 to 5 minute rest and the total stimuli number for one condition is 270.

Results and discussion

The data of the first six subjects are presented here. The full data set with more subjects will be presented at the conference. As expected, the proportion of rightward response increased as the heading directions of the cues moved from −16 degrees to +16 degrees (c.f., Figure 1). Following the methodology used in previous studies, data were fitted with a cumulative normal distribution before extracting the heading deviation (in degrees) responsible for a 84% rightward response (Fetsch et al. 2009; Fründ et al. 2011). This was needed in case some subjects did not reach the 84% rightward response level even with 16 degrees heading deviation to the right. Using the above methodology, visual HDTs and the three bi-modal HDTs (visual and inertial cues of 9, 14, and 19 mg) were calculated for each subject. For the three conditions with inertial cues only (9 14, and 19 mg) however, five out of the six subjects were not able to discriminate headings of +16° from −16° and their estimated vestibular HDTs approached infinity. This was understandable as the inertial cues used in this study were near the magnitude perception threshold and the near infinity vestibular HDTs indicated that subject were not able to percept the inertial headings cues with acceleration levels of 9, 14, and 19 mg. But all subjects verbally confirmed that they could detect the movement of motion simulator.

The near infinity vestibular HDTs suggest that H1 could not be supported. In this study, even though the visual heading cues had 8% coherency, the associated visual HDTs were much smaller than those of vestibular HDTs. Consequently, subsequent analyses could only test H2 in situations where visual heading cues were much stronger than the vestibular cues. In the rest of the analyses of HDTs, we will focus on visual HDTs and bimodal HDTs.

The mean visual HDT and the three bimodal HDTs were as follows: (i) visual HDT = 12.57° ± 4.29° (standard errors of mean); (ii) bimodal HDT with 9 mg inertial cue = 22.81° ± 5.95°; (iii) bimodal HDT with 14 mg = 24.58° ± 8.68°; and (iv) bimodal HDT with 19 mg = 13.69° ± 3.29°. There were no significant differences among the three bimodal HDTs (Wilcoxon signed-rank test, p > 0.05) and they were pooled into a new data set called bimodal HDTs. Wilcoxon tests were used because they were not normally distributed (Shapiro-Wilk test). The visual HDTs were found to be marginally significantly smaller than the pooled bimodal HDTs (Wilcoxon, p = 0.06). As these results were only based on the first six subjects, as more subjects' data are available, the differences are likely to be more significant and the full data set will be presented at the conference.

The finding that bimodal HDTs were higher than visual HDTs supports hypothesis H2. The same phenomenon was also reported in previous research (Fetsch et al. 2009; de Winkel et al. 2010) and was explained by vestibular overweighting. But this explanation could not be applied in this study because the vestibular cues in this study were so weak (with near-infinity vestibular HDTs). In other words, the vestibular cues were so weak that they should not be able to cause the phenomenon of vestibular overweighting. A review of literature indicates that in the presence of vestibular stimuli, the internal visual response could be suppressed (visual-vestibular reciprocal cortical inhibition: Deutschla et al. 2002). According to cortical inhibition theory, the presence of inertial heading cues could suppress the visual cortex and resulted in weaker heading perception and hence, higher bimodal HDTs.

We acknowledge that the visual heading stimuli used in this study were with 8% coherency (i.e., relatively weak heading stimuli) and whether the findings are confounded with this 8% is a valid question. Following the Bayesian model methodology in Gu et al. (2008), we had calculated that the model weighting for visual heading perception (W_{vis}) and most of them were above 0.98. This suggests that the visual heading cues, though with 8% coherency only, had been strong stimuli. Consequently, we hypothesize that if the study was repeated with visual heading cues of 100% coherency, similar findings will be obtained.

Conclusions and future work

When exposed to visual motion, adding non-conflicting near-threshold inertial motion cues can cause marginal significant increases in heading perception thresholds (HDTs). In other words, it can degrade heading perception. This result contradicts past findings and could be explained by attention distraction or visual-vestibular reciprocal inhibition in the cortices. Potential applications include predicting the limitations in heading discrimination of a pilot who remotely controls a vehicle from another moving platform. For example, an operator controls an Unmanned Aerial Vehicle while he is carried by a moving car. Furthermore, the designers could optimize steering interfaces for remote vehicle based on the HDT data. Future work to follow this up is on-going. Bimodal HDTs under 100% visual coherence and weak inertial motion will be measured to verify the results of this paper. The H1 which was not supported in this experiment will also be tested in the future work.

Acknowledgement

This research is partially supported by the Hong Kong Research Grants Council through 619210 and 618812.

References

Benson, A. J., Spencer, M. B. and Stott, J. R. 1986. "Thresholds for the detection of the direction of whole-body, linear movement in the horizontal plane." *Aviation, Space, and Environmental Medicine* 57: 1088–1096.

Deutschla, A., Bense, S., Stephan, T., Schwaiger, M., Brandt, T. and Dieterich, M. 2002. "Sensory System Interactions During Simultaneous Vestibular and Visual Stimulation in PET." *Human Brain Mapping* 16: 92–103.

Butler, J. S., Smith, S. T., Campos, J. L. and Bülthoff, H. H. 2010. "Bayesian integration of visual and vestibular signals for heading." *Journal of Vision* 10(11):23, 1–13.

de Winkel, K. N., Weesie, J., Werkhoven, P. J. and Groen, E. L. 2010. "Integration of visual and inertial cues in perceived heading of self-motion." *Journal of Vision* 10(12):1, 1–10.

Fetsch, C. R., Turner, A. H., DeAngelis, G. C. and Angelaki, D. E. 2009. "Dynamic reweighting of visual and vestibular cues during self-motion perception." *The Journal of neuroscience* 29(49): 15601–15612.

Fründ, I., Haenel, N. V. and Wichmann, F. A. 2011. "Inference for psychometric functions in the presence of nonstationary behavior." *Journal of vision* 11(6):16, 1–19.

Gu, Y., DeAngelis, G. C. and Angelaki, D. E. 2007. "A functional link between area MSTd and heading perception based on vestibular signals." *Nature neuroscience* 10(8): 1038–47.

Gu, Y., Angelaki, D. E. and DeAngelis, G. C. 2008. "Neural correlates of multisensory cue integration in macaque MSTd." *Nature neuroscience* 11(10): 1201–1210.

Kwok, K. C. S., Hitchcock, P. a. and Burton, M. D. 2009. "Perception of vibration and occupant comfort in wind-excited tall buildings." *Journal of Wind Engineering and Industrial Aerodynamics* 97(7–8): 368–380.

FACTORS CONTRIBUTING TO THE PERCEIVED SLIPPERINESS RATING

Wen-Ruey Chang, Mary F. Lesch, Chien-Chi Chang & Simon Matz

Liberty Mutual Research Institute for Safety, Hopkinton, MA 01748, USA

Perceived slipperiness rating (PSR) has been widely used to assess walkway safety. In this experiment, 29 participants were exposed to 15 different floor conditions. The relationship between their perceived slipperiness rating and all other objective measurements, such as utilized coefficient of friction (UCOF), kinematics and available coefficient of friction (ACOF), was explored with multiple linear regression analysis. The results show that the UCOF and ACOF were the major predictors of the PSR.

Introduction

Data from the Liberty Mutual Workplace Safety Index (Liberty Mutual Research Institute for Safety, 2012) show that the costs for disabling workplace injuries in 2010 due to falls on the same level in the U.S. were estimated to be approximately 8.61 billion US dollars or 16.9% of the total cost burden. Slippery floors, typically caused by contaminants, are a critical factor for falls on the same level (Chang et al., 2001). Bell et al. (2008) indicated that liquid contamination was the most common cause (24%) of slip, trip and fall incidents in healthcare workers. Various approaches have been used to assess slipperiness. Chang et al. (2003) outlined biomechanics, human-centered approaches, available coefficient of friction (ACOF) and surface roughness as major elements in the measurement of slipperiness.

Perceptions can be considered measurements of slipperiness. Perceptions, based on both visual cues and proprioceptive feedback, can be used prospectively or retrospectively to assess slipperiness and can supplement objective measurements of slipperiness.

In this experiment, the participants were exposed to walkways of different slipperiness and were asked to avoid a slip while walking on these surfaces. The relationship between perceived slipperiness ratings and different objective measures, including ACOF, kinetics and kinematics that might affect their outcome, was explored using multiple linear regression analysis. This relationship could identify critical factors that help form the final outcome of perceived slipperiness rating.

Methods

The five floor types used in this experiment, referred to as floor types A to E, were: (A) a standard quarry tile with raised-profiled tread lines perpendicular to the walking direction, (B) standard flat quarry tile, (C) vinyl composition sheet, (D) marble tile and (E) glazed porcelain tile. Five walkways were constructed, one for each floor type. Each walkway was 6.08 m in length and 0.81 m wide. Two force plates (Model 9281C, Kistler Instrument Corporation, Amherst, New York, USA) were installed underneath the tiles in the middle of each walkway.

Surface conditions were dry, water and glycerol. A fine water mist was applied with a garden sprayer. A mixture of 45% glycerol with water by weight was used as another contaminant and was applied with a paint roller.

Thirteen males and 16 females took part in this experiment. The protocol used was approved by an institutional review board and written, informed consent was obtained from each participant. Leather loafers were used as the standard footwear in this experiment. A safety harness was used. Prior to the start of the first trial, the contaminant, if required by the condition, was applied to the entire walkway. The surface condition started from dry and then wet, and was completed with glycerol. Under each surface condition, the five different floor types were randomized. The participants had a clear view of the entire walkway prior to walking on it. They started from one end of the walkway and went to the other end, then turned around and walked back to the beginning of the walkway. This was considered one trial. Data were collected only on the first leg of this trial. They repeated walking back and forth in the walkway sections four times. After each fifth trial, participants rated perceived slipperiness.

The sampling rate for the force plates was 1000 Hz. All the force plate data were processed with the fourth order zero-lag Butterworth low pass filter with a cut-off frequency of 24 Hz. Utilized coefficient of friction (UCOF), obtained from the vector sum of the transverse and longitudinal components of the ground reaction force divided by the normal force at the same instant, was identified and extracted from each successful strike.

Kinematics were collected with a motion tracking system (Motion Analysis Corp., Santa Rosa, CA, USA) with a sampling rate of 200 Hz. Seventeen passive reflective markers were attached to each participant on the sacral vertebrae, and right and left of the os calcani, ankle, anterior superior iliac spine, heel, knee, shank, thigh and toe. All the kinematic data were processed with the fourth order zero-lag Butterworth low pass filter with a cut-off frequency of 12 Hz. The following kinematic parameters were extracted from the data: step length, walking speed, stance duration and, at heel contact, heel angle with the floor, horizontal heel velocity, vertical heel velocity and ankle angle.

The participants rated their perceptions of slipperiness immediately after the fifth trial for each walkway and surface condition (for a total of 15 times) by drawing a horizontal line across a vertical line representing the range of slipperiness from 'not at all slippery' at one end to 'extremely slippery' at the other end. A ratio was obtained by dividing the distance from the line drawn by the participants to 'not at all slippery' by the total distance from 'extremely slippery' to 'not at all slippery' on the survey form. This ratio was then multiplied by 100 to obtain the final rating used in the subsequent analyses. Based on these calculations, the perception rating of 100 represented 'extremely slippery', while a rating of zero represented 'not at all slippery'.

A portable inclinable articulated strut slip tester (PIAST), also known as the Brungraber Mark II, was used in this experiment to measure ACOF. A total of 100 friction measurements were carried out for each walkway under each surface condition.

The biomechanical measurements, including UCOF and seven kinematic variables, of the fifth trial and the perceived ratings for each condition were extracted at the individual participant level. The ACOF values measured on each walkway under each surface condition were averaged across all the measurements to reflect an overall condition of the walkway under that particular surface condition. This ACOF value was assigned to each individual participant under the same conditions. A multiple linear regression analysis with the perceived slipperiness rating as the dependent variable, and UCOF, ACOF and all kinematic variables as the independent variables was used to explore their relationship.

Results and discussion

The averaged ACOF for each floor type and surface condition had a maximum of 0.907 and a minimum of 0.001 while the averaged perceived slipperiness rating had a maximum of 94.0 and a minimum of 6.6.

The adjusted R^2 value for the multiple linear regression equation was 0.52. Among the independent variables, UCOF and ACOF were the only variables with statistical significance in the regression equation with p values of <0.001 for UCOF and 0.0018 for ACOF.

The UCOF and ACOF were the major contributors to the outcome of perceived slipperiness rating. These two variables also determined whether a slip would happen at the shoe and floor interface. A slip is more likely to occur when the UCOF exceeds the ACOF. It has been reported in the literature that increases in stance and stride times, and step width, as well as decreases in stride length, walking speed, heel horizontal velocity, heel horizontal and vertical accelerations, heel and floor angle and UCOF are used to avoid a slip on slippery surfaces (Swensen et al., 1992; Bunterngchit et al., 2000; Fong et al., 2005; Lockhart et al., 2007; Menant et al., 2009; Cappellini et al., 2010). The results appear to suggest that the participants in this experiment relied on foot slip to form the perceived slipperiness rating.

Conclusion

The results of the linear regression analysis indicated that the UCOF and ACOF were the major contributors to the outcome of perceived slipperiness rating. These two variables also determined whether a slip would happen at the shoe and floor interface. Therefore, the results appear to suggest that the participants in this experiment relied on foot slip to form the perceived slipperiness rating.

References

Bell, J. L., Collins, J. W., Wolf, L., Grönqvist, R., Chiou, S. S., Chang, W. R., Sorock, G. S., Courtney, T. K., Lombardi, D. A. and Evanoff, B., 2008, Evaluation of a comprehensive slip, trip, and fall prevention program for hospital employees, *Ergonomics* 51 (12), 1906–1925.

Bunterngchit, Y., Lockhart, T., Woldstad, J. C. and Smith, J. L., 2000, Age related effects of transitional floor surfaces and obstruction of view on gait characteristics related to slips and falls, *International Journal of Industrial Ergonomics* 25, 223–232.

Cappellini, G., Ivanenko, Y. P., Dominici, N., Poppele, R. E. and Lacquaniti, F., 2010, Motor patterns during walking on a slippery walkway, *Journal of Neurophysiology* 103 (2), 746–760.

Chang, W. R., Grönqvist, G., Leclercq, S., Myung, R., Makkonen, L., Strandberg, L., Brungraber, R. B., Mattke, U. and Thorpe, S. C., 2001, The role of friction in the measurement of slipperiness, Part 1: friction mechanisms and definition of test condition, *Ergonomics* 44, 1217–1232.

Chang, W. R., Courtney, T. K., Grönqvist, R., Redfern, M. S., 2003, *Measuring Slipperiness: Human Locomotion and Surface Factors*, (Taylor & Francis, London), ISBN 0-415-29828-8.

Fong, D. T. P., Hong, Y. and Li, J. X., 2005, Lower-extremity gait kinematics on slippery surfaces in construction worksites, *Medicine & Science in Sports & Exercise* 37 (3), 447–454.

Liberty Mutual Research Institute for Safety, 2012, 2012 Workplace Safety Index. *From Research to Reality*, 15 (3) available at http://www.libertymutual.com/researchinstitute.

Lockhart, T. E., Spaulding, J. M. and Park, S. H., 2007, Age-related slip avoidance strategy while walking over a known slippery floor surface, *Gait and Posture* 26, 142–149.

Menant, J. C., Steele, J. R., Menz, H. B., Munro, B. J. and Lord, S. R., 2009, Effects of walking surfaces and footwear on temporo-spatial gait parameters in young and older people, *Gait & Posture* 29, 392–397.

Swensen, E. E., Purswell, J. L., Schlegel R. E. and Stanevich, R. L., 1992, Coefficient of friction and subjective assessment of slippery work surface, *Human Factors* 34, 67–77.

THE STUDY ON EFFECTIVENESS OF BAMBOO CHARCOAL FOR IMPROVING WORK EFFICIENCY IN LEARNING ENVIRONMENT

Hiroki Nishimura[1], Zelong Wang[2], Atsushi Endo[2],
Akihiko Goto[3] & Noriaki Kuwahara[2]

[1]*ARC EDU CO. Ltd., Osaka, Japan*
[2]*Kyoto Institute of Technology, Kyoto, Japan*
[3]*Osaka Sangyo University, Osaka, Japan*

Because bamboo charcoal has high porosity, it is widely used in products to keep the room environment comfortable; e.g. wall material. We investigated the effectiveness of bamboo charcoal board for improving work efficiency in a learning environment. We set up two booths; one surrounded by bamboo charcoal boards on three sides, and the other surrounded by normal boards. In each booth, 11 subjects solved a numerical calculation, and we measured the time required and brain waves. The results suggest that in the booth of bamboo charcoal boards, subjects' brain activity became more active, but there were no significant differences in the task time or accuracy between the control and experimental conditions.

Introduction

It is known that Bamboo charcoal has outstanding characteristics, e.g. in eliminating odour, an absorption effect, far-infrared discharge, an electromagnetic wave interception effect, water quality purification, soil improvement, etc. Therefore, the usage of bamboo charcoal has expanded in various fields, such as environment and garments. In recent years, a product called the bamboo charcoal board, which is bamboo charcoal molded to a flat panel, was developed. We presented the effectiveness of bamboo charcoal boards on eliminating bad odor in a room by measuring physiological data and an inquiry survey (Takao et al. 2012). Also, we clarified that bamboo charcoal boads achieved much better impression for learners in a learning environment than normal boards from the inquiry survey (Isoda et al. 2013). In this study, we investigated the effectiveness of bamboo charcoal board for improving work efficiency in a learning environment by comparing to normal boards.

Method

Experimental protocol

Figure 1 shows the experimental protocol. We set up two booths; one was surrounded by bamboo charcoal boards on three sides, and the other was surrounded

(A)	Attach EEG Monitor	Move to Booth 1	Rest 1 in Booth1	Task 1 in Booth1	Move to Booth 2	Rest 2 in Booth2	Task 2 in Booth2
(B)	Attach EEG Monitor	Move to Booth 2	Rest 2 in Booth2	Task 2 in Booth2	Move to Booth 1	Rest 1 in Booth1	Task 1 in Booth1

| | 3 min. | Arbitrary time length | | 3 min. | Arbitrary time length |

Booth1: Control condition (No bamboo charcoal)
Booth2: Experimental condition (Bamboo charcoal)

Figure 1.   Experimental Protocol.

Control condition   Experimental condition

Figure 2.   Booths for the experiment.

by normal boards in the same way. In each booth, eleven subjects solved a numerical calculation, and we measured the time required and brain waves (EEG) of subjects. Half of the subjects followed the procedure (A) and the rest of the subjects followed the procedure (B) in order to eliminate order effects.

Booths

Figure 2 shows the booths that we used in the experiment. The size of each booth was 120 cm width × 120 cm depth × 180 cm height. A small table and chair were placed in the centre of each booth. The inside of one of the booths was covered by bamboo charcoal paper board produced by DAIWA ITAGAMI Co. Ltd. This paper board contained 5% bamboo charcoal micro-particles by weight.

Task

One sheet of the test paper was put on the table in each booth for the task of the subject. The task in each booth was 30 problems of two-digit number addition.

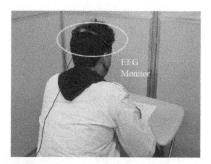

Figure 3. Portable EEG monitor used in the experiment.

We asked subjects to finish the task as quickly as possible. The time required for solving all problems was measured for the evaluation.

EEG monitor

The brain activities were measured by a portable EEG monitor developed and distributed by Digital Medic Inc. This portable EEG monitor obtained a correlation coefficient of 0.94 in a comparative study of measurement results with a full medical EEG monitor (NEC SanEi Synafit 1000). Figure 3 shows a subject with the portable EEG monitor attached.

Subjects

The number of subjects was 11. All subjects were male. Eight subjects were Japanese. Two subjects were Chinese. One subject was Korean. All subjects understood the experimental protocol and had ability to complete the tasks. Their age was from 23 years old to 47 years old. The average age of all subjects was 31.4 years old. The standard deviation of their age was 7.9 years.

Results

EEG monitoring

We compared α wave occupancy rate between the rest period and the task period. α wave occupancy rate was calculated by dividing the power of α wave by the sum of the powers of α wave and β wave during each period. Figure 4 shows the comparison of α wave occurrence rate between the rest period and the task period in the control condition. We found the tendency that α wave occupancy rate became lower in the task period than in the rest period. However, there was no significant difference between them. Figure 5 shows the comparison of α wave occurrence rate between the rest period and the task period in experimental condition. We found

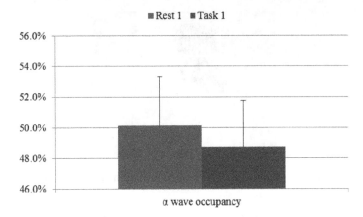

Figure 4.   Comparison of α wave ocupancy rate in control condition.

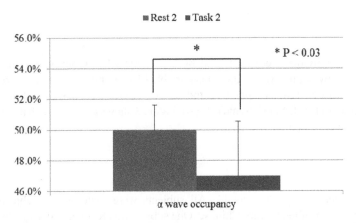

Figure 5.   Comparison of α wave ocupancy rate in experimental condition.

the tendency that α wave occupancy rate became lower in the task period than that in the rest period. There was the significant difference between them.

Accuracy rate of tasks

Figure 6 shows the comparison of the accuracy rate of the task in control condition (Task 1) and experimental condition (Task 2). In both conditions, the accuracy rate was 99.1%.

Time required to complete the task

Figure 7 shows the comparison of the time required to complete the task in control condition (Task 1) and experimental condition (Task 2). The time required to

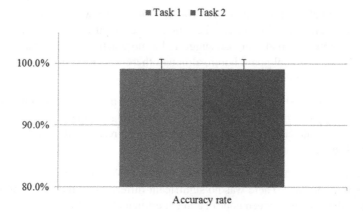

Figure 6.   Comparison of the accuracy rate of tasks in control condition and experimental condition.

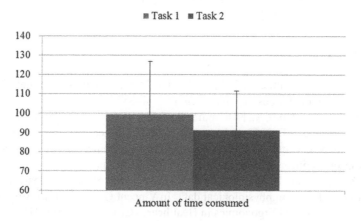

Figure 7.   Comparison of the time required to complete the task in control condition and experimental condition.

complete the task was shorter in the experimental condition than that in the control condition. However, there was no significant difference between them.

Conclusion

After discovering α blocking phenomenon (Adrian and Matthews, 1934), decreasing α wave occupancy rate is considered to increase the brain activity. Therefore, the result of EEG monitoring indicated that the subjects' brain activity was more active in bamboo charcoal booth than that in normal board because α wave occupancy rate became significantly lower in the task period than in the rest period in the experimental condition. In the control condition, there was the same tendency, but with no significant difference. The shorter times required to complete the task in the experimental condition also suggested a more active state of brain activity in the bamboo charcoal board booth, although there was no significant difference between the two conditions.

Isoda et al. (2013) showed that students in a private tutoring school preferred to study by themselves in a room with walls covered by black bamboo charcoal boards because they could concentrate more effectively. The result of this study dovetails with Isoda et al.'s work.

The accuracy rate was the same in both conditions. We think that this is because the task was too easy. There was no significant difference in the time required to complete the task in between control and experimental condition.

As the conclusion, we might say that the bamboo charcoal booth was effective for improving work efficiency in learning environment. However, we need to investigate the whether it comes from good chemical properties of bamboo charcoal, a colour effect, or something else.

References

Adrian, E.D., and Matthews, H.C., 1934, The Berger, rhythm: Potential changes from the occipital lobes in man. *Brain* 57, 354–385.

Isoda, S., Takai, Y., Goto, A., Nishimura, H., Takao, K., Ota, T., Kida, N., Kuwahara, N., and Hamada, H., 2013, Subjective Evaluation in Learning Environment with Different Wall Material, In Proceedings of Spring Conference of Japan Society of Kansei Engineering, Vol. 8th, Paper No. 6-3 (In Japanese).

Takao, K., Kuwahara N., and Kida N., 2012, Evaluation of bamboo charcoal effect to comfort based on physiological data, In Vincent G. Duffy (Ed.) Advances in Human Factors and Ergonomics in Healthcare (CRC Press), 209–216.

PREDICTORS OF INJURY AMONGST WHITE-WATER RAFT GUIDES WORKING IN THE UK

Iain Wilson, Hilary McDermott & Fehmidah Munir

Loughborough University, UK

Workers in the Outdoor Industry may be at risk of sustaining work-related injuries. This study aimed to establish a prevalence of injury and identify predictors of such injuries experienced by raft guides working in the UK. An online survey was distributed across providers based in the UK. Responses were gained from 126 participants. Lower back pain was the most prevalent chronic injury reported. Binary logistic regression identified that the need for recovery, river type and bilateral guiding were predictors of these injuries. However, none of the variables predicted the presence of knee pain. Specific working practices and conditions may need to be examined in future research.

Introduction

Work-related injury and ill-health are problems across a range of industries in the UK with musculoskeletal disorders (MSDs) being the most prevalent type of work-related injury or ill-health experienced in the UK (Health and Safety Executive, 2012). Anecdotal evidence suggests that there is a high prevalence of work-related injury and ill-health among those working in the Outdoor Industry (Adventure Activities Industry Advisory Committee [AAIAC], 2006) whereby activity leaders have reported worn knees, bad backs and surfer's ear. This is supported with empirical evidence whereby research examining the work-related health of Mountain Leaders in the UK has identified that work-related injury is problematic in the industry (McDermott & Munir, 2012). In addition, McDermott and Munir (2012) identified that work-related injuries were perceived as part of the job and did not necessarily prevent respondents from working.

Although injuries associated with white-water activities have been identified, a distinction between work-related and recreation-related injury has not been established (Wilson, McDermott, Munir & Hogervorst, 2013). The only known study which specifically looked at work-related injury examined back pain in raft guides working in the US (Jackson & Verscheure, 2006). This study identified manual handling as a primary contributor to back pain. However, it did not examine other injury sites such as the shoulder or knee which were identified as vulnerable areas of injury in a small qualitative study exploring work-related health amongst UK raft guides (Wilson, Munir & McDermott, 2013). Very little is known about the extent to which work-related injury is a problem in the white-water industry in the UK.

This study therefore aimed to establish a prevalence of injury amongst white-water raft guides working in the UK and identify predictors of such injuries.

Methods

The study used an online survey to collect data from a geographically diverse population. The research was subject to and in compliance with the requirements of the Loughborough University Ethical Advisory Committee in relation to research with human participants.

Utilising data from an initial qualitative study, an on-line, self-completed questionnaire comprising ten sections was developed using Likert-scale questions. Open ended questions and dichotomous questions were used to obtain data on a number of variables. Several validated instruments were also used to gain information on employee health and well-being. These included the Recovery Experience Questionnaire (Sonnentag & Fritz, 2007), the Need for Recovery Survey (Veldhoven & Broersen, 2003), Hypermasculinity Inventory (Mosher & Sirkin, 1984), the Utrecht Work Engagement Scale (Schaufeli & Bakker, 2004), the Outcome Exercise Expectations Scale-2 (Resnick, 2005), and the Nordic Musculoskeletal Questionnaire (Kuorinka et al., 1987).

The survey was distributed to all 577 (357 male) qualified raft guides registered in the UK via the British Canoe Union internal email. The email included information about the study, a link to the online survey and contact details for the researcher. In addition, all 45 providers identified through the Adventure Activities Licensing Authority were contacted by email. Of these the researcher was invited to visit eleven of the centres to recruit participants. This involved an initial email sent by the provider to their staff with details of the research, followed by a site visit by the researcher with paper copies of the survey.

Data were analysed using the Statistical Program for Social Sciences version 20. Tests of relationship were conducted using Chi Square and Spearman's Rank Correlation correlations. Binary Logistic Regression analyses were utilised to predict the outcome of injury or not.

Results

A total of 126 (90.7% male) completed surveys were included in analyses. The participants were most frequently Level 1 raft guides (N = 59) and were employed either full-time (N = 59) or freelance (N = 47). Demographics of the participants are presented in Table 1.

The prevalence of chronic injury reported was high with 89.7% of the sample reporting at least one chronic injury within the four weeks prior to completing the survey. The most frequently reported injury sites were the lower back

Table 1. Demographics of Participants.

	Mean ± Standard Deviation
Age	30.1 ± 9.7
Body Max Index	24.5 ± 3.8
Years of Experience as a Raft Guide	5.5 ± 6.2

(N = 87 [69.1%]), shoulder (N = 82 [65.1%]), upper back (N = 61 [48.4%]) and knee (N = 60 [47.6%]). The extent to which these injuries limited normal activity varied depending on the location of the injury. The lower back was the most limiting (N = 18 [20.7%]), whilst the upper back was least limiting (N = 2 [3.3%]). The shoulder (N = 10 [12.2%]) and knee (N = 7 [11.7%]) were comparable in the amount they limited daily activity.

Shoulder

The chi-square analyses identified a significant relationship between shoulder pain and river type with guiding on natural rivers only being associated with more injuries than guiding on man-made courses or a combination of the two ($\chi^2_{(2,n=126)} = 7.02$, p = 0.03). Positive relationships were also observed between shoulder pain and hours worked as a raft guide ($r_s = 0.18$, n = 126, p = 0.04), hours of physical activity ($r_s = 0.21$, n = 126, p = 0.02) and the need for recovery ($r_s = 0.20$, n = 126, p = 0.03). A negative relationship between age and shoulder pain was identified ($r_s = -0.20$, n = 126, p = 0.03).

The binary logistic regression analysis was performed with sex, age, BMI, years of experience, Need for Recovery, river type, hours worked in a physically active job, hours of physical activity and the vigor component of Work Engagement as predictor variables. The model significantly predicted the status of chronic shoulder injury ($\chi^2_{(10,n=126)} = 38.72$, p < 0.0005). Between 26.8% and 37.0% of the variance in chronic neck pain was accounted for in this model. Overall 75.8% of predictions were correct in this model. The model successfully predicted 91.4% of cases of shoulder injury occurring. However, it only successfully predicted 46.5% of no shoulder injury cases. Age, years of experience, Need for Recovery and guiding on both natural rivers and man-made courses, hours of physical activity and the vigor component of Work Engagement were significant predictors of chronic neck pain. The coefficients showed that increases in age (95% CI 1.03–1.59) and guiding on a mixture of natural rivers and man-made courses (95% CI 1.31–11.84) increase the odds of no shoulder injury occurring. However, increases in years of experience (95% CI 0.79–0.96), need for recovery (95% CI 0.66–0.98), hours of physical activity (95% CI 0.95–1.00) and vigor (95% CI 0.27–0.96) are associated with an increased chance of injury occurring.

Upper back

Using the chi-square and Spearman's rank correlation analyses, two significant associations were identified. Age was negatively associated with upper back pain ($r_s = -0.19$, n = 126, p = 0.03). The Need for Recovery and upper back pain was positively related ($r_s = 0.20$, n = 126, p = 0.03).

The binary logistic regression analysis was performed with sex, age, BMI, years of experience, Need for Recovery, the vigor component of WE, hours of non-physically active job, warm-up, personality and whether raft guides paddle on a single side (unilateral guiding) or paddle on both sides (bilateral guiding) as predictor variables. The model significantly predicted the status of chronic neck injury ($\chi^2_{(10,n=126)} = 30.44$, p = 0.001). The model accounted for between 21.8% and 29.0 % of variance in chronic upper back injury. The model successfully predicted 70.5% cases of upper back injuries and 71.4% of no upper back injury cases. Overall 71.0% of predictions were correct. Significant predictor variables of upper back pain from this model include sex, need for recovery and bilateral guiding. Female raft guides are associated with experiencing more upper back injuries (95% CI 0.04–0.77). An increase of one on the need for recovery scale is associated with an increased chance of injury being reported by a factor of 0.757 (95% CI 0.64–0.90). Finally, unilateral guiding is associated with an increased chance of upper back injury being reported (95% CI 0.162–0.927).

Lower back

The Need for Recovery was identified as being positively associated with chronic lower back pain ($r_s = 0.35$, n = 126, p < 0.01). This was the only variable found to be significantly associated with lower back pain in the chi-square and Spearman's rank correlation analyses.

The binary logistic regression analysis was performed with sex, age, BMI, years of experience, Need for Recovery, informed of the benefits of warming up, the dedication and absorption components of work engagement, river type, hours of physical activity and preferred side to guide as predictor variables. The model significantly predicted the status of chronic lower back injury ($\chi^2_{(12,n=126)} = 25.84$, p = 0.01). Between 18.8% and 26.4% of the variance in chronic upper back injury was explained in this model. A total of 88.2% of predictions for the occurrence of upper back injury were correct. However, only 43.6% of predictions for no injury were successful. Overall, 74.2% of predictions were successful. Need for recovery, guiding on man-made courses and the number of hours of physical activity were all significant predictors of chronic lower back pain. The coefficients showed that an increase in one unit of need for recovery and hours participating in physical activity are associated with decreased odds of no injury being reported by a factor of 0.72 (95% CI 0.59–0.88) and 0.98 (95% CI 0.96–1.00) respectively. Guiding on a man-made course may protect against lower back pain (95% CI 1.02–11.10).

Knee

The chi-square and Spearman's rank correlation analyses identified no significant relationships between any of the independent variables and chronic pain in the knee.

The binary logistic regression analysis was performed with sex, age, BMI, years of experience and hours worked as an raft guide as predictor variables. The model did not significantly predict the status of chronic knee injury ($\chi^2_{(5,n=126)} = 4.46$, $p = 0.485$). None of the variables significantly predicted chronic injury of the knee.

Discussion

The response rate was encouraging with 116 of 577 (20.1%) of qualified raft guides and 10 trainee raft guides completing the survey. The paper surveys appeared to be more successful, this could be due to the presence of the researcher which facilitated completion of the surveys.

There appears to be a high prevalence of chronic injuries experienced by raft guides. The prevalence of lower back pain is more than double (69% compared to 31%) than the global estimates (Hoy et al., 2012). It is possible that the prevalence of chronic injuries may be inflated as a result of the sampling technique. As participants are self-selecting, there is a chance that raft guides who had suffered a recent injury may have been more likely to complete the survey than their peers who had not experienced any issues.

Although the frequency of self-reported lower back pain is high, the prevalence of activity limiting lower back pain is similar to that reported globally (Hoy et al., 2012). However, this may be the nature of the sample population. McDermott and Munir (2012) suggested that mountain leaders in the UK continue working regardless of injury. Raft guides may also work through their injuries. This may be due to there being no paid sick leave because of the nature of freelance work. There is also an attitude which has been observed in workers in the outdoor industry whereby injury is perceived as just part of the job (McDermott & Munir, 2012; Wilson, Munir & McDermott, 2013).

As expected, the need for recovery was a significant predictor of chronic injury. This suggests that raft guides may be overworking themselves and should rest more frequently. Bilateral guiding contributed to upper back pain. This was surprising as previous work suggested that unilateral guiding was related to chronic back pain and in severe cases can lead to curvature of the spine (Wilson, Munir & McDermott, 2013). It may be that there is a risk of back pain whilst working as a raft guide, regardless of whether unilateral or bilateral guiding is practiced.

Working on a man-made course may reduce the risk of lower back injury because of the Statutory Operating Procedures in place to reduce manual handling conducted by the guide, for example, the use of trollies to move rafts and equipment. This is similar to what was reported by Jackson and Verscheure (2006) in the US whereby loading and unloading equipment onto a vehicle for transportation was related to back pain. At a man-made course, loading vehicles with equipment is not a common practice when compared to rafting on natural rivers.

Although this survey examined a variety of variables, it is clear that there are more to be measured. This is particularly evident in chronic knee pain, whereby almost half the sample population reported a chronic knee injury yet none of the included variables were significant predictors. Knee pain could be related to general wear and tear on the body from an active lifestyle, however, it could also be foot positioning in the raft which may result in rotational forces on the knee. This has not yet been empirically tested.

Although survey data is reliant on respondents providing accurate information, this was the most appropriate method to collect a large data set. This is the first study which has assessed the prevalence of work-related injury in the white-water rafting industry in the UK. This may be a representative sample of male guides in the UK rafting industry as almost a third of the 357 registered male guides have completed the survey. Although work in the outdoor industry is primarily male dominated, a very small sample of female raft guides has been represented in this study. The 9.5% female sample is much lower than the estimated 44% of the working population in the Outdoor Industry being female (SkillsActive, 2011).

Conclusions

It is clear that there is a high prevalence of injury amongst raft guides working in the UK, with 89% of participants reporting at least one recent chronic injury, however these injuries don't appear to be activity limiting. This study is consistent with other literature, identifying lower back pain as the most prominent issue experienced by raft guides. Insufficient recovery and guiding on natural rivers contribute to this pain. Specifically why working on natural rivers contributes to back pain requires further investigation, as it could be related to manual handling practices on the river bank as well as the nature of the water worked on. Although various variables were included in analysis, knee pain was not successfully predicted. This requires further investigation. The next step of this research is to continue collecting data across the working season to identify patterns of injury and working practices longitudinally. The findings of the longitudinal study will be used to improve current training advice and guidelines for raft guides in the UK. In addition, this longitudinal data may be beneficial to other seasonal occupations in the outdoor industry. Other work examining the forces on raft guides would be beneficial.

Acknowledgments

Many thanks go to the English White-Water Rafting Committee for their coopera-tion in this work.

References

Adventure Activities Industry Advisory Committee [AAIAC]. 2006, "Surviving a career in adventure activities." (AAIAC, UK), from: www.docs.hss.ed.ac.uk/education/outdoored/surviving_a_career_in_outdoor_activities.pdf.

Health and Safety Executive [HSE]. 2012, "Health and Safety Executive annual statistics report 2011/2012." (HSE, UK), from: www.hse.gov.uk/statistics/overall/hssh1112.pdf.

Hoy, D., Bain, C., Williams, G., March, L., Brooks, P., Blyth, F., . . . Buchbinder, R. 2012, "A systematic review of the global prevalence of low back pain." *Arthritis & Rheumatism* 68(6): 2028–2037.

Jackson, D. M., & Verscheure, S. K. 2006, "Back pain in whitewater rafting guides." *Wilderness & Environmental Medicine* 17(3): 162–170.

Kuorinka, I., Jonsson, B., Kilbom, A., Vinterberg, H., Bieringsorensen, F., Andersson, G., & Jorgensen, K. 1987, "Standardized Nordic questionnaires for the analysis of musculoskeletal symptoms." *Applied Ergonomics* 18(3): 233–237. doi:10.1016/0003-6870(87)90010-X

McDermott, H., & Munir, F. 2012, "Work-related injury and ill-health among mountain instructors in the UK." *Safety Science* 50(4): 1104–1111. doi:10.1016/j.ssci.2011.11.014

Mosher, D. L., & Sirkin, M. 1984, "Measuring a macho personality constellation." *Journal of Research in Personality* 18(2): 150–163.

Resnick, B. 2005, "Reliability and validity of the outcome expectations for exercise scale-2." *Journal of Aging and Physical Activity* 13(4): 382–394.

Schaufeli, W., & Bakker, A. 2004, "UWES Utrecht Work Engagement Scale preliminary manual (1.1st ed.)." (Occupational Health Psychology Unit, Utrecht University) from: www.wilmarschaufeli.nl/publications/Schaufeli/Test%20Manuals/Test_manual_UWES_English.pdf.

SkillsActive. (2011). The UK outdoor sector. A guide. Retrieved Sept 2, 2013, from www.skillsactive.com/images/stories/PDF/SkillsActive_OutdoorsSectorGuide_Final_06July2011.pdf.

Sonnentag, S., & Fritz, C. 2007, "The recovery experience questionnaire: Development and validation of a measure for assessing recuperation and unwinding from work." *Journal of Occupational Health Psychology* 12(3): 204–221. doi:10.1037/1076-8998.12.3.204

Veldhoven, M. V., & Broersen, S. 2003, "Measurement quality and validity of the "need for recovery scale." *Occupational and Environmental Medicine* 60(Suppl 1): 3–9. doi:10.1136/oem.60.suppl_1.i3

Wilson, I., McDermott, H., Munir, F., & Hogervorst, E. 2013, "Injuries, ill-health and fatalities in white water rafting and white water paddling." *Sports Medicine* 43(1): 65–75.

Wilson, I., Munir, F., & McDermott, H. 2013, "It's just part of the job!" Raft guides working with back pain. In Anderson, M. (ed.) *Contempory Ergonomics and Human Factors 2013*, (Taylor and Francis, London), 77–78.

HF IN HIGH VALUE MANUFACTURING SYSTEMS

USING SHERPA TO INFORM THE DESIGN OF INTELLIGENT AUTOMATION FOR TIG WELDING

William Baker

Industrial Ergonomics and Human Factors Group, School of Engineering, Cranfield University, UK

Due to the inability of automation to adapt to unanticipated situations, designers have tended to automate the simple task components, while leaving the more complex processes under human control. The study presented in this paper used a Systematic Human Error Reduction and Prediction Approach (SHERPA) to identify human errors associated with a complex manual tungsten inert gas (TIG) welding task, in order to provide design recommendations for an 'intelligent' automated system. Task performance was assessed across a range of different skill levels to understand the adaptive capability of individuals to overcome potential errors and maintain performance. Results showed that recommendations for automation design could be developed through human error analysis.

Introduction

Many high-value manufacturing processes continue to be performed manually by highly-skilled operators (for example, tungsten inert gas (TIG) welding). However, as the manufacturing industry continues to face a shortage of skilled operators, organisations are regularly considering automated alternatives as a means to increase productivity without compromising quality. Rapid developments in technology have enabled automation to become increasingly more complex, capable and 'intelligent'; performing functions that previously could only be carried out by human operators (Parasuraman, 2000). Automation promises greater efficiency and a reduction in human errors, and yet in many cases, these benefits are not delivered (Lee and Seppelt, 2009).

Historically, automation has been designed and implemented with a focus on the technical aspects (for example, algorithms and actuators), without fully considering the human factors issues associated with such systems. Such a technology-centred approach has led to greater system autonomy and authority, with almost no progression in feedback mechanisms for system status and behaviour (Woods, 1996). The problem with this type of 'strong and silent' automation (Norman, 1990) is that under normal operating conditions the task functions are performed appropriately; however, it lacks the flexibility to handle unanticipated situations. During manual task performance, operators rely of feedback for monitoring actions, as well as detecting and correcting for potential errors; yet the importance of providing adequate feedback in automation has not been recognised in the design of automated systems.

Often automation does not simply relieve the operator of the tasks, but rather it changes the nature and structure of the task components (Lee and Seppelt, 2009). Designers tend to automate the simple task components while leaving the complex aspects of the task to the operators (Bainbridge, 1983). For industry to realise the potential benefits of automation, a greater understanding is required of how these complex manual tasks are performed so that appropriate, well-applied and intelligent automated systems can be developed. The aim of this research was to use a Systematic Human Error Reduction and Prediction Approach (SHERPA) to identify and classify the human errors associated with a complex manual task, in order to provide design recommendations for an intelligent automated alternative.

Methodology

A tungsten inert gas (TIG) welding task lasting approximately 5 minutes was selected for this study. This selection was made on the basis that TIG welding represents one of the most versatile and extensively used welding techniques in many manufacturing industries (for example, aero-structure assembly).

Ten participants were recruited to take part in the study. Based on the level of welding experience, the participants were categorised accordingly:

- 'Skilled' (n = 3): formal training and routinely performed TIG welding tasks as part of work practices.
- 'Semi-skilled' (n = 4): formal training but did not routinely perform TIG welding tasks.
- 'Unskilled' (n = 3): no formal training or experience of TIG welding (training provided prior to data collection).

An additional independent 'skilled' welder served as a SME to assess the reliability and validity of the task analysis. All participants received information concerning experimental procedure and informed consent was obtained prior to data collection. The study was approved by the Cranfield University Science and Engineering Research Ethics Committee.

Prior to the data collection, the participants were provided the opportunity to familiarise themselves with the welding equipment and experimental procedures. During this time, 'unskilled' participants received additional training, consisting of video-based instruction and practical demonstration. Data collection sessions lasted approximately two hours, during which the participants were observed while performing the welding task. Once the task was completed, participants were asked to describe their actions in recurrent interview sessions. Data collection and analysis was progressive and iterative, with observations and interviews conducted until redundancy was achieved (i.e. until it was evident that no further or new insights

Table 1.   Hierarchical Task Analysis (HTA) for manual TIG welding task.

1   Setup equipment.
2   Simulate laying a weld.
3   Lay tacks.
4   Lay weld.
....
4.6   Manipulate the filler rod.
4.6.1*   Move (stroke) the filler rod in and out of the weld pool.
....
5   Make post weld inspections.

*Task component represented in the SHERPA analysis (Table 2).

were emerging). The data were analysed using a Systematic Human Error Reduction and Prediction Approach (SHERPA) technique. The first step in conducting the SHERPA analysis was to produce a Hierarchical Task Analysis (HTA), from which error classifications were assigned to individual task steps in order to identify credible errors associated with human task performance (Stanton *et al.,* 2005).

Results

The HTA represented the overall task as a hierarchy of goals, operations (actions executed by the individual to achieve the goals) and plans (the sequence in which the operations are performed). For the manual TIG welding task, the HTA consisted of five sub-goals and 47 operational task steps. Each sub-goal represented a distinct task process and these have been summarised in Table 1.

For the SHERPA analysis, each of the lower level task steps (operations) identified during the HTA were assigned individual error modes (Table 2). These error modes were categorised according to the following SHERPA error taxonomy:

- Action: e.g. depressing the welding pedal, moving weld torch
- Retrieval: e.g. obtaining information from a manual or task cues
- Checking: e.g. conducting a procedural check
- Selection: e.g. choosing one alternative over another
- Information communication: e.g. talking to another party

Each credible error (i.e. judged by a subject matter expert to be possible) was described fully within the SHERPA, including the immediate consequences for task performance should the error occur (Stanton and Baber, 2002). The remaining stages of the SHERPA established the possibility for error recovery, the probability of the error occurring, the criticality of the error and illustrative remedies (Stanton and Baber, 2002). The error probability was categorised according to the skill classification of the participants (skilled, semi-skilled and unskilled) and the illustrative remedies were used to provide design considerations relating to automation. These remedial strategies included changes to equipment design, procedural recommendations for error recovery and technology alternatives for automation (for example, thermal imaging).

Table 2. Extract of SHERPA analysis applied to a manual TIG welding task.

Task Step	Error Mode	Error Description	Consequence	Recovery	Probability			Criticality	Considerations for Automation
					Skilled	Semi-Skilled	Unskilled		
4.6.1	A5	Filler rod moved into the weld pool at the incorrect angle.	Filler rod contacts electrode = weld contamination.	None.	M	M	H	!	- Selection of gas nozzle (ensure adequate access to weld). - Filler rod attached to torch (maintain correct filler-to-electrode angle). - Adjustable orientation of filler rod attachment (depending on weld specifications).
R3;	A2i	Incomplete retrieval of information regarding weld characteristics. Filler rod introduced too early.	Weld pool has not properly formed (insufficient weld heat), filler material causes weld pool to cool = lack of fusion / penetration.	Immediate.	L	L	M		- Sensor to monitor weld pool and heat distribution (e.g. thermal imaging). - Filler rod feed mechanism based on state of weld pool, not current. - Increase current to re-form weld pool (based on state of weld pool).
R3;	A2ii	Incomplete retrieval of information regarding weld characteristics. Filler rod introduced too late.	Weld area becomes too hot, causing weld pool to 'fall' through the weld joint = excessive penetration (burn-through).	None.	L	M	H	!	- Sensor to monitor weld pool and heat distribution (e.g. thermal imaging). - Filler rod feed mechanism based on state of weld pool, not current.

Discussion

The HTA not only formed the basis to the SHERPA but it was also invaluable in removing some of the uncertainty from the complex task processes. This is particularly important as there is often a limited understanding of how or why particular task processes are conducted and failures of automation may occur because the ability of the human operator has been underestimated (Lee and Seppelt, 2009). Using the HTA as the initial step of the SHERPA provided a comprehensive and structured representation of the task. This would enable designers to consider the human performance associated with each task component and ensure appropriate types and levels of automation are considered. In addition to shifting responsibility for task components away from the human, automation may change the nature of the task by adding or removing task steps (Miller and Parasuraman, 2003). Consequently, multiple alternative hierarchies of the original task should be considered when designing automated systems. A SHERPA analysis could be used to form a baseline against which the error probability and criticality associated with alternative task hierarchies could be compared to determine the relative impact of additional or fewer tasks steps. Furthermore, groups of operations identified in the HTA could be used to establish which task components are inherently more error-prone than others.

Considering the TIG welding task, the information provided by the SHERPA has been summarised in Figure 1 to illustrate the error criticality and the types of error associated with different task components. A total of 105 potential errors were identified, consisting of: 66 action errors, 15 checking errors, 10 retrieval errors and 14 selection errors. The absence of communication errors was due to the nature of the task (no requirement for the participants to engage with other personnel). The SHERPA error description (Table 2) provided a detailed account of the individual errors and the potential causal links between different error modes (for example, 'selection' errors were often not recovered due to 'checking' errors). This was evident during the setup of the welding equipment (Figure 1), where many selections were performed however, there were minimal checks to confirm these selections. The consequences of 'selection' errors were often only realised during the later stages of the task (for example, when laying the weld), at which point there was little opportunity for recovery. Using the SHERPA to identify the different types of errors could inform the HTAs of alternative automated systems. For example, adding task steps to confirm equipment selections could reduce the probability of errors occurring and improve the recovery potential for errors associated with other task components.

Additionally, the inability to accurately identify and retrieve information regarding the current state of the weld ('retrieval' errors) often resulted in the incorrect execution of an action ('action' errors). The relationship between the retrieval of information and the execution of actions, highlights one of the main limitations of current automation; the inability to adapt to system variations. Understanding type of information that forms the basis for deciding a particular action could be used to establish design recommendations for automation. Intelligent systems should

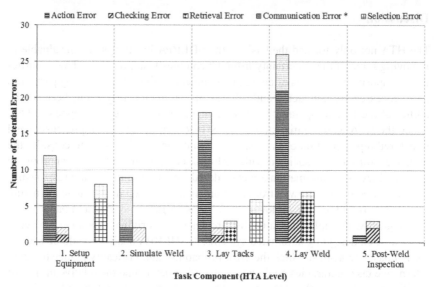

Figure 1.   Types of potential errors associated with manual TIG welding *(black pattern = critical; grey pattern = non-critical)*.

not only be able to perform the required actions but should implement the most appropriate response in order to maintain task performance and minimise errors.

For the TIG welding task, the error consequences were grouped into six main categories (for example, weld contamination or excessive penetration). Using the error analysis provided in the SHERPA, the reasons behind each consequence were collated (for example, weld contamination could result from incorrect grinding of the electrode, unclean parent metal, incorrect filler rod angle). This information could be used as a troubleshooting guide to provide feedback for human operators acting in a supervisory role to automation as well as for designers trying to control specific task parameters to improve system performance. During the TIG welding task, the majority of errors (68%) were considered to have critical consequences many of these demonstrated little opportunity for recovery once the error had occurred (for example, once the filler rod touches the electrode, the weld is contaminated and the process has to be stopped). When considering the design of automation, task components with critical errors should be prioritised, especially those associated with a 'high' probability. Automation design should consider whether the error could be eliminated by removing the task step from the HTA, and if not; what measures could be introduced to minimise the residual error probability after automation (for example, performing checks to minimise 'selection' errors).

Perhaps unsurprisingly, the error probability associated with the welding task was lowest for the 'skilled' participants and increased consistently with decreasing levels of skill (from the 'semi-skilled' to the 'unskilled' participants). Unskilled task performance was associated with the high probability of 'selection', 'retrieval' and 'action' errors, representing an inability of 'unskilled' participants to obtain the necessary information required to make informed decisions as well as a limited capacity to accurately perform co-ordinated actions. The 'semi-skilled' participants however, were able to perform the required actions but lacked the ability to adequately retrieve the necessary information to make appropriate adjustments and maintain task performance. Finally, the 'skilled' participants were able to make continual adjustments in response to subtle changes in the system characteristics to minimise the potential for errors. As mentioned by Norman (1990), automation seems to represent an intermediate level of intelligence where systems are able to perform actions previously controlled by humans but cannot adapt to abnormalities and maintain performance. This would best describe the level of performance associated with the 'semi-skilled' participants. Previous studies have found that the use of 'unskilled' participants can be useful in eliciting additional information from 'skilled' participants (Caird-Daley *et al.*, 2013). By investigating task performance across a range of skill levels, the current study was able to identify the full extent of potential errors associated with the welding task ('unskilled' participants). Additionally, the 'skilled' participants provided insight into how potential errors could be avoided, and demonstrated the compensatory ability of human operators.

Conclusions

This study demonstrated that a SHERPA analysis can be used to provide design recommendations for the automation of complex manual tasks. Although time-consuming, the initial HTA provided a baseline to compare the error probability and criticality of proposed automation alternatives. The analyst applying the SHERPA should have knowledge of the skills and procedures of the task (in this context, TIG welding) and potential automation technologies. The SHERPA was expanded to include different classifications of skill levels. Error probability was found to increase as the level of participant skill decreased (the highest error probability was associated with 'unskilled' performance). Each level of skill classification enabled different aspects of performance to be assessed using the SHERPA (for example, the 'unskilled' participants enabled all the potential errors to be captured, while the 'skilled' participants provided insight into error recovery and adaptability). Future work will consider the application of the SHERPA throughout the development of an automated TIG welding system to establish the practical validity of the initial design recommendations. Additional human error methodologies will be considered as a comparison, to quantify the benefits and limitations of the SHERPA for automation design.

Acknowledgements

This was an EPSRC Centre for Innovative Manufacturing in Intelligent Automation study. Many thanks to John Thrower for additional technical support.

References

Bainbridge, L. 1983, Ironies of Automation. *Automatica,* 19 (6), p. 775–779.

Caird-Daley, A., Fletcher, S. R. and Baker, W. D. R. 2013, *Automating Human Skill: Preliminary Development of a Human Factors Methodology to Capture Tacit Cognitive Skills.* Proceedings of the 11th International Conference on Manufacturing Research (ICMR), Cranfield, UK.

Lee, J. D. and Seppelt, B. D. 2009, Human Factors in Automation Design. In Nof, S. Y. (eds). *Handbook of Automation, 2009,* Springer, p. 417–136.

Miller, C. A. and Parasuraman, R. 2003, *Beyond Levels of Automation: An Architecture for more Flexible Human-Automation Collaboration.* Proceedings of the Human Factors and Ergonomics Society 47th Annual Meeting, Santa Monica, CA.

Norman, D. A. 1990, The Problem with Automation: Inappropriate Feedback and Interaction, not 'Over-Automation'. *Philosophical Transactions of the Royal Society of London,* B327, p. 585–593.

Parasuraman, R. 2000, Designing Automation for Human Use: Empirical Studies and Quantitative Models. *Ergonomics,* 43 (7), p. 931–951.

Stanton, N. A., Salmon, P. M., Walker, G. H., Baber, C. and Jenkins, D. P. 2005, *Human Factors Methods: A Practical Guide for Engieering and Design.* (Ashgate Publishing, England).

Stanton, N. A. and Baber, C. 2002, Error by Design: Methods for Predicting Device Usability. *Design Studies,* 23 (4), p. 363–384.

Woods, D. D. 1996, Decomposing Automatin: Apparent Simplicity, Real Complexity. In Parasuraman, R. and Mouloua, M. (eds). *Automation and Human Performance: Theory and Applications, 1996,* Erlbaum, p. 3–17.

HUMAN FACTORS ANALYSIS IN RISK ASSESSMENT: A SURVEY OF METHODS AND TOOLS USED IN INDUSTRY

Nora Balfe & M. Chiara Leva

Centre for Innovative Human Systems, Trinity College Dublin, Ireland

This paper outlines results of 15 interviews conducted to establish the methods and tools currently used to support risk assessment in industry. The interviews covered general risk assessment and looked at human factors (HF) tools and methods in use, considering the representation of the system under analysis and tools to identify and analyse human error. The interview results show that only five companies used any form of structured technique to analyse HF, and two of these companies had specific human factors teams. This highlights a gap in risk assessment, in stark contrast to the high attribution of major accidents to human error. Possible reasons for this gap and the need to better include guidance on HF assessment in applicable standards are discussed.

Introduction

The contribution of human performance, which has its roots in human and organisational factors (HOF), to major accidents is often cited as between 70% and 80% (e.g. Shappell & Wiegmann, 1997) and human and organisational factors are understood to have a dominant influence on safety in the offshore oil and gas industry with approximately 80% of accidents attributable to unanticipated actions of people during operations and maintenance activities (Bea, 1998). These statistics make clear the importance of accounting for human and organisational factors in risk assessments in order to manage safety. Major accidents, including Piper Alpha, Chernobyl and Three Mile Island have all had human factors described as a root cause (Gordon, 1998).

Direct human factors contributions to accidents and incidents tend to be described as human errors (Reason, 1990). Various factors including biological factors (e.g. age, gender, size, handicaps), physical condition, mental condition, competence, and personality may all contribute to human performance. All of these can be managed to some degree by the organisation, and the failure to do so can be regarded as an organisational contribution to an accident. Organisational factors may not be as readily identified during an accident investigation, as they tend to require a deeper level of analysis than is always performed following an accident/incident. Leveson (2011) suggests that organisational factors are frequently not identified because traditional accident investigation models stop once an immediate causal event has been identified. Therefore, the factors leading an operator towards an error are not analysed. Organisational factors can include management commitment to safety, training, communications, stability of the workforce, supervison and teamwork, among others (Gordon, 1998).

The primary approach to risk assessment uses established tools and methods developed to identify and quantify technical risks, such as HAZOP, risk matrices, FMEA, etc. These tools were originally developed to account for technical risks, and not for human performance, although some attempts have been made to incorporate human factors (e.g. human HAZOP; Kirwan, 2005). The limitations of traditional quantitative risk assessments methods are illustrated by Einarsson & Brynjarsson (2008) who describe an actual fire at a chemical storage facility; the facility was retrospectively analysed through a quantitative risk assessment (QRA) including a HAZOP for the entire plant. The results of the QRA indicated that the plant was safe by international standards. Einarsson & Brynjarsson suggest that this result could only have occurred if the significant human and organisational factors at work in the accident were not sufficiently captured by the risk assessment.

Human factors approaches to risk assessment tend to start with an analysis of the human activities, typically using a form of task analysis. Without a good representation of the human tasks and activities within the system it is difficult to identify with any confidence where human errors can occur. There are then a number of tools for identification of human errors, including SHERPA (Embrey, 1986), TRACEr (Shorrock & Kirwan, 2002), and human HAZOPs (Kirwan, 2005). Some analyses may base control measures and design changes on the results of the human error identification phase, but human reliability assessments (HRA) go on to quantify the probability of an error. The benefit of HRA is that human error mechanisms can be analysed and prioritised within a systems context, allowing human factors specialists to identify the critical areas for improvement (Kirwan, 2005). Again, there are a number of techniques available to conduct a HRA, including THERP (Swain & Guttman, 1983), HEART (Williams, 1985), CREAM (Hollnagel, 1998), and a wide range of nuclear specific tools (e.g. SPAR-H (Gertman et al, 2004), ATHEANA (USNRC, 2004), HFW (Embrey & Zaed, 2010) and MERMOS (Pesme et al., 2007).

Human factors specialists may also use a wide variety of methods relating to mental workload assessment, equipment and control room design, fatigue management, user interface design, etc. to design systems for optimal human performance, but these are not considered to be part of the risk assessment process here.

The approaches and tools briefly outlined above are known within the human factors domain, and HRA in particular is widely used in the nuclear industry, but the level of adoption of techniques to identify and mitigate human and organisational risks in wider industry (i.e. those who may not have internal human factors teams) is not known. These interviews sought to establish the risk assessment processes of the companies surveyed and the incorporation of human and organisational factors within that.

Study methodology

The interviews were undertaken as a collaborative effort as part of the Innovation through Human Factors in Risk Analysis and Management (InnHF) project. The 15 semi-structured interviews were conducted across five different European countries using a standard interview template. The majority of the interviewees worked in the processing industry, but interviews were also undertaken in the transport, energy production, and manufacturing domains. Participants were primarily chosen from the partner companies in the project and other contacts. 11 of the interviewees held positions in risk analysis or safety management, two were human factors specialists, and two were production managers.

The interview questions were drafted based on the risk assessment process as described in ISO: 31000 and covered both design and operational/in-service risk assessments. The objectives were:

* To collect information about the risk assessment process used in the companies
* To collect information about the data and tools available to companies to perform the risk assessment
* To identify how human and organisational factors are accounted for in the process and tools

Only the data relevant to the third objective is presented in this paper.

The interview was divided in to three sections:

* The background to the risk assessment
* The risk assessment process
* Management of risk

Companies were approached to determine their willingness to take part, usually either the safety department or, where it existed, the human factors department were contacted. The interviews were undertaken face-to-face if possible, run by the researchers on the InnHF project, all of whom had previously been trained on interview techniques. If face-to-face interviews were not possible, the interviews were conducted over the telephone. All interviews were recorded. The countries in which the interviews were undertaken included Italy (eight interviews), Ireland (three interviews), Serbia (two interviews), UK (one interview) and France (one interview). Following the interviews, the recording was transcribed or detailed notes were drawn up. These documents were used for the analysis, which was a thematic-led qualitative analysis.

Results

The topics of relevance to human factors within the interviews were:

- Standards used to support risk assessment
- Representation of human tasks and activities within the system
- Human factors methods and tools used in risk assessment

The companies used a wide variety of standards to support risk assessment and safety management, including internal standards, industry standards, national standards (or regulations), and international standards. OHSAS 18001, ISO 9001, and ISO 14001 were the most common international standards used by the companies surveyed. However, none of the interviewed companies mentioned any human factors standards as supporting their risk assessment and, despite stating that human factors are important, there is a paucity of guidance in the standards on incorporating human factors within safety assessments. This shows a gap in the standards, which could either be filled by a specific human factors risk assessment standard, or perhaps more beneficially by providing more and better human factors guidance in the existing risk assessment standards which have already been adopted by companies. The key results for task representation and human factors (HF) risk assessment are shown in Figure 1.

Of the 15 companies surveyed, only six had some form of representation of human tasks and activities. All of these were based around Task Analysis. This illustrates

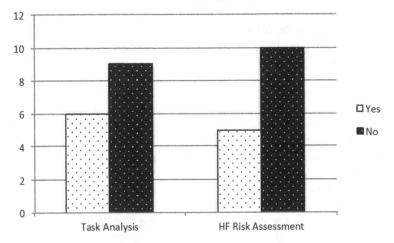

Figure 1. Numbers of companies undertaking HF assessments.

that Task Analysis is a broadly accepted approach to representation of human tasks and activities within a system, but that the general idea of using this kind of information in a risk assessment is not as common as might be hoped. Without a formal representation of the human tasks, the risk assessment cannot have a firm structure on which to base its assessment of the risk associated with human factors. Although not using a structured technique, several organisations did include operations and maintenance staff in HAZOP or other risk assessment meetings in order to try to bring their expertise and knowledge to the assessment.

"The main tool is attendance of the people to the risk assessment meeting, they way they can provide their experience to the process"

However, this method of including operational expertise to cover human factors is limited for several reasons, First, as mentioned by one of the interviewees, the input of the operators may be hit and miss:

"We encourage the supervisors to visit the site and they do risk assessment and discuss it with the people involved, but in practice we suspect that this is done in an office and is copied from the previous one"

This may be because, as in the experience of this organisation, the operators are not always consulted as expected, or because the operational experiences of the operators themselves are not sufficient to identify all possible failure modes. The second reason that this approach is limited is the lack of structure involved. It relies entirely on the operators raising issues as they occur to them during the meeting without any attempt to ensure that the full range of potential human activities is accounted for.

All companies interviewed used a risk assessment method of some type, with risk matrices being the most common (used by eight companies). This is likely because risk matrices are relatively simple to understand and apply. More complex methods, such as FMEA and Fault Trees or Event Trees, were used by only four companies. In contrast to the wide (if varied) usage of risk assessment tools, only five companies reported using specific human factors tools for risk assessment. The analysis tools used in this sample included:

- A second generation HRA method
- TRACEr
- HEART
- Method Statements – modified to capture human factors risks
- Incident reporting/recording
- World Class Manufacturing approaches for human factors
- A company specific human performance modelling tool

Only the two companies with human factors specialists used known human factors tools; all the remaining companies used modified or general tools to address human factors risks. One company also stated that they were too focused on occupational health and safety risks related to manual handling:

"That's all we seem to do, and that's all every company seems to do, is concentrate on manual handling whereas there's a lot more going on. We miss an awful lot."

In contrast to the representation methods, there were a wide variety of tools and methods used for human factors risk assessment (seven approaches across five organisations). This could be due to differing requirements of different industries or it could represent a level of dis-satisfaction with current available analysis methods, which means that no single approach has yet emerged as best practice in the area. Another possible reason is that risk assessment companies do not have specific expertise in human factors and therefore are not comfortable with its inclusion:

"If you are expert in natural or technological risks you often need the expertise from someone else to deal with human factors"

This again points to a need for more guidance for risk professionals in how to incorporate human factors in their assessments. It is not realistic that all companies will have access to human factors expertise to conduct their risk assessment, although companies with a particularly high level of risk linked to human error should continue to use specialists. The high use of risk matrices as a risk assessment tool suggests that many companies are most comfortable with a simplistic approach to risk assessment. There is therefore a need for simple methods or approaches that can be easily described and applied as part of the risk assessment process.

Conclusions

In terms of human and organisational factors, there was very low usage of human factors tools and methodologies among the interviewed companies. Only five companies specifically included HOF in their risk assessment process, and since two of the interviewees were specialists in this area, only three companies without a specialist team included HOF in their analysis. Despite the small sample size, this points to a clear gap in risk assessment across all industry domains, particularly given the high prevalence of human and organisational factors in major accidents. This result reinforces the view that human factors are still not adequately addressed in risk assessment and safety management (e.g. Taylor, 2012). The recommendation on the basis of this research is to support better analysis of human and organisational factors risks by providing more, better, and simple guidance on the identification and analysis of human error in the most relevant standards, building on the existing guidance available (e.g. Henderson & Embrey, 2012; Widdowson & Carr, 2002). The IEHF can assist in this by collecting together the current best practices, developing or identifying appropriate training, and promoting these in collaboration with other professional institutions.

Acknowledgements

This research was funded by Marie Curie Initial Training Networks under the FP7 Agreement, project ID 289837. The data was collected by a number of researchers on this project based in Ireland, Italy and Serbia as part of the Innovation through Human Factors in Risk Analysis and Management, Work Package 4 (InnHF, 2013). The data presented in this paper was partially analysed by Geng Jie and Evanthia Giagouglou.

References

Bea, R. 1998. Human and organisational factors: engineering operating safety into offshore structures. *Reliability Engineering & System Safety* 61(1): 109–126.

Einarsson, S. & Brynjarsson, B. 2008. Improving human factors, incident and accident reporting and safety management systems in the Seveso industry. *Journal of Loss Prevention in the Process Industries*, 21(5): 550–554.

Embrey, D. E. 1986. *SHERPA: A systematic human error reduction and prediction approach.* International Meeting on Advances in Nuclear Power Systems: Knoxville, Tennessee.

Embrey, D. E. and Zaed, S. 2010. *A set of computer-based tools for identifying and preventing human error in plant operations.* Proceedings of the 6th Global Conference on Process Safety: American Institute of Chemical Engineers, San Antonio USA.

Gertman, D., Blackman, H., Marble, J., Byers, and Smith, C. 2004. *The SPAR-H human reliability analysis method.* NUREG/CR-6883. Idaho National Laboratory, prepared for U. S. Nuclear Regulatory Commission Office of Nuclear Regulatory Research Washington, DC 205555-0001.

Gordon, R. P. E. 1998. The contribution of human factors to accidents in the offshore oil industry. *Reliability Engineering and System Safety*, 61(1): 95–108.

Henderson, J. & Embrey, D. 2012. Guidance on quantified human reliability analysis. (Energy Institute, London).

Hollnagel, E. 1998. Cognitive Reliability and Error Analysis Method. (Elsevier, Oxford).

InnHF. 2013. Categorisation of current issues – Human and organizational factors. Internal project report. www.innhf.eu.

Kirwan, B. 2005. Human reliability assessment. In J.R. Wilson & N. Corlett (Eds.) *Evaluation of Human Work*, 3rd edition, (CRC Press, Boca Raton): 833–876.

Leveson, N. 2011. Engineering a safer world: Systems thinking applied to safety. (MIT Press, Cambridge MA).

Pesme, H., Le Bot, P., & Meyer, P. 2007. *Little stories to explain Human Reliability Assessment: A practical approach of the MERMOS method.* In Proceeding of the Human Factors and Power Plants and HPRCT 13th Annual Meeting, Monterey, CA, USA.

Reason, J. 1990. Human Error. (Cambridge University Press, New York).

Shappell, S. A., & Wiegmann, D. A. 1997. A human error approach to accident investigation: The taxonomy of unsafe operations. *The International Journal of Aviation Psychology*, 7(4): 269–291.

Shorrock, S. T., & Kirwan, B. 2002. Development and application of a human error identification tool for air traffic control. *Applied Ergonomics* 33(4): 319–336.

Swain A. D. and Guttmann H. E. 1983. *Handbook of human reliability analysis with emphasis on nuclear power plant applications.* US Nuclear Regulatory Commission), Washington, DC. NUREG/CR-1278.

Taylor, J. R. 2012. Lessons learned from forty years of HAZOP. *Loss Prevention Bulletin,* 227: 20–27.

US Nuclear Regulatory Commission (USNRC). 2000. *Technical Basis and Implementation Guidelines for A Technique for Human Event Analysis (ATHEANA).* NUREG-1624. Division of Risk Analysis and Applications. Office of Nuclear Regulatory Research, Washington DC.

Widdowson, A. & Carr, D. 2002. Human factors integration: Implementation in the onshore and offshore industries. (HSE, Norwich, UK).

Williams, J.C. 1985. *HEART – A Proposed Method for Achieving High Reliability in Process Operation by means of Human Factors Engineering Technology.* In Proceedings of a Symposium on the Achievement of Reliability in Operating Plant, Safety and Reliability Society, Southport, UK.

INVESTIGATING THE IMPACT OF PRODUCT ORIENTATION ON MUSCULOSKELETAL RISK IN AEROSPACE MANUFACTURING

Sarah R. Fletcher & Teegan L. Johnson

Industrial Ergonomics and Human Factors Group
Centre for Advanced Systems
School of Engineering
Cranfield University

Like most high value aerospace components, aircraft wings are held in relatively fixed positions during manufacture. Their internal systems are installed by human operators entering into the wing box cavity as it is held at a static horizontal position. These are confined and cramped conditions that force workers to adopt awkward working postures which increase the risk of physical illness and hamper task performance. To address this problem, a study has been conducted to investigate a twofold idea for changing product orientation and accessibility: a) keeping one side of the structure open until a later stage of the production process and b) turning the product to improve access.

Introduction

It is always preferable to avoid moving very large or heavy components during manufacture given that sizeable structures require greater physical space and logistical efforts. In safety-critical industries such as aerospace manufacture it is also important to avoid surplus movements of the product as these might increase the risk of structural stress or damage. So, in these circumstances, human operators need to bring their tools, materials and smaller sub-assembly components to the product and work on it in a static position which often brings ergonomic challenges as it forces operators to adopt awkward physical postures and positions. In the aerospace manufacturing sector, a prime example occurs in the long and labour-intensive installation of internal systems into aircraft wings (fuel, hydraulic, electrical and electronic) as traditional structural designs and production methods restrict physical access and mobility. Before systems installation work begins wings are already built as sealed units, with the inside area subdivided into small cavities by spars and rib to create confined spaces. Entry into each of the small cavities is only possible via a number of horizontal access holes (purpose-built for manufacture and post-production maintenance) which are located along the bottom side of the wing box. Internal access across the cavities is then available via through vertical access holes positioned in the internal ribs. As the wing boxes are held at a fixed height and horizontal orientation throughout this phase of work operators need to perform their systems installation tasks with the 'closed box' in this static position.

Confined space is obviously more of a problem in the shallower wing boxes of smaller aircraft but is a typical condition to which assembly operators are exposed. It is therefore likely to increase their risk of musculoskeletal injury (Veasey, Craft McCormick, Hilyer, Oldfield, Hansen, and Krayer, 2005; Gallagher, 2005) and, at least in the short term, consistently raise levels of discomfort and postural stress (Mozrall, Drury, Sharit, and Cerny, 2009). The risks are further augmented when systems installations tasks involve forceful and repetitive exertions (Punnett, Gold, Katz, Gore and Wegman, 2004).

Musculoskeletal disorders (MSDs) have been more highly associated with unhealthy conditions in the manufacturing sector than in other sectors (Punnett and Wegman, 2004). Despite a general downward trend in the development of work-related MSDs *per se*, the rate of new cases diagnosed between 2009 and 2011 in manufacturing was almost double that for all industries (670 per 100,000 workers; HSE 2012a). As aircraft wing systems installation work involves such restrictive working conditions it is likely to be contributing to these MSD rates and this is inevitably a costly problem for industry (Dunning, Davis, Cook, Kotowski, Hamrick, Jewell, and Lockey, 2010). On average, organisations must cope with 17 lost work days per MSD case (HSE 2012b) and untold costs for the rehabilitation and redeployment of returning personnel.

An obvious solution would be to adapt the designs and production methods that have been described. Altering the wing structure and positions of access panels or internal components (spars and ribs) would require such major redesign it is currently not feasible. However, greater access might instead be achieved by altering production methods, namely: a) if the wing box was not already sealed/a closed box during systems installations and b) if the traditional static horizontal orientation was adapted. If one side (the top) was left unattached until a later stage of production operators would be able to reach straight into the open wing box. However, due to the large size of the product, it would still not be possible to reach very far across the open top with the wing held in the fixed horizontal position so operators would still need to use the underside access holes to reach any central areas that are not reachable from the sides. Thus, to maximise access into the open box and eliminate the need to use the purpose-built access panels the wing would also need to be turned to a different orientation.

Ergonomic benefits from adapting product orientation have recently been demon-strated in the automotive manufacturing sector (Ferguson, Marras, Allread, Knapik, Vandlen, Splittstoesser and Yang, 2010) and the general importance of integrating ergonomics with design engineering to optimise work conditions and output is increasingly being understood (Colombini and Occhipinti, 2006; Cimino, Longo and Mirabelli, 2009; Battini, Faccio, Persona and Sgarbossa, 2011). The study described in this paper explored the potential ergonomic benefits of adapting the orientation of an open wing box for systems installations work. Specifically, the aim of the study was to test the impact of different wing orientations and dichotomous conditions of accessibility (open versus closed box) on MSD risk.

Methodology

Design

A repeated measures design was used to test the effect of wing box orientation and task position (independent variables) on working postures (dependent variables) at a controlled environment test facility. Participants wore a non-optical motion capture suit while performing a series of simulated installation tasks in a wing box section model positioned at different angles for systematic collection of physical activity data in two conditions: in Condition 1 the 'open box' had no top panel; in Condition 2 the top panel remained sealed as a 'closed box' (as in current production).

Tasks

Based on previous analysis of real shop floor systems installations the tasks were designed to represent those of real assembly work in wing production, involving simple insertion and screw fixing of pipes and small components (although for reasons of commercial sensitivity they cannot be described further here). The tasks were positioned in different regions of the model wing box (Figure 1) to ensure accessibility and postural concerns were sufficiently observed. The sequence in which tasks and orientations were presented to participants was randomised to counter any potential order effects.

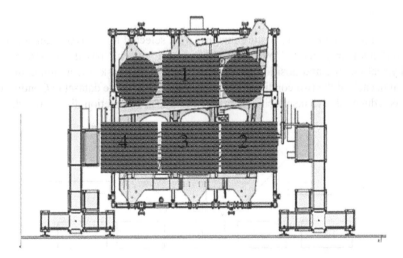

Figure 1. Selected tasks of the wing box structure.

Orientations

Different orientations were tested using a set of predefined angles, as shown in Figure 2 where the dashed line denotes the top cover, and the adjacent solid line represents the bottom side where access holes are located. Note the 180° angle represents the current horizontal wing position for systems installation tasks.

Sample

Participants (n = 10) were local office-based workforce volunteers; none were assembly operators and only one had experience of systems installations work.

Equipment

In both conditions the participants wore a full body non-optical inertial motion capture suit (Animazoo IGS-180i) With 17 sensors to record body segment and joint movements, even when occluded by the wing box structure, and generate both metric and visual image data. The visual output was examined and exemplar 'key frames' of extreme postures.

Participants were able to self-select any supporting apparatus to perform tasks: staging, ladders or wheeled office chair. Although this introduced an additional confounding variable it was considered important to allow participants the freedom to decide how to reach tasks as this is true to real shop floor conditions, and to prevent participants exceeding their personal limitations and comfort. Personal Protection Equipment was not necessary given that the tasks were innocuous, plus it would have impeded the motion capture.

Analysis

Key frames (single static posture images) were selected for analysis from an enormous number generated by the motion capture system, based on extremes of physical posture and position to represent each person-task-orientation combination in each of the two conditions. However, the final usable dataset in Condition 2 was reduced due to insufficient data caused by task completion difficulties during

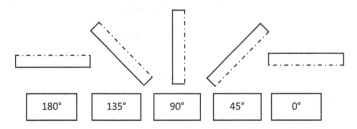

| 180° | 135° | 90° | 45° | 0° |

Figure 2.   Selected orientations of the wing box structure.

the trials: across all tasks the closed box in Condition 2 proved too difficult at the 0° and 45° angles and task 3 was generally found too difficult to complete at ANY orientation.

The Rapid Entire Body Assessment (REBA) scoring method developed by Hignett and McAtamney (2000) was used as this has been found accurate for capturing very awkward postures that included squatting, sitting on the ground, lumbar flexion greater than 90 and torso twisting (Chiasson, Imbeau, Aubry & Delisle, 2012) and found to elicit acceptable inter-rater reliability levels (Jacobs, 2008). The ability to accurately score awkward positions is a key factor within this study because of the nature of the postures adopted while working within the wings. The analysis produced individual scores for six body segments – neck, trunk, legs, upper arm, lower arm and wrist – in each of these combinations. However, as the objective for this study was to assess static postures according to wing orientation rather than dynamic activity the supplementary weighting factors that are usually included in a REBA (such as loads, couplings or duration and dynamics) were not added. Inter-rater reliability analysis on 10% of the data (randomly selected) confirmed good concordance between ratings (ICC = 0.778).

Results

The full data analysis conducted in this study included multifarious individual scoring for each participant's six body segments across each task, orientation and condition. However, for the scope of this paper the results presented are only the overall mean average total REBA scores for each experimental condition.

REBA risk and action levels (Table 1) were used to determine the criticality of postures as an impact of wing orientations. It should be noted that overall levels will be higher for real flow line work given that our analysis does not include the REBA adjustment weightings for dynamic activities and loads.

Condition 1 'open box' results

Table 2 summarises average unweighted mean scores for each task, orientation and condition combination. According to these scores the lowest risk orientation in the

Table 1.   REBA method risk levels.

Action level	REBA score	Risk Level	Action
0	1	Negligible Risk	None Necessary
1	2–3	Low Risk	Maybe Necessary
2	4–7	Medium Risk	Necessary
3	8–10	High Risk	Necessary Soon
4	11+	Very High Risk	Necessary Now

Table 2.   Condition 1 'open box' overall mean average results.

	0°	45°	90°	135°	180°
Task 1	8.7 (1.49)	8 (2)	6 (1.73)	5.3 (1.34)	8 (1.33)
Task 2	8.2 (1.13)	8.2 (1.55)	8 (1.76)	6 (1.49)	8.2 (1.69)
Task 3	7.7 (1.64)	6 (1.56)	7.1 (1.1)	9.13 (0.99)	6.33 (2.18)
Task 4	7.78 (1.99)	6.71 (1.9)	5.6 (1.6)	7.7 (1.42)	9 (1.07)

Table 3.   Condition 2 'closed box' mean average results.

	90°	135°	180°
Task 1	6.9 (2.42)	9.1 (1.97)	6.88 (2.30)
Task 2	7.8 (2.62)	9.2 (1.93)	8.3 (2.41)
Task 4	7.88 (2.03)	8.33 (1.97)	6.67 (1.97)

open box condition is 90°, and the highest risk presented at the 180° angle. No mean average score falls below the medium risk level and all orientations present a high risk for performance of at least one of the tasks.

Condition 2 'closed box' results

As described, the dataset was reduced due to participants finding that, in some positions, it was not possible to perform the task. In Condition 2 the 0° and 45° angles were not included as it was soon found that with the top cover attached the access holes were only available on top of the structure so participants could not reach. Task 3 is also excluded because most participants felt unable to reach and declined to complete the task; consequently, there were not enough data sets to analyse. Thus, Table 3 presents a summary of the mean average unweighted scores pertaining to the remaining task, orientation and condition combinations.

Once again, scores show that mean average MSD risk levels do not fall below a medium level of risk. The lowest risk orientation was 90° and the highest 135°, as it presented high risk for all three tasks.

Comparison of Conditions 1 and 2

Table 4 below provides mean averages and standard deviations for the collective scores of the six body segments across all tasks, orientations and conditions (missing data reflects where insufficient participants attempted task 3 in Condition 2 and the 0° and 45° angles in Condition 2.

Table 4.   Summary of Scores Across Tasks, Orientations and
Conditions.

Angle	Condition	Task 1	Task 2	Task 3	Task 4
0°	Open	8.7 (1.49)	8.2 (1.13)	7.7 (1.64)	7.78 (1.99)
45°	Open	8 (2)	8.2 (1.55)	6 (1.56)	6.71 (1.9)
90°	Open	6 (1.73)	8 (1.76)	7.1 (1.1)	5.6 (1.6)
	Closed	6.90 (2.42)	7.80 (2.62)	–	7.88 (2.03)
135°	Open	5.3 (1.34)	6 (1.49)	9.125 (0.99)	7.7 (1.42)
	Closed	9.1 (1.97)	9.2 (1.93)	–	8.33 (1.97)
180°	Open	8 (1.33)	8.2 (1.69)	6.33 (2.18)	9 (1.07)
	Closed	7.88 (2.03)	8.33 (1.97)	–	6.67 (1.97)

Conclusions

Given that all tasks were generally found too difficult when the wing box was closed, when it was positioned at a 0° or 45° angle, and as task 3 was unachievable at all other orientations in Condition 2, this shows that the closed box condition is considerably less accessible than the open box condition in general. Overall, the lowest MSD risk was associated with working on the open box at the 90° and 135° orientations whilst the highest risk of developing musculoskeletal damage is associated with working on a closed box at the 135° orientation. However, with respect to the individual body segment scores there was not one apparent ideal orientation as risk and accessibility would vary depending on the particular task location and participant's ability (and to some extent their individual preference).

Introducing a rotating fixture that would hold the entire wing in the open box condition would be a substantial engineering endeavour for wing manufacturing. Further ergonomic studies are needed to examine the effects of individual capabilities and support apparatus in relation to product orientations in order to assess whether this is a viable venture. Although also a significant endeavour, it may also be important to reconfigure the internal systems to make assembly tasks a little easier for operators to perform inside the wing, or at least to reduce the duration needed for task completion. Given that the access holes are also the only means of non-invasive entry for post-production maintenance this would also help in the wider life cycle of the product. Work is already underway to explore potential system redesign options towards this long term objective. In the short term, the findings from this study are to be used to develop more immediate remedial solutions to improve work processes and reduce MSD risk.

Acknowledgements

The authors would like to particularly thank John Thrower, Tim Hall and Antoinette Caird-Daley for their invaluable technical support during this research.

References

Battini, D., Faccio, M., Persona, A. & Sgarbossa, F. 2011, New methodological framework to improve productivity and ergonomics in assembly system design. *International Journal of Industrial Ergonomics*, 41, 1, 30–42

Chiasson, M., Imbeau, D., Aubry, K. & Delisle, A. 2012, Comparing the results of eight methods used to evaluate risk factors associated with musculoskeletal disorders. *International Journal of Industrial Ergonomics*, 42, 478–488

Cimino, A., Longo F. & Mirabelli, G. 2009, A multimeasure-based methodology for the ergonomic effective design of manufacturing system workstations. *International Journal of Industrial Ergonomics*, 39, 2, 447–455

Colombini, D. & Occhipinti, E. 2006, Preventing upper limb work-related musculoskeletal disorders (UL-WMSDS): New approaches in job (re)design and current trends in standardization. *Applied Ergonomics*, 37, 4, 441–450

Dunning, K.K., Davis, K.G., Cook, D., Kotowski, S.E., Hamrick, C., Jewell, G. & Lockey, J. 2010, Costs by industry and diagnosis among musculoskeletal claims in a state workers compensation system: 1999-2004. *American Journal of Industrial Medicine*, 53, 276–284

Ferguson, S.A., Marras, W.S., Allread, W.G., Knapik, G.G., Vandlen, K.A., Splittstoesser, R.E. & Yang, G. 2010, Musculoskeletal disorder risk as a function of vehicle rotation angle during assembly tasks, *Applied Ergonomics*, 42, 5, 699–709

Gallagher, S. 2005, Physical limitations and musculoskeletal complaints associated with work in unusual or restricted postures: A literature review. *Journal of Safety Research*, 36, 1, 51–61

Hignett, S. & McAtamney, L. 2000, Rapid Entire Body Assessment: REBA. *Applied Ergonomics*, 31, 201–5

HSE 2012a. Retrieved 30th September, 2013, from: http://www.hse.gov.uk/statistics/industry/manufacturing/manufacturing.pdf

HSE 2012b. Retrieved 30th September, 2013, from: http://www.hse.gov.uk/statistics/dayslost.htm

Mozrall, J.R., Drury, C.G., Sharit, J. & Cerny, F. 2000, The effects of whole-body restriction on task performance. *Ergonomics*, 43, 11, 1805–1823

Punnett, L. & Wegman, D.H. 2004, Work-related musculoskeletal disorders: the epidemiologic evidence and the debate. *Journal of Electromyography and Kinesiology*, 14, 13–23

Punnett, L., Gold, J.E., Katz, J.N. & Gore, R. 2004, Ergonomic stressors and upper extremity disorders in automotive manufacturing: a one-year follow-up study. *Occupational and Environmental Medicine*, 61, 668–674

Veasey, D.A., Craft McCormick, L., Hilyer, B.M., Oldfield, K.W., Hansen, S. & Krayer, T.H. 2005, *Confined Space Entry and Emergency Response*. Hoboken, N.J.:Wiley

A COMPUTER SOFTWARE METHOD FOR ERGONOMIC ANALYSIS UTILISING NON-OPTICAL MOTION CAPTURE

Teegan L. Johnson & Sarah R. Fletcher

Industrial Ergonomics and Human Factors Group, School of Engineering, Cranfield University, Bedford, MK43 0AL

This study assesses whether postures of individuals captured using non-optical motion-capture technology and analysed using an ergonomics computer program can be used with confidence. Three postures from a simple task were captured using a Motion Capture suit; these postures were analysed by participants and the computer program. The concordance between the participant's scores and the computer program were measured using an Intra-Class Correlation. The resulting scores between the software and human scores were in the poor-to-fair level of agreement, however further analysis of the results is required to gain a greater understanding of these results.

Introduction

The ability to capture and assess the entire posture of an individual in a manufacturing environment using computer technology would improve the replicability for postural scoring in the field i.e. on the shop floor. Some tasks undertaken within industrial environments are performed in areas that cannot easily be observed, or are completely out of view within structures making it difficult to accurately capture the postural information of operators. Traditional methods of posture capture in these environments involve in-person observations along with video capture (Chiasson, Imbeau, Aubry & Delisle, 2012), and although multiple cameras may be used to increase the proportion of activity captured surrounding structures and parts often prevent a full capture of the activity. A key consequence of this occlusion is that ergonomists carrying out a postural assessment may need to make estimations about the exact position of body parts (Chang & Wang, 2007) and critical postures that play an important role in the development of Musculoskeletal Disorders (MSDs) may be missed (Klippert, Gudehus & Zick, 2012). This situation is exacerbated when operators work in confined conditions, as operators often need to adopt particularly awkward postures but these cannot be observed. Even though the operators may be asked to replicate the postures they adopt, the re-enactment is unlikely to be accurate when the physical structures are not available for support and balance.

An answer to this problem is the use of a non-optical Motion Capture (MoCap) suit which captures and presents a 360° interactive visual output of the postures adopted by the individual within the suit within its associated computer program

IGS Bio (IGS-Bio v1.8 Animazoo Ltd.). The data this sort of system can produce may then be translated into ergonomic software programs or computer aided design (CAD) programs with internal ergonomics tools where the data can be analysed. These programs can significantly reduce the time expended by ergonomists scoring postures; all frames are scored by the system and the posture with the highest risk of MSD can be identified. Although these capabilities are readily available very few ergonomic studies were found to have applied this technique. One such study by Klippert, Gudehus and Zick (2012) compared the results of two experienced ergonomists with a specific postural assessment software system using a non-optical MoCap suit and found a sufficient correlation between the scores completed by the humans and the software system.

CAD software was used in this study as opposed to a postural assessment specific program as used in the Klippert *et al* (2012) study. This is due to the advantage of creating a virtual environment around the visual output of individual within the motion capture suit. This virtual environment can be used to give context to the postures adopted and can be manipulated to investigate the impact of structural changes on postures before design changes are put in place.

There has been little previous research investigating the use of non-optical motion capture systems and the translation of their data into CAD modelling programs where their associated ergonomic assessment tools can be utilised. This investigation aims to address this dearth of research and highlight the advantages of using such a system.

This paper presents the findings of one component study from a wider project that aims to investigate the level of correlation between a larger rater population than that used in the Klippert *et al* (2012) study and looks to validate the postural assessment tool within a CAD software tool. This validation will include static postures and postures captured during a dynamic activity to identify whether the CAD software can be used with confidence to analyse postural data captured using the MoCap suit. This paper presents an analysis of the Inter-Rater Reliability between the three types of media: video stills from a simple dynamic activity; and the same postures in IGS Bio (the software output for the suit); the resulting scores will be compared to the postural analysis completed in the CAD software ergonomics suit. It is hypothesised that there will be a high correlation between human scores from IGS Bio and the CAD scores. A second and final hypothesis is that there will be a correlation between IGS Bio (IGS-Bio v1.8 Animazoo Ltd.) scores and the scoring in the CAD program. The null hypothesis is that there will be low correlations between the video and the scoring form the CAD program.

Method

Participants

One volunteer participant wore the MoCap suit and completed a simple pick and place task to create the data. 15 participants, 8 female and 7 male, were recruited to act as postural raters using availability sampling; all were naïve to RULA analysis (age range 18–55 years, mean: 33, SD: 12.4).

Design

The study was a semi-experimental 3 × 3 mixed design to investigate the effect of the output of the postures (independent variables) on postural assessment scores (dependent variables). The three conditions are three postures created from a simple pick-up and place activity: pick-up, transition and place. All participants scored a posture from each activity presented in video stills and IGS Bio (IGS-Bio v1.8 Animazoo Ltd.) images. The same three postures were produced in the CAD software and analysed using its postural analysis tool.

The pick-up and place activity data created for analysis involved a single participant wearing the MoCap suit completing a simple pick-up and place task with a torso twist during placement. A box weighing 2 kg with handle holes was lifted from the ground and placed on a raised surface.

Equipment

Data was electronically captured using an fully body non-optical inertial motion capture suit (Animazoo Ltd. IGS-180i) that records body segment and joint movements, with no restriction to motion. 17 Inertial Measurement Unit (IMU) sensors are used to capture small changes in position generating metric and visual output data. The system provides a wireless real-time data transmission to a computer program that translates the data and provides a visual output in a software program called IGS-Bio (version 1.8, Animazoo Ltd). This visual output gives a 360° interactive representation of the suit's position (see figure 1), allowing all movements to be recorded into motion data files. This data can then be translated into DELMIA (version 5, Dassalt system) a CAD modelling program which provides the ability to create 3D environments with realistic human manikins. The manikins are editable using anthropometric data and analysed using DELMIA's internal ergonomics tools.

Manual scoring

The postural analysis scoring method used was the Rapid Upper Limb Assessment (RULA) (McAtamney & Corlett, 1993). RULA is one of the postural analysis tools available within DELMIA, and has been shown to have a high construct validity and inter-rater reliability (Jacobs, 2008).

The RULA analysis carried out by the participants involved analysing the body in three sections. Section one involves scoring the right arm: upper arm, lower arm and wrist are scored based on their position. An overall arm score in then identified to which weightings for force and muscle use are added. The same is done for the left arm. The third section involves analysis of the neck, trunk and legs, an overall score is identified and weightings are included. The overall scores from the first, second and third sections are then used to identify a global or whole body score; in this case there are two global scores due to the two arm scores.

Figure 1.   IGS Bio output (Place).

Procedure

Performance of the pick-up and place task was simultaneously recorded using a video camera in a fixed position using a tripod and the IGS Bio system; the data was then exported to DELMIA. The task was broken down to three sections: pick-up, transition and place. Each section was isolated and analysed in DELMIA using RULA to identify a posture with the highest whole body score in each of the three sections. The corresponding video stills and IGS Bio frames were identified and isolated. Two training images were additionally isolated from IGS Bio and the video.

The participants were trained to use RULA using two images, a video still and an IGS Bio frame. The posture showed the back of the participant standing straight with their hands by their sides. Participants were presented with a RULA worksheet and the process of analysis was explained. Participants were asked to analyse the video posture first and then the IGS Bio posture with help given as needed.

Figure 2.   Video frame of task (transition).

The participants were presented with the 6 frames (3 video, 3 IGS Bio) and asked to score the postures using RULA. The order in which the participants were presented with the video frames and IGS bio frames were identified using the random number generator function in Excel (Boa *et al*, 2009) and cyclical counterbalancing was used to prevent practice effects (Keppel, Saufley & Tokunaga, 1992).

Analysis

The data in this paper consists of the left arm, right arm, the second section (neck, trunk and leg), right overall and left overall scores. DELMIA scores consisted of a single score for each of the sections, leading to a single data set compared to the 15 sets of results for IGS Bio and Video stills. Therefore the DELMIA scores were repeated 15 times creating data set of the same size.

Data was analysed using Intra Class Correlation (ICC), this method is used to compute inter-rater reliability for observational ratings (Hallgren, 2012). This is a suitable method of analysis for ordinal level data and studies with two or more coders (Portney and Watkins, 2009). Two types of ICC were used to analyse the data, overall inter-rater agreement was analysed using a two way mixed, absolute agreement, single measures ICC. The levels of agreement between the different posture outputs (IGS Bio, DELMIA and video still) were measured using two-way random, absolute agreement, single measures ICCs. Two-way ICCs are used when postures are assessed by the same set of raters. Absolute agreement was chosen over a consistency agreement, due to the importance of raters providing scores similar in value (Hallgren, 2012) and single measures because the postures scores were not averaged (Hallgren, 2012).

A mixed model was used for the overall inter-rater agreement because there was no intention to generalize the findings to a wider population of raters (Portney

Table 1.   Inter-rater Reliability for Pick-up, Transition
and Place.

		IGS Bio	DELMIA
Pick-up	Video Still	.72	.34
	DELMIA	.54	
Transition	Video Still	.05	.10
	DELMIA	.31	
Place	Video Still	.46	.24
	DELMIA	.57	

and Watkins, 2009). Whereas random models are used when raters are randomly chosen and there is an intention to generalize results to a wider audience (Portney and Watkins, 2009).

Results

Overall inter-rater agreement

The resulting ICC for overall inter-rater agreement (between all raters across both image and IGS Bio scores) was 0.7; which would be judged as excellent according to Cicchetti's (2004) recommendations. Less than 0.4 is considered poor, between 0.40–0.59 fair, 0.60 to 0.74 good and greater than 0.75 is considered excellent (Cicchetti, 1994).

The degree to which postural scoring was consistent between Video stills, IGS Bio and DELMIA in the different conditions is presented in Table 1.

Video stills and IGS Bio

The highest of all ICCs across all conditions was for the pick-up activity and was in the excellent range (Cicchetti, 1994). This indicates a high level of agreement between the Video Stills and IGS Bio for this posture. The ICC for the place condition would be considered in the fair level of agreement range (Cicchetti, 1994), and in the poor range for the transition condition was (Cicchetti, 1994) indicating a low level of agreement.

IGS Bio and DELMIA

IGS Bio and DELMIA ICCs for the pick-up and place conditions would be considered in the fair range (Cicchetti, 1994), indicating a fair level of agreement and some measurement error between the scores given by participants. However for the transition condition the ICC would be considered poor with a high level of disagreement between the scores for IGS Bio and the computer program.

Video stills and DELMIA

The resultant ICC scores for the DELMIA output and the Video Stills is in the poor range (Cicchetti, 1994) for all three conditions indicating a level of disagreement between the scores. However the highest of the low ICC scores was for the pick-up condition and the lowest for the transition posture, this score was the lowest of all conditions. The transition condition ICCs were consistently low between the three output types.

Discussion

It was hypothesised that the highest levels of agreement would be between IGS Bio and DELMIA, with the lowest levels of agreement between the Video Stills and both IGS Bio and DELMIA. Of the three hypotheses outlined, the null hypothesis that there is no relationship between the video stills and DELMIA is supported.

The poor ICC results for the transition and place conditions between the Video Stills and both IGS Bio and DELMIA may be explained by the occlusion of the left arm within the image due to the angle of recording which can be seen in figure 2 causing scorers to miss score the arm. Further analysis of the data will investigate the levels of agreement at the body part level of RULA analysis between raters and between the three media to identify whether the scores for the left side of the body are different from those in IGS Bio and DELMIA. This may lead to the identification of similar agreement levels to those found by Klippert *et al* (2012) whose analysis was at the individual body part level.

The poor to fair levels of agreement between IGS Bio and DELMIA cannot be explain by occluded sections of the body as both sides should have been captured. Therefore additional analysis will include an investigation into static postures using a goniometer to measure the physical placement of an individual, recording the postures created and measuring them within IGS Bio using a protractor to identify whether the angles are the same.

At this point of the study it is not possible to suggest which method should be used to analyse the postures of individuals, however further investigation of the data captured is expected to develop an understanding of the results so far and take steps towards developing a quantified method of postural capture and analysis.

Acknowledgements

The authors would like to thank John Thrower and Wil Baker for their support during the research. This work was supported by the UK Engineering and Physical Sciences Research Council as part of the Centre for Innovative Manufacturing in Intelligent Automation.

References

Bao, S., Howard, N., Spielholz, P., Silverstein, B., & Polissar, N. 2009, "Interrater Reliability of Posture Observations." *Human Factors: The Journal of the Human Factors and Ergonomics Society.* 51, 292–309.

Chang, S., & Wang, M. 2007, "Digital Human Modeling and Workplace Evaluation: Using an Automobile Assembly Task as an Example." *Human Factors and Ergonomics in Manufacturing.* 17(5): 445–544.

Chiasson, M., Imbeau, D., Aubry, K., & Delisle, A. 2012, "Comparing the results of eight methods used to evaluate risk factors associated with musculoskeletal disorders." *International Journal of Industrial Ergonomics.* 42: 478–488.

Hallgren, K. A. 2012, "Computing Inter-Rater Reliability for Observational Data: An Overview and Tutorial." *Tutorials in Quantitative Methods for Psychology.* 8(1): 23–34.

Jacobs, K. 2008, *Ergonomics for Therapists.* (Elsevier Health Sciences, Missouri).

Keppe, G., Saufley, W. H., & Tokunaga, H. 1992, *Introduction to Design and Analysis A students Handbook.* (Freeman, New York).

Klippert, J., Gudehus, T., & Zick, J. 2012, "A Software-Based Method for Ergonomic Posture Assessmet in Auromotive Preproduction Planning: Concordance and Differenc in Using Software and Personal Observation for Assessment." *Human Factors and Ergonomics in Manufacturing and Service Industries.* 22(2): 156–175.

McAtamney, L., & Corlett, E. N. 1993, "RULA: a survey method for the investigation of work-related upper limb disorders." *Applied Ergonomics.* 24(2): 91–99.

Portney, L. G., & Watkins, M. P. 2009, *Foundations of Clinical Research Applications to Practice.* (Pearson Education International, New Jersey).

RELATIONSHIPS BETWEEN DEGREE OF SKILL, DIMENSION STABILITY, AND MECHANICAL PROPERTIES OF COMPOSITE STRUCTURE IN HAND LAY-UP FABRICATION METHOD

Tetsuo Kikuchi[1,3], Yuichiro Tani[3], Yuka Takai[2],
Akihiko Goto[2] & Hiroyuki Hamada[3]

[1]TOYUGIKEN CO., Ltd., Kanagawa, Japan
[2]Osaka Sangyo University, Osaka, Japan
[3]Kyoto Institute of Technology, Kyoto, Japan

Hand lay-up fabrication has been used composite structure forming since ancient times. This method is able to handle different production volumes and product sizes due to inexpensive moulds and low facility costs. However, hand lay-up fabrication work relies on human skills, resulting in variation in finishes, product quality and occurrence of defects. The authors conducted a motion analysis with hand lay-up fabrication experts. This quantified techniques that are not visibly apparent and considered 'tacit knowledge'. Mechanical properties and dimension stability of samples and their relationships with the motions of experts were measured. It is suggested that specialised control techniques, the training of non-experts, and technical tradition are possible.

Introduction

In hand lay-up (HLU) fabrication work, the characteristics of the composite material will be usually the same regardless of which forming method is used as long as the reinforced substrate, reinforcement morphology, matrix resin, and volume content of reinforcing material are the same. Composite materials, particularly fibre reinforced plastics (FRP) made of fibres and resins, are basically formed by impregnating fibres with resin, i.e., replacing the air contained in fibres with resin. With the HLU technique, rollers are used to impregnate reinforced fibres with resin. Consequently, the impregnation method is expected to contribute to changes in the properties of the interface formed. To review the effects of different roller use in the HLU technique on the mechanical properties of the composite structure, an experiment was conducted to analyse the process of work and investigate the relationships with mechanical strength and dimension stability of the structures built in craftsmen (experts, intermediates, and non-experts) specializing in making bathtubs using the HLU technique, who were asked to create FRP structures.

Method

Participants

The participants consisted of a total of five craftsmen ranging from an expert with 25 years of experience to one with only one year of experience. In this experiment, the craftsman with 25 years of experience is defined as expert, one with only one year of experience as non-expert, and those in between as intermediates. Table 1 shows the demographic data of the participants.

Experimental protocol

FRP laminate moulding using the HLU technique was analysed. Figure 1 shows the size of the mould used in this experiment: length of 415 mm, width of 1,600 mm, and height of 50 mm.

Materials

The mould surface was pre-coated with a gel coat (white) of 0.5 mm in thickness, after which lamination was carried out using the HLU technique. Glass fibre chopped strand mat (380 g/m^2) was used as the reinforced substrate, and isophthalic unsaturated polyester resin was used as the matrix. Cutting agent (MEKPO) was added to the resin at the ratio of 100 to 1.0. Two sheets of glass fibre chopped strand mats were used in the forming process.

Table 1.   Demographic data of participants.

No.	Subject	Age	Career (year)	Height (cm)	Weight (kg)	Dominant-Hand
1	Expert	48	25	171	52	Right
2	Intermediate-1	41	18	163	61	Right
3	Intermediate-2	45	15	175	63	Right
4	Intermediate-3	28	4	163	60	Right
5	Non-expert	29	1	164	52	Right

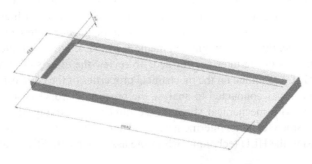

Figure 1.   Mould used in this study.

Motion analysis & eye movement analysis

To measure the motion of the participants, the MAC3D System (Motion Analysis Corporation) was used. The left side of Figure 2 shows the motion analysis system. The infrared markers attached to the body of the participants were captured using five cameras, and the collected information was sent to the PC as the 3D positional data of all the markers. The sampling frequency was set at 60 Hz. For eye movement, eye movement detector Talk Eye II (Takei Scientific Instruments Co., Ltd.) was used. The sampling frequency was set at 30 Hz. The right side of Figure 2 shows the eye movement analysis system. The centre camera was used to capture the field of vision of the participants, while the cameras on both sides were used to capture eye movements.

Dimension stability

To compare dimension stability between the expert and non-experts, the planar section and cross-sectional surface of the moulded product obtained were observed (Figure 3). A micrometer was used to measure the thickness. To compare the roughness of the surface of planar section, the 300 mm × 300 mm area in the centre of the

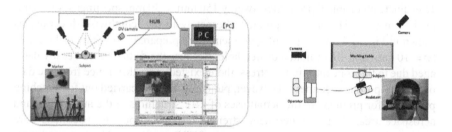

Figure 2.   Motion analysis and eye movement measurement systems in this experiment.

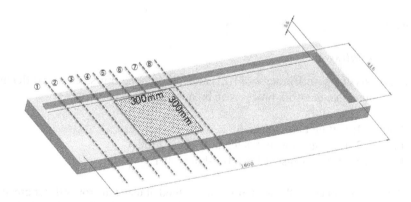

Figure 3.   Observation of dimension stability.

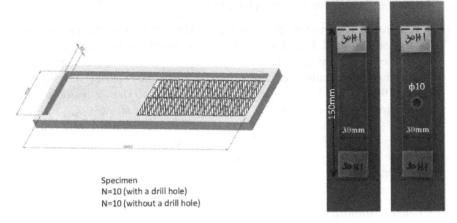

Figure 4.   Sampling of specimens for tensile testing.

sample was measured. Subsequently, eight sections on the short side of the samples were cut out to observe the thickness distribution of the cross-section.

Mechanical property

To evaluate mechanical properties, twenty 150 mm × 30 mm specimens were sampled from the FRP moulded product alternately in the lengthwise direction, as shown in Figure 4. Two types of specimens were prepared: 10 with a drilled hole ($\varphi = 10$ mm) and 10 without a drilled hole. Because the drilled hole is easily damaged due to the concentration of stress, there existed some difference from the data of the non-porous specimens. However, punching must be carried out for attaching parts, etc. for products in the actual uses of FRP structures in the automotive and aerospace areas. In this experiment, strength tests using porous specimens were also conducted to evaluate the actual uses of FRP structures. The test conditions were inter-chuck distance of 80 mm Cand crosshead speed of 1 mm/min.

Results

Process analysis

The analysis of the work process of HLU moulding among craftsmen with different careers found that the work process can broadly be divided into three:

    Process 1: Setting glass substrate
    Process 2: Impregnation and defoaming using a roller
    Process 3: Surface finish using a finishing roller

Particular for process 3, the work time and method of using the roller differ greatly according to the craftsman, which is suggested to characterise the HLU technique. For this reason, comparison was made focusing on process 3.

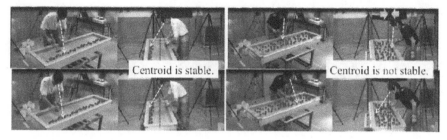

Expert   Non-expert

Figure 5.   How to use a roller between expert (left) and non-expert (right): short-side direction.

Motion analysis & eye movement analysis

The results of the motion and eye movement analyses carried out on each subject for process 3 are shown below. First, the point of regard (POR) (right eye) and operations were compared among participants in process 3. Figure 5 compares how the expert and non-expert use the roller in the short-side direction. The left photos represent the expert while the right photos represent a non-expert. During the movement in the short-side direction, POR and the center of gravity is consistent regardless of the movements of the roller in the expert (left), suggesting the movements of the roller and the load are stable. On the other hand, the POR of the non-expert (right) moves when the roller moves, and the center of gravity also moves to and fro. This suggests that the stability of the roller movement and load are poorer in the non-expert than the expert.

Figure 6 traces the movements of the right eye POR of different participants. The charts clearly demonstrate that only the POR locus of the expert moves in a narrow range.

The POR variation of the expert is small. The stable-POR movement is unique to the expert, independent of career. The expert is estimated to check the finish of the moulded product through experience, touch, sound, etc.

Figure 7 compares how the expert and non-expert use the roller in the long-side direction. The left side photos represent the expert while the right side photos represent the non-expert. The movements in the long-side direction were observed to be similar to those in the short-side direction. Instead of following the movement of the roller, the expert (left) constantly looked ahead, and the to and fro variations of the centre of gravity were limited. In contrast, the POR of the non-expert (right) moved together with the roller, and the to and fro fluctuations of the centre of gravity were about 2.5-fold larger than the expert. During the to-and-fro movements of the roller, the right elbow of the expert extended and bended in the range of 60–120 degrees. On the other hand, the elbow of the non-expert virtually did not extend or bend (about 100–110 degrees).

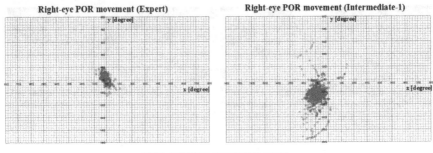

Figure 6.   Right-eye POR movements in participants (process 3: during the motion in the short-side direction).

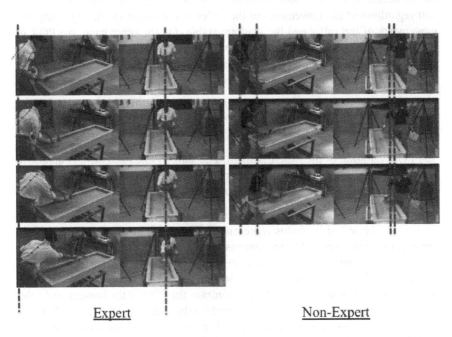

Expert   Non-Expert

Figure 7.   How to use a roller between expert (left) and non-expert (right): long-side direction.

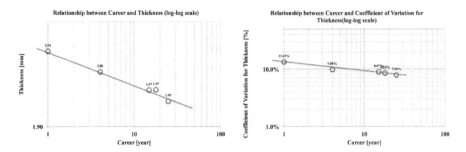

Figure 8.   Relationships between career and thickness or the coefficient of variation.

The roller speed of the expert was about two times faster than that of the non-expert. However, the acceleration rate was almost the same between the two, presumptively because of the smooth extension and bending movements of the right elbow in the expert.

Furthermore, the expert held the handle of the roller from the top so that force is applied when pulling the roller. In fact, it was found later in interviews with the participants that the expert tend to concentrate more on pulling than pushing the roller. On the other hand, the non-expert was found to hold the handle of the roller from below upwards so that force is applied when pushing the roller. This is thought to have resulted in the marked difference in the extension/bending of the right elbow and speed of the roller between the expert and non-expert.

Dimension stability

Distribution of the thickness in the short-side direction demonstrated that thickness is consistent (1.9 mm) for the expert even in the elevation surface and R convex section, while it is not consistent for the non-expert in the convex section and its vicinity, which are thought to be difficult to form. The convex section was about 15% thinner, while the vicinity of the convex section was about 15% thicker than the average thickness.

These results suggest that how the roller is used clearly affects the thickness of the planar, elevation, and R convex sections, in other words the stability of the shape. In addition, it is evident that how the roller is used is closely related to operations that define "career," as shown in Figure 8.

Mechanical property

The effects of career on mechanical property were investigated.

Figure 9 shows the tensile strength and coefficient of variation, with respect to career, in specimens with and without drilled holes. First, comparison of the tensile strength of specimens between the expert and non-expert indicates that it is about 15% higher in the expert's specimen. In addition, a strong correlation was

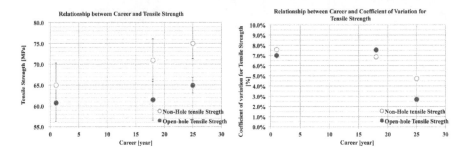

Figure 9.   Relationships between career and tensile strength or the coefficient of variation.

found between career and tensile strength (coefficient of correlation: 0.99(NHT), 0.92(OHT)). On the other hand, correlation was weak in the coefficient of variation (coefficient of correlation: −0.87(NHT), −0.65(OHT)).

Conclusion

The findings suggest that the craftsman's career has a large effect on mechanical property in the HLU technique because FRP demonstrated different characteristics even if the same reinforced substrate, reinforcement morphology, matrix resin, and volume content were used. The results also suggest a close relationship between how the roller is used (e.g., direction, frequency, and load), which is a measure of career, and mechanical property.

Characteristics of the expert can be summarised as follows:

1) Smooth movement of the right elbow during roller movements*.
2) POR is consistent and is independent of the movements of the hand.
3) Centre of gravity is consistent and stable.
4) Holding the roller handle from the top and focusing on pulling movements.

* As force is evenly distributed, the coefficient of variation of the thickness and mechanical strength is estimated to be small.

Furthermore, it was found that the above four points can be incorporated into the educational tools of non-experts and intermediates, sharply reducing the skill acquirement time. The inclusion has also increased the fun and joy of skill acquirement among them.

References

T. Kikuchi, T. Koyanagi, H. Hamada, A. Nakai, Y. Takai, A. Goto, Y. Fujii, C. Narita, A. Endo and T. Koshino "Biomechanics investigation of skillful technician in hand lay up fabrication method". *The ASME 2012 International Mechanical*

Engineering Congress & Exposition, Houston, Texas, USA, IMECE2012-86270, pp. 288, 2012.

Z. Zhang, Y. Yang and H. Hamada "Mechanical property of glass mat composite with open hole". *The ASME 2012 International Mechanical Engineering Congress & Exposition, Houston*, Texas, USA, IMECE2012-87270, pp. 79, 2012.

M. Shirato, M. Kume, A. Ohnishi, A. Nakai, H. Sato, M. Maki and T. Yoshida "Analysis of daubing motion in the clay wall craftsman: Influence on several kinds of clay with different fermentation period on the daubing motion properties". *Symposium on sports engineering: symposium on human dynamics 2007*, JAPAN, pp. 254–257, 2007.

ERGONOMIC ISSUES ARISING FROM THE 'NEXT MANUFACTURING REVOLUTION'

M.A. Sinclair & C.E. Siemieniuch

*ESoS Research Group, School of Electrical,
Electronic & System Engineering,
Loughborough University of Technology*

The paper outlines briefly the contents of the government-sponsored document, 'The Next Manufacturing Revolution', with its emphasis on practices appropriate to the demands for sustainability required by population growth and emissions control, and then explores some of the implications of the contents for the practice of ergonomics, particularly in relation to job design. It is clear that there are some significant extensions required to the knowledge classes, processes and practices of ergonomists; there are also some implications for the Institute's role as a source of advice.

Introduction

The government-endorsed report, 'The Next Manufacturing Revolution' (http://www.nextmanufacturingrevolution.org) is an attempt to make Britain's manufacturing sector more competitive in the world of 2030 and beyond; a world in which environmental sustainability will be a key issue, in addition to the usual competitive and technical challenges (Sulston 2012)

The document describes the benefits that will accrue from this revolution:

"This study presents opportunities to improve non-labour resource productivity which could enable a revolution in manufacturing and are estimated, conservatively, to be worth for the UK:

- *£10 billion p.a. in additional profits for manufacturers – a 12% increase in average annual profits.*
- *314,000 new manufacturing jobs – a 12% increase in manufacturing employment.*
- *27 million tonnes of CO_2 equivalent p.a. greenhouse gas emissions reduction – 4.5% of the UK's total greenhouse gas emissions in 2010."*

The document states that the following side-benefits will also accrue:

- *improved security of resources (energy, food and raw materials)*
- *reductions in waste and landfill*

- *less traffic congestion and reduced load on energy and transport infrastructures (requiring less investment)*
- *improved prosperity in UK agriculture (mainly through waste reductions)*
- *economic development in those developing nations supplying UK manufacturers*

The document specifically excludes discussion of 'labour productivity' (labour productivity improvements are said to have reduced costs since 2001 at 3% p.a. to £75bn in 2011, a loss of about 1,000,000 jobs).

However, as Air Chief Marshal Sir Sidney Dalton said, "But it is people who turn technology into capability'; the document implicitly recognises the truth of this when it discusses eight barriers to the delivery of these benefits:

- senior executive leadership (strategy, allocation of resources and investments, leadership and impetus for change, etc.)
- information (awareness, futurology, technical developments, etc.)
- skills (servitisation, making linear companies cyclic, etc.)
- resources (financial, relevant manpower, etc.)
- design (better design processes & tools, design for recycling, better utilisation of fewer materials, etc.)
- infrastructure (e.g. energy efficiencies, recovery and reuse of materials)
- legal constraints (e.g. bans on remanufactured parts in new goods)
- collaboration (unwillingness to talk to suppliers, security issues, etc)

Amelioration of these barriers involves the direct participation of people, and it here where the knowledge and skills of the ergonomics/human factors profession become significant and useful.

Some process issues emerging from the document

To avoid an abstract discussion, we consider 'Circular manufacturing' (the rest of us might call it 'recycling and re-use', as a very important means to address sustainability. Figure 1 below illustrates what is meant by this.

The intention is both to reduce impact on the environment through mining and landfill, by re-use of materials as much as possible, following the right-to-left arrows above. Triage is of critical importance in this; note that the design of the product has a big impact on the difficulty of materials separation.

More specifically, recycling/re-use includes the following:

- re-use – redeploying a product without the need for refurbishment – e.g. reselling mobile phones in less developed regions.
- remanufacturing – restoring a product to its original performance. Caterpillar has a successful engine remanufacturing business.

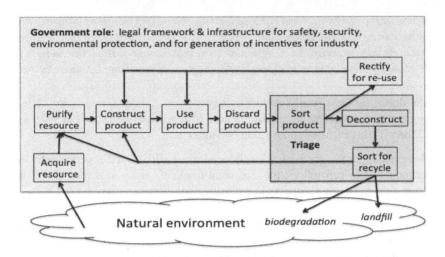

Figure 1.   Illustration of 'circular manufacturing'. See text below for discussion.

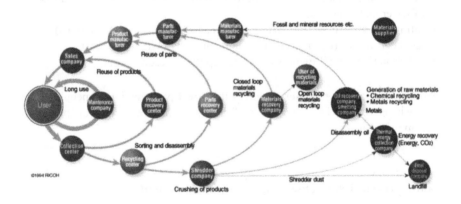

Figure 2.   The Ricoh COMET Circle to recycle/ reuse photocopiers. This has resulted in a '90% reduction in emissions and waste'.

- cascaded use – using a product for a lower value purpose – e.g. turning used clothes into pillow stuffing.
- recycling – extracting a product's raw materials and using them for new products – e.g. aluminium and steel are widely recycled.
- recovery – using a product's materials for a basic, low value purpose such as road base or combustion to produce heat.

As a concrete example of this, consider the Ricoh COMET Circle, shown in figure 2.

Ricoh has generated new companies and processes, and a new business model in which they charge by the copy rather than by the machine, since the copiers they supply may be remanufactured rather than new. In fact, the report states that some companies in the UK have been able to reduce waste for landfill to zero.

Implications for the ergonomics of job design

A recent review paper (Zink 2013) summarises a number of changes that will be necessary to meet the new working environment illustrated above, in particular the sustainability aspects. A number of points are made; firstly, the changes in design of the product to use sustainable materials will equally need job redesign that also minimises waste in the job; entropy in the form of heat, emissions, materials waste, tool redesign, task removal or simplification, and so on. These will be in addition to the usual job criteria that ergonomists use.

Secondly, because of the ever-increasing pace of change, driven both by shorter-term drives of competitive advantage and by the longer-term needs of the sustainability agenda, there is an issue of 'human workability' (Ilmarinen 2006, Docherty, Kira et al. 2009), defined as having the health, physical, psychological and social functional capacity to do designated work, including professional competence, values and ethics, and understanding of work conditions, work community, supervision, and the characteristics and processes of change.

Thirdly, 'workability' in relation to changes entails considerations of 'employability'. This includes an individual's ability to gain and maintain employment, to move between roles within an organisation, to obtain new employment in an equivalent role, and to be equipped with the knowledge and skills to do so. Since the latter applies across all employees, employers have a role to play in ensuring collective learning as well; it becomes an organisational skill, particularly in imparting the skills of sustainability. While people may learn these skills in the workplace, the thinking involved will be useful in other aspects of their lives.

Finally, from an ergonomics perspective, all the issues mentioned above come covered with the complexities arising from the fact that all the people are individual in their hopes, aspirations, and personal histories. Clearly, then, what we as ergonomists normally call 'job design' with its emphasis on health, safety satisfaction and performance will need some reconsideration. It seems that there is a role here for the IEHF to bring about such reconsideration through recommendations for initial professional education and through subsequent Continued Professional Development.

Other implications for ergonomics emerging from this

Whether or not the recycling/re-use routes indicated in Figs. 1 and 2 happen within one company or within a constellation of smaller companies, they imply a linked set of systems that constitute a 'system of systems', in which people will play an important part. While ergonomics practitioners are skilled in systems ergonomics, there is a requirement for extension of this into systems of systems ergonomics, since these systems of systems have extra characteristics not evident at the systems level (Maier 1998, Dahmann and Baldwin 2008, Firesmith 2010, Barot, Henson et al. 2012, Henshaw, Barot et al. 2013).

Secondly, as implied in the discussion above, there is a large degree of organisational re-arrangement and redistribution of roles, authority, responsibilities and skills involved. Inevitably there will be associated organisational culture changes to be carried through as well; as a manager commented when involved in an organisational shift to a service orientation (similar to Ricoh), "Basically [we] spend [time] explaining to people some of the new skills that we need and some of the new mindset issues about, for example, spares always used to be a profit opportunity, now they're a cost, and that's a completely 180 degree turn for us" (quoted in Johnstone, Dainty et al. 2008). There is an evident role for ergonomics skills in accomplishing this, but these skills need the extra knowledge base about this area to make the skills most effective.

Thirdly, there is likely to be a knowledge and culture upheaval in the design and engineering parts of the organisation, both to redesign products, to incorporate recycling considerations, to creating new processes and their associated technology, and creating new jobs. For all of these, there is a role for ergonomics knowledge and skills, especially in the light of creating 314,000 new ones in the UK alone.

Fourthly, given that government involvement will be necessary, both to create a business environment in which private investment will be encouraged and to provide a regulatory environment to protect people, the general environment and the businesses involved, there may be an important role for the IEHF as an institution to provide sound advice to the government of the day.

Finally, for the IEHF to be in a position to fulfil its role in these new aspects of business, it must ensure that its practitioners are equipped to provide the expertise to their clients. There is a role here for Continuous Professional Development and for initial skills training by the educational providers, and for research to develop the new knowledge required. It would be timely for the IEHF Council to consider how this role might be fulfilled.

It would be nice to think that the IEHF will be ensconced comfortably on the sustainability bandwagon of the future, rather than trailing in its dust, catching up with it.

References

Barot, V., S. Henson, C. E. Siemieniuch, M. J. d. C. Henshaw, M. A. Sinclair, S. L. Lim, C. Ncube, M. Jamshidi and D. DeLaurentis. 2012. *State of the art report: Trans-Atlantic Research and Education Agenda in Systems of Systems (T-AREA-SoS)*. p. n. EU Framework Programme 7. Loughborough, UK, Loughborough University,: 126.

Dahmann, J. and K. Baldwin 2008. Understanding the Current State of US Defense Systems of Systems and the Implications for Systems Engineering. *2nd Annual IEEE Systems Conference*. Montreal.

Docherty, P., M. Kira and A. B. A.B. Shani, Eds. 2009. *Creating sustainable work systems*. London, Routledge.

Firesmith, D. 2010. *Profiling systems using the defining characteristics of Systems of Systems (SoS)*, SEI, Carnegie-Mellon University.

Henshaw, M. J. d., V. Barot, M. A. Sinclair, C. E. Siemieniuch and S. Henson 2013. *Common Vision and Strategic Research Agenda (Work Package 5, Deliverable D5.1)*. Loughborough, UK, Loughborough University.

Ilmarinen, J. 2006. *Towards a longer work life: aging and the quality of work life in the European Union*. Helsinki, National Institute for Occupational Health.

Johnstone, S., A. Dainty and A. Wilkinson 2008. *Delivering 'product-service': key challenges for Human Resource Management*. Loughborough, UK, Loughborough University.

Maier, M. W. 1998. "Architecting principles for systems-of-systems." *Systems Engineering* 1(4): 267–284.

Sulston, J. 2012. *People and the planet*. London, The Royal Society.

Zink, K. 2013. "Designing sustainable work systems: The need for a systems approach." *Applied Ergonomics* 45: 126–132.

USERS' UNDERSTANDING OF INDUSTRIAL ROBOT GESTURE MOTIONS AND EFFECTS ON TRUST

Gilbert Tang, George Charalambous, Philip Webb & Sarah R. Fletcher

Aerostructure Assembly and Systems Installations Group/Industrial Ergonomics and Human Factors Group,
Centre for Advanced Systems, School of Engineering, Cranfield University

Industrial human-robot collaboration has been identified as a promising concept for optimising manufacturing output. One aspect of this concept which can have positive impact on the overall cooperation is the concept of robot to human communication through robot gestures. Although robotic gestures have been investigated for social robots, the impact of such gestures on humans has received little attention for industrial applications. This research aims to investigate the impact of industrial robot gestures on: (i) human understanding of robot gestures and (ii) effect of robot gestures on human trust in the robotic assistant.

Introduction

Industrial robots have been identified as an effective solution for optimising a manufacturing plant's output (Walton, Webb and Poad 2011). However, there are a number of manufacturing applications involving complex tasks which prohibit the use of fully automated solutions in the foreseeable future. Health and safety regulations are being advanced and updated to reflect that in some circumstances it is now safe and viable for humans to work more closely with industrial robots (ISO 10218-2-2011). Robots in such systems will operate as "intelligent assistants" in a shared workspace and will require a high level of cooperation with effective communication to avoid unwanted conflicts. Understanding robotic gestures has been a major topic of interest in human-robot interaction (HRI) (Nakagawa, Shinozawa, Ishiguro, Akimoto and Hagita 2009). There has been little focus on investigating the implications of industrial robots and environments.

Understanding of robot gestures by the human partner

Industrial robots vary greatly in appearance and capability with different features such as number of arms and articulation, degrees of anthropomorphism, size and payload, different types of end-effector, etc. These features can pose a significant challenge when attempting to incorporate robot gestures for communicating to the human partner during a collaborative task Gesture communication has been investigated with humanoid robots (e.g. Riek, Rabinowitchy, Bremnerz, Pipez and Fraserz 2010) but as industrial robots are typically less anthropomorphic interpretation of similar gestures can be a significant challenge. Ende, Haddadin, Parusel,

Wusthoff, Hassenzahl and Albu-Schaffer (2011) studied gestures in human-human interaction and transferred a selection to a single arm robotic system demonstrating that some are better recognised than others. Gleeson, MacLean, Haddadi, Croft and Alcazar (2013) investigated collaborative human gestural communication in assembly task performance for the development of a lexicon for industrial human-robot collaboration and demonstrated that successful cooperation relies on gestures being understood by the human. Although industrial contexts can vary significantly, the aforementioned studies support that human understanding of robotic gestures is one aspect of successful cooperation in collaborative tasks. Recognising and understanding gestures is likely to determine human operator trust

Effect of robot gestures on human trust

Human trust in a robotic teammate *"plays a critical role when operating a robotic system in terms of both acceptance and usage"* (Yagoda and Gillan 2012, p. 235). It is a multi-faceted condition determined by the human's mental model of a robot's capabilities and the given task context (Ososky, Schuster, Phillips and Jentsch 2013). In a meta-analytic review of factors affecting trust in HRI, it was found that robot performance-related factors (e.g. robot behaviour, predictability etc.) had the highest influence on trust (Sanders, Oleson, Billings, Chen and Hancock 2011). Also, non-predictable robot motions have been found to be hard to understand and subsequently have a negative impact on human well-being (Bortot, Born and Bengler, 2013). However, although trust has been the topic of many recent studies in the domain of HRI, very little research has been directed towards understanding the impact of industrial robotic gestures on human trust.

Although research has started to place focus on industrial robotic gestural communication, it is still at an infancy level in terms of developing a set of gestures suitable for industrial human-robot collaborative tasks. To ensure effective industrial human-robot cooperation, it is important to investigate the effects of industrial robot gestures on: (i) human understanding of the gestures and (ii) impact of gestures on users' trust in the robotic teammate.

Method

A study was designed to test human understanding of robot gestures and the effects of the gestures on their trust in the robot during a collaborative task.

Participants

Sixteen students and staff of Cranfield University (12 male and 4 females, age: $M = 28.6$, $SD = 7.1$) took part. Five participants were classified as having no prior experience with robots/automation, six participants were classified as having intermediate experience with robots/automation (took part in previous robot experiments) and five participants classified as having high involvement with

Table 1.   Experiments summary table.

Robot gesture	Gesture meaning
Robot gesture No. 1	"Come here"
Robot gesture No. 2	"Step back"
Robot gesture No. 3	"Stop"

robots/automation (involved in research projects/used automated machines, such as computer numerically controlled machined).

Materials

The robot system used was a single arm robot system incorporating a two-clamp gripper. Four plastic drain pipes were used approximately 15 cm each.

Design

A repeated measures design approach was followed. In the first part participants observed three robot gestures and answered what they think each of the gestures meant. In the second part participants took part in a human-robot collaborative task where the robot utilised identical gestures. The gesture order in the interaction task was random to reduce the possibility of order effects.

To identify suitable robotic gestures, previous studies were reviewed (Ende *et al.* 2011). For this study, three gestures were selected. Gesture selection was based on two criteria: (i) gestures that received high recognition rate without any contextual information provided to participants and (ii) gestures applicable for the needs of the task to be performed in this study. The gestures selected are shown in (Table 1).

Procedure

This study aims to investigate the effects of industrial robot gestures (independent variable) on: (i) human understanding of the gestures (dependent variable 1) and (ii) impact of gestures on users' trust in the robotic teammate (dependent variable 2).

Part A: Participants observed three robot gestures, one at a time, in a video format on a computer screen. No task-related information was provided to participants. A text box was provided underneath each of the videos for participants to write what the intended message of each of the gestures was. Each of the gestures was shown on a separate page and participants were allowed to view each gesture three times. Upon completing this, participants were taken to the robot work cell.

Part B: Participants took part in a human-robot interaction task. The task involved the robot positioning four small pipes, one at a time, in a pipe holder. Two pipes had a pink sticker, one had a green sticker and one had a blue sticker. A pink and green

Table 2.   Complete categorisation for comprehension score.

Correctness	Weight	Frequency	Comprehension Score
Correct	1	A	[(A * Weight)/Total answers] * 100
Likely	0.75	B	[(B * Weight)/Total answers] * 100
Arguable	0.5	C	[(C * Weight)/Total answers] * 100
Suspect	0.25	D	[(D * Weight)/Total answers] * 100
Incorrect	0	E	[(E * Weight)/Total answers] * 100

colour area was created next to the pipe holder. Participants had to select only the pink and green pipes and position them in the appropriate area. When the blue pipe was placed in the holder, the robot utilised the "stop" gesture. When the pink and green pipes were positioned the robot utilised the "come here" gesture followed by the "step back" gesture. At the end of the interaction task, a short semi-structured interview took place. The order of the pipes presented was random to reduce the possibility of order effects. The interview started by asking participants to give their thoughts regarding the interaction and what they think the gestures meant. Then to identify the impact of gestures on users' trust, participants were asked how the gestures influenced their trust and if they could rely on the robot gestures to complete the task.

Analysis

Written responses from the first part of the study were grouped and interviews from the second part were fully transcribed. A comprehension score for each gesture was obtained from the written responses and the interviews for each participant. Also, through the interviews we were able to identify the impact of gestures on users' trust.

Comprehension score: Participant responses were initially categorised in two categories: correct and incorrect. Correct responses received a weight of 1 and incorrect a weight of 0. To assess whether a response was correct or incorrect researchers generated a correct response for each of the gesture. The correct meaning for each gesture is shown in table 1. Any positive answer not considered 100 per cent correct was subjected to a different algorithm to evaluate the comprehension score. The algorithm was based on a study conducted by (Corbett, McLean and Cosper, 2008) and includes three more categories: likely, arguable and suspect, each of which carries a different weight. The complete categorisation and weight for each category is shown in table 2. Once the responses were tabulated, the comprehension score was calculated as shown in table 2. Total comprehension score was obtained by summing the individual scores of each category.

Impact of gestures on trust: Transcripts were analysed by coding volume of text frequently discussed among participants into common themes. This approach led to the development of a coding template. The template structure was revised iteratively to ensure it reflected the data in the most suitable manner.

Table 3. Comprehension score table.

Experiment part	Gesture 1 (Come here)	Gesture 2 (Step back)	Gesture 3 (Stop)
Part A	51.6%	4.7%	3.1%
Part B	87.5%	29.7%	43.8%

Inter-rater reliability: Analysis of inter-rater reliability was carried out to confirm the level of consensus between raters and, therefore, the suitability of the measures used to measure gesture understanding and trust.

For the comprehension scores, one of the researchers categorised the responses for parts A and B of this study and obtained a comprehension score. A second researcher categorised the responses individually. Results were then tabulated for calculation of the Cohen's kappa statistic, for this data, chosen because it shows the level of concordance between ratings corrected for the probability of agreement by chance thus giving a more conservative result when compared to simple agreement percentage. The Cohen's kappa for part A was 0.72 and for part B was 0.63 indicating there was 'substantial agreement' among raters (Landis and Koch 1977).

The coding template developed as described above was used by an independent rater to code the interview transcripts. Results were then tabulated for calculation of the Cohen's kappa statistic. The Cohen's kappa was 0.616 suggesting 'substantial agreement' among raters (Landis and Koch 1977).

Results

The aim of the user study was to investigate human understanding of robotic gesture and their impact on users' trust in the robot.

Gestures comprehension scores

Results indicated that comprehension scores were lower in part A when compared to part B (Table 3).

With the exception of gesture 1, gestures 2 and 3 received very low comprehension scores for part A (4.7% and 3.1%). For part B on the other hand, all gestures received higher comprehension scores (87.5%, 29.7% and 43.8%).

Impact of robot gestures on users' trust

Impact of robotic gestures on user's trust pointed four major themes (Table 4).

The majority of participants felt that the impact of robot gestures on their trust was mainly influenced by two factors: (i) gesture understanding and intuitiveness and (ii) human-likeness of the robot gestures. Another common aspect influencing trust in the robot was the development of familiarisation with the gestures and the

Table 4.   Major trust related themes.

Trust related theme	Frequency
Gesture understanding and intuitiveness	14
Robot gestures motion	12
Experience and familiarisation development	5
Robot programmer	5

robot. Some participants reported that their trust in the robotic assistant was due to trusting the person who developed the program rather than the gestures.

Discussion

In part A, with the exception of the "come here" gesture (51.7%), the "step back" and "stop" gestures received very low comprehension scores (4.7% and 3.1%). This is congruent with a previous study (Ende *et al.* 2011) where the "stop" and "step back" gestures received a much higher identification rate (92% and 84% respectively). A possible explanation for the low comprehension score in our study is that no context information was provided in the videos. This appears to have had some influence since all gestures received a higher score (87.5% gesture 1, 29.7% gesture 2 and 43.8% gesture 3) in part B. However, for part B gesture 2 ("step back") still appears to receive a low comprehension score (29.7%) when compared to gesture 1 ("Come here") and gesture 3 ("Stop") (87.5% and 43.8%). This indicates the lack of gesture 2 to convey the appropriate message.

In relation to the impact of robotic gestures on users' trust, it was found that if they could not understand the gestures their trust decreased. In addition, participants found trust in the robot increasing with human-like gestures. These two themes appear to be interrelated. It was discussed that human-like gestures are more easily understood and are considered more trustworthy. To this end, most participants found gestures 1 ("Come here") and 3 ("Stop") human-like and some participants described them as "universally gestures" in human-human non-verbal communication. Gesture 2 on the other hand, was the hardest to interpret and this had a negative impact on their trust. This is reflected in the comprehension scores obtained for part B and can potentially explain why gesture 2 received the lowest comprehensive score. This appears to be consistent with the notion that human-like motion enhances social acceptance and comprehensibility of the gestures (Gielniak, Liu and Thomaz 2013).

Also, some participants suggested that developing familiarisation with the gestures was important in order to trust the robot. Initially participants felt unsure approaching the robot, particularly when the "Come here" gesture was initiated. This was because they could not predict what the robot would However, after familiarising themselves with the gestures, participants reported having higher trust in the robot that enabled them to overcome the initial surprise effect. This is consistent with the notion that experience and familiarisation with robotic teammates foster trust in human-robot interaction (Ososky *et al*. 2013).

Some participants reported that their trust in the robot was due to trusting the person who installed program rather than the gestures. Interestingly, these participants were classified as having high exposure to robots. These participants achieved a lower comprehension score when compared to participants with intermediate or low experience. It appears that participants with higher exposure to robots appear to already have formed certain expectations towards robots.

Conclusion

In this study we investigated the implications of three gesture motions performed by a small industrial robot on: (i) human understanding of the gestures and (ii) user's trust in the robotic teammate.

An important result is that human-like robot gestures, such as "come here" and "stop" were found intuitive and can convey the intended message more accurately. The "step back" gesture on the other hand was not well understood which was reflected in the comprehension scores in both parts of the study. Thus, when considering robot gestures for industrial applications, more research is required to identify a selection of gestures that can be applied across a variety of robots. The investigation on the impact of robot gestures on users' trust identified that being able to correctly comprehend the intended message of robot gesture has a positive impact on trust. Subsequently, human-liken robot gestures appear to foster trust in the robot when humans are collaborating with a robot to complete a task. Also, it was found that experience and familisarisation with robotic gestures can foster trust in the robotic assistant. At the same time, participants with higher exposure to robots were found to place more trust in the robot programmer rather than the robotic gestures.

The study shows that robot gesture communication requires additional research to identify a set of gestures that can be applied across a variety of robots. Therefore, some gestures which are not easily interpreted can be improved when coupled with other indication methods such as lamp signal or audio notification.

Acknowledgements

This work was supported by the UK EPSRC Centre for Innovative Manufacturing in Intelligent Automation. Particular thanks to John Thrower for his technical support for this study.

References

Bortot, D., Born, M. and Bengler, K. (2013). Directly or detours? How should industrial robots approach humans? *International Conference on Human-Robot Interaction*. Tokyo: IEEE.

Corbett, C. L., McLean, G. A. and Cosper, D. K. (2008). Effective presentation media for passenger safety I: Comprehension of safecard pictorials and *pictograms*. FAA.

Ende, T., Haddadin, S., Parusel, S., Wusthoff, T., Hassenzahl, M. and Albu-Schaffer, A. (2011). A human-centered approach to robot gesture based communication within collaborative working processes. *IEEE/RSJ International Conference.* San Francisco: IEEE.

Gielniak, M. J., Liu, C. K. and Thomaz, A. L. (2013). Generating human-like motions for robots. *International Journal of Robotics Research*, 1275–1301.

Gleeson, B., MacLean, K., Haddadi, A., Croft, E. and Alcazar, J. (2013). Gestures for industry: Intuitive human-robot communication from human observation. *IEEE International Conference on Human-Robot Interaction* (pp. 349–356). Tokyo, Japan: IEEE.

International Standards Organisation. (2011). *Robots and robotic devices – Safety requirements for industrial robots-Pt 2: Robot systems and integration.* ISO.

Landis, J. R. and Koch, G. G. (1977). The measurement of observer agreement for categorical data. *Biometrics*, 159–174.

Nakagawa, K., Shinozawa, K., Ishiguro, H., Akimoto, T. and Hagita, N. (2009). Motion modification method to control affective nuances for robots. *In Proceedings of the 2009 International Conference on Intelligent Robots and Systems* (pp. 3727–3734). IEEE.

Ososky, S., Schuster, D., Phillips, E. and Jentsch, F. (2013). Building appropriate trust in human-robot teams. *AAAI Spring Symposia Series.* Stanford University: AAAI.

Riek, L. D., Rabinowitchy, T. C., Bremnerz, P., Pipez, A. G., Fraserz, M. and Robinson, P. (2010). Cooperative gestures: Effective signalling for humanoid robots. *5th International Conference on Human-Robot Interaction* (pp. 61–68). Osaka: IEEE.

Sanders, T., Oleson, K. E., Billings, D. R., Chen, J. Y. and Hancock, P. A. (2011). A model of human-robot trust: Theoretical model development. *Proceedings of the Human Factors and Ergonomics Society Annual Meeting* (pp. 1432–1436). SAGE.

Walton, M., Webb, P. and Poad, M. (2011). Applying a concept for robot-human cooperation to aerospace equipping processes. *SAE International.*

Yagoda, R. E. and Gillan, D. J. (2012). You want me to trust a ROBOT? The development of a human-robot trust scale. *International Journal of Social Robots*, 235–248.

AN EVALUATION OF THE ERGONOMIC PHYSICAL DEMANDS IN ASSEMBLY WORK STATIONS: CASE STUDY IN SCANIA PRODUCTION, ANGERS

Mohsen Zare[1], Michel Croq[2], Agnès Malinge-Oudenot[2] & Yves Roquelaure[1]

[1]*LUNAM Université, Université d'Angers, Laboratoire LEEST, Angers, France*
[2]*XSXEA, Safety, Health and Work Environment Support,*
Scania Production Angers SAS

This study assessed the physical workloads among assembly line operators. The SCANIA manufacturing assembly Angers was our experimentation plant. The in-house Ergonomic Standard method for assembly line was used to evaluate ergonomic workloads in assembly work stations. We performed an intervention for the job station in which the ergonomic workloads were high, and repeated the ergonomic evaluation after modification. The results showed that a large variability of ergonomic workload existed among different workstations. Also, we observed that risk factors reduced after ergonomic intervention. We discuss the different ergonomic risk factors that might influence on the probability of musculoskeletal disorders.

Introduction

Industrial companies and manufacturers have to be competitive due to being faced with new challenges in the industrial world. Higher productivity, better quality as well as reduced workplace disease particularly work-related musculoskeletal disorders (WMSDs) which are related to ergonomic are some important factors in order to reach organizational goals (Falck & Rosenqvist, 2012; Törnström, Amprazis, Christmansson, & Eklund, 2008). WMSDs are main causes of disability in many workplaces, especially assembly lines. More than 40% of occupational costs are related to musculoskeletal disorders. Medical costs, lost work times, workers' compensations and absenteeism are some apparent expenditures of WMSDs for industries (Speklé et al., 2010). Automotive industries are subject to a prevalence of WMSDs due to the characteristics of assembly line. Thus, corporation of ergonomic in assembly line and evaluation of adverse risk factors not only might reduce the WMSDs but also could improve productivity and quality (Morken et al., 2003; Törnström et al., 2008). Assemblers have large variety of tasks including tightening, picking pieces and parts, lifting with tools or hands and material handlings. Physical risk factors among mentioned tasks are awkward postures, force demands, repetition, and duration of exposures. In addition, there are different kinds of materials should be carried which the most of them have the weight between 3–7 kg, although a few pieces are heavier than 14 kg in some stations. The static forces for whole body are rather high among assemblers because of pushing and pulling. The assessments of these factors have indicated their relations with incidence of WMSDs among a variety of occupations (David, 2005). Ergonomists, occupational health practitioners, and employers are eager to perform accurate assessments of workers' exposures to those ergonomic risk factors that might cause WMSDs. Exposure to potential ergonomic risk can be reduced by ergonomic interventions, before accurate and precise ergonomic evaluation is necessary (van der Beek, Erik Mathiassen, Windhorst, & Burdorf, 2005). Physical risk factors could be assessed by observational exposure assessment techniques which are widely used for posture assessment and determining physical workload in manufacturing. Using observational methods, they could give us the opportunity to find preventive ergonomic interventions. A large numbers of observational methods were developed as well as the in-house methods were built up by industrial companies particularly assembly line corporations. There are a number of studies about workplace analysis with the methods like RULA, Strain Index, REBA, OCRA etc. However, few papers address ergonomic workloads with corporate-internal ergonomics methods (Berlin, Örtengren, Lämkull, & Hanson, 2009). Our purposes in this study are to assess exposures to potential ergonomic risk factors in truck manufacturing assembly line with corporate-internal method and prioritise the risks for further ergonomic intervention. In addition, we compared the results of modification for the station with high ergonomic risks.

Materials and methods

This is a case study that was carried out in SCANIA assembly plant in Angers. In this study, we chose 12 assembly workstations to evaluate with the Ergonomic Standard Method, which is an internal-corporate technique. Various kinds of trucks are assembled in each station according to the clients' demands. The most frequent kinds of trucks are the standard one, while the other options of trucks compose moderate percentage of assembly rate per day. Therefore, we chose the standard and significant variants and 28 evaluations were done in function of truck variants. All of the research population who participated in our study were men. The operators changed their stations rotationally after 2 hours. Lacks of the acute diseases, pains or accidents in musculoskeletal system were the inclusion criteria. To have an accurate ergonomic assessment, we recorded all of our evaluation with camera, so station analyses were done according to the movies. SCANIA in-house Ergonomic Standard method (SES) for production has been used for identifying potential ergonomic risks. This method is adapted with the ergonomic risk requirements in assembly manufacturing which could evaluate multi task work stations. SES includes 20 parameters which are classified in 5 categories. The categorizations of these parameters are shown in table 2. For prioritization of ach assessment, we used traffic light model and the results are sorted in the following order: Green or normal zone which show the minimal risk of WMSDs and these kinds of risks are acceptable. Yellow shows the zone which has a moderate WMSDs risk. Yellow

Table 1.   Categorization of risk factors in the observational method used in this study for ergonomic evaluation.

Category	Examples
Repetition	Parameters ask about the number of repetition or qualifying repetitive movement
Posture	Questions review whole body posture, Obstruction in the workspace, Distance for manually fitting parts, workspace for hands during operation, handgrip, contact surface on a part or mounting part
Lifting	Questions about manual material handling with two and one hand
Force	Questions about pushing and pulling with whole body, hands and fingers as well as using tightening tools
Activity	Questions about Movement and Climbing

Table 2.   Prioritization of risk factors in the observational method used in this study.

Category	Quantity	Station Colour
Number of Yellow	0–8	Green
	9–16	Yellow
	≥17	Red
Number of Red	0–6	Green
	7–9	Yellow
	≥10	Red
Number of Yellow + Red	0–16	Green
	≥17	Red
Number of DR	0	Green
	1–32	Double Red

tasks and job stations might need some improvement actions in the future. Red is action zone and there are considerable WMSDs risks for people, so changes require as soon as possible. Finally, double red (DR) shows the potential for excessive ergonomic risks. Tasks that are assessed as DR should be stopped immediately and solutions found.

The number of yellow, red and DR were then totalled and determine the colour of workstation (Table 3). After identification of major ergonomic risks, especially those tasks with double red colour, we proposed some interventions. Then, the proposed modifications were discussed among technicians and finally the best action was selected to perform. Executing the modifications, we repeated the evaluation for the modified stations and changing in the ergonomic workloads were assessed after intervention. All the data were inserted in the Excel file and analysis were done.

Results & discussion

The colour zone ratings are tabulated and show the ergonomic workloads in the evaluated workstations for standard trucks (Table 4). As shown in Table 3 there is large

Table 3.   Ergonomic evaluation of each station for standards trucks: the colour of each ergonomic risk factors in different stations.

Position	Repetition	Work posture	Hidden assembly	Clearance	hand workspace	Handling Grip	Area for pressure	component size	back posture	Neck posture	shoulder posture	wrist posture	2 handed lift	1 handed lift	push/pull whole body	push/pull hand	push/pull finger	movement	Climbing	tightening torque
Air filter	Y	R	R	R	R	Y	G	Y	Y	G	Y	R	R	Y	R	G	G	G	G	Y
boarding steps	G	R	R	R	G	Y	G	G	Y	Y	R	R	G	G	G	R	Y	G	G	Y
Mudguards	G	Y	Y	G	R	Y	G	R	R	Y	R	R	Y	G	Y	R	R	Y	G	R
PA air filter	G	Y	G	G	R	R	G	R	Y	R	G	R	R	Y	R	G	R	Y	G	Y
Bumper	Y	G	Y	R	G	Y	G	Y	Y	Y	Y	R	G	Y	R	G	Y	Y	G	Y
PA Bumper	G	G	Y	R	R	Y	G	G	Y	Y	R	R	G	G	Y	G	Y	G	G	Y
bumper the line	G	R	R	G	R	Y	G	G	Y	Y	R	G	G	Y	Y	G	DR	G	G	R
Picking	G	G	R	G	G	Y	G	R	Y	Y	Y	G	G	Y	Y	G	G	G	G	G
Bumper,	G	G	Y	G	R	G	G	R	Y	Y	R	R	G	Y	Y	G	G	G	G	Y
preparation sun	Y	G	G	G	G	R	R	R	R	G	Y	Y	R	R	G	G	Y	R	G	G
visor	G	G	G	G	R	Y	G	G	Y	G	R	G	Y	Y	Y	Y	Y	G	G	G
PA SCR	G	G	G	R	G	Y	G	Y	Y	Y	Y	R	G	Y	G	R	Y	G	G	Y
SCR line	G	R	Y	R	G	Y	G	Y	Y	Y	Y	R	R	Y	R	DR	Y	G	G	R

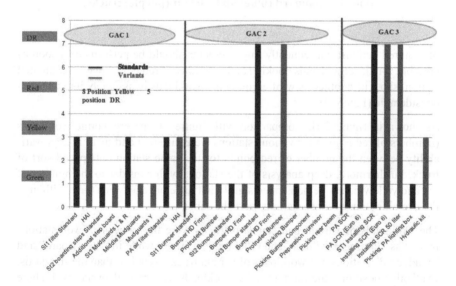

Figure 1.   The overall colour for all assessed stations regarding to standard & variants truck.

variability of workloads between different stations. The assessments of all stations in the cluster regarding to various kinds of trucks gave us the overall ergonomic scores for each station. The stations' colours and their comparison were shown in Figure 1. The overall ergonomic workloads of the stations (the overall colour of each station) either standard trucks or variant trucks are the same. Assembly of Bumper and Anti-air pollution Tank (SCR) for standard and prioritised in the zone

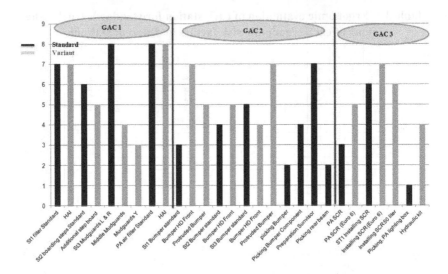

Figure 2. The distribution of the number of red points in different workstations for standard (blue) and variant (purple) trucks.

of required actions; consequently improvements should be performed as soon as possible. On the other hands, looking at the number of red points in each evaluated station, show us another view of ergonomics workloads in assembly line which is considerable (Figure 2).

As shown in figure 2 (in comparison with figure 1), the ergonomic workloads (numbers of red points) for various stations were different and there is large variability between the number of red points for the same station but special sort of truck. Furthermore, deep analysis of the stations with consider to the number of red and yellow points demonstrated significant difference between them, although the overall colour of those stations were the same.

The workloads of variant and standard trucks were the same in 2 workstations named as "Putting Air Filter" and "Picking Air Filter" while in "Step Board" and "Mudguard" stations, the workloads of variant trucks were less than the standards. In all other positions, the variant trucks' workloads were more than standard. There were clearly more ergonomic workloads (more red points) for variant compared to standard trucks. Therefore, our findings prove that we have to take into account the other variant of trucks in assembly line and work on them. Currently, most of assembly manufacturers believe that assessing the potential of ergonomic risk factors for more frequent sorts of their products are sufficient nevertheless we observed the risk factors would change during the 8 hours of working for operators. It would be necessary to consider in further research developing a visual tool that not only could show the ergonomic risks for most frequent products but also could illustrate changes in the patterns of ergonomic workloads among assemblers during a typical workdays in the function of different products.

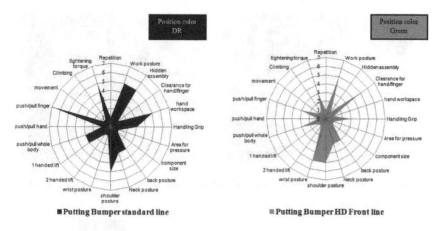

Figure 3. The visualization of the pattern of risk factors at "Putting Bumper on the Line" station for standard (blue) and variants (pink) trucks.

Ergonomic intervention

Figure 3 shows the patterns of ergonomic workloads in the station named as "Putting Bumper on the Line" for two different kinds of trucks (Standard & Variant). The overall workloads are different in this station for different types of trucks as the colour for standard truck is double red while the variant truck is green (Figure 3). The main reason of high ergonomic workload for standard trucks is the task unlocking lifting tool for displacing the bumper (Figure 4a) which causes extra forces for fingers. Our measurement with dynamometer showed that unlocking lifting tool required the force approximately 200N. To unlock lifting tool the thumb and index finger are involved as it is so-called finger grip or finger-top grip.

Dynamometer measurement was not enough accurate because the access was not sufficient to have a right method for using dynamometer. Assembling a variant truck did not require this task, so the ergonomic workload for the station is green (Figure 3). Therefore, we decide to eliminate the double red task in this station for standard truck. Designing a new unlocking system, we changed the overall colour of the station from double red to green and the extra required force for finger-top grip was eliminated (Figure 4). The new unlocking system did not need the involvement of the fingers as we used a cord to unlock the lifting tools. The ergonomic evaluation for redesign station is shown in figure 5. As it is illustrated in the pattern of this station, the colour for push/pull with finger is green and overall colour of this station for standard truck is green as well. The ergonomic evaluation for redesign station is shown in figure 5. As it is illustrated in the pattern of this station, the colour for push/pull with finger is green and overall colour of this station for standard truck is green as well.

Conclusion

In conclusion, our study showed that there are relatively important ergonomic risks among various stations of assembly line. The comparisons between standard and

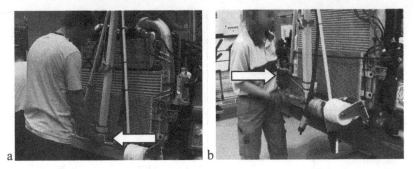

Figure 4.   The task "Unlocking Lifting Tool" a- before intervention b- after.

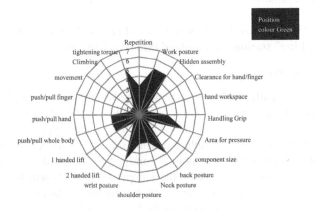

Figure 5.   The pattern of ergonomic workloads for the station "Putting Bumper in Line" after intervention.

variant trucks were done and show a large variability of the physical workloads for the same station. Therefore, unlike the current believing of the manufacturer, it is essential to consider not only the ergonomic risks of the frequent kinds of products but also the other types of production have to be considered. Furthermore, our intervention successfully influenced the ergonomic workload.

References

Berlin, C., Örtengren, R., Lämkull, D., & Hanson, L. (2009). Corporate-internal vs. national standard–A comparison study of two ergonomics evaluation procedures used in automotive manufacturing. *International Journal of Industrial Ergonomics, 39*(6), 940–946.

David, G.C. (2005). Ergonomic methods for assessing exposure to risk factors for work-related musculoskeletal disorders. *Occupational Medicine, 55*(3), 190–199.

Falck, A.-C. & Rosenqvist, M. (2012). What are the obstacles and needs of proactive ergonomics measures at early product development stages?–An interview study in five Swedish companies. *International Journal of Industrial Ergonomics, 42*(5), 406–415.

Morken, T., Riise, T., Moen, B., Hauge, S.H.V., Holien, S., Langedrag, A. Thoppil, V. (2003). Low back pain and widespread pain predict sickness absence among industrial workers. *BMC Musculoskeletal disorders, 4*(1), 21.

Speklé, E., Heinrich, J., Hoozemans, M., Blatter, B., van der Beek, A., van Dieën, J., & van Tulder, M. (2010). The cost-effectiveness of the RSI QuickScan intervention programme for computer workers: Results of an economic evaluation alongside a randomised controlled trial. *BMC musculoskeletal disorders, 11*(1), 259.

Törnström, L., Amprazis, J., Christmansson, M., & Eklund, J. (2008). A corporate workplace model for ergonomic assessments and improvements. *Applied ergonomics, 39*(2), 219–228.

van der Beek, A.J, Erik Mathiassen, S., Windhorst, J., & Burdorf, A. (2005). An evaluation of methods assessing the physical demands of manual lifting in scaffolding. *Applied Ergonomics, 36*(2), 213–222.

HF IN PRODUCT DESIGN

REFERABILITY: MAKING THINGS EASY TO REFER TO

Gisle Andresen & Øystein Veland

Department of Product Design, Norwegian University of Science and Technology

Referring, the act of drawing another person's attention towards some entity through words or actions, is an essential part of collaboration. In this paper we propose a new design principle: referability. This principle should aid designers in making the contents of collaborative tools easy to refer to. The principle is backed up by a framework that outlines key elements of the practice of referring: what vocabularies people use when they refer, what elements they refer to, and how they do the actual referring. The framework should give designers a better understanding of how it is possible improve the referability of a tool and what pitfalls to avoid.

Introduction

Referring is the act of directing a person's attention towards some entity through words or actions (Yule, 2010). Several studies indicate that for collaborative tools to be successful, they need to enable efficient referring (e.g., Tatar, Foster and Bobrow, 1991; Gergle, Kraut and Fussell, 2003; Whittaker, 2003). No studies have compiled this knowledge into a form that is easy for designers to utilise, however.

We propose a new design principle: referability. It says that to support collaboration, a collaborative tool's content should be easy to refer to. On its own the principle is not very informative, but it is backed up by a framework that outlines the vocabularies, design elements, and methods people use when they refer to the contents of collaborative tools.

The main contribution of the paper is to provide practical guidance on how collaborative tools can be enhanced through better support for referring. The paper also identifies areas in which it is necessary to conduct further research.

Background

In this paper we are primarily interested in definite referring; i.e., referring to specific elements in a collaborative tool. There are three ways of doing this (Cruse, 2010). You can describe the target element using a noun phrase ("the grey button at the top of the screen"). You can use a name that uniquely identifies the specific element ("the menu button"). Or you can use deixis, i.e., point to the element via language ("that button").

Referring is a collaborative process (Clark and Wilkes-Gibbs, 1986). In other words, both the person initiating the referring, the 'speaker', and the person which the speaker adresses, the 'listener', must contribute to the process. For example, if the speaker says "that button", and the listener finds the reference ambiguous and responds "do you mean the white?", then the speaker must make additional contributions ("no, the grey one") to complete the referring process. This collaborative view of language use is not unique for referring, however. Many researchers on collaboration and collaborative tools have adopted this view. It will also be the theoretical foundation of this paper. That is, we will build on Clark's theory of language (Clark, 1996).

It is beyond the scope of this paper to give a thorough description of Clark's theory, but we need to introduce a few concepts. Let us begin with *common ground*. Common ground is the knowledge, experiences, and beliefs we assume we have in common with other people. With our focus on collaborative tools we are both interested in things that are shared in the physical world and the vocabularies we use for referring to the physical world.

People try to minimise the effort they spend on collaborative processes. Clark has called this the principle of least collaborative effort. It is then natural to ask: what can we as designers of collaborative tools do to make the referring more efficient?

To begin with, there is substantial evidence that so-called shared workspaces are critical for referring (Whittaker, 2003). This has been demonstrated both in experiments and ethnographic studies (Gergle et al., 2003; Hindmarsh and Heath, 2000). There are also studies that provide hints to what factors may hamper efficient referring. Tatar, Foster and Bobrow (1991) showed that that flexible layout of interface windows may cause difficulties with referring to locations ("What window in the upper left corner?"). Gergle, Kraut and Fussell (2003) demonstrated similar effects when they displayed information upside-down. What happens in these situations is that one strategy of referring, deixis, is rendered more or less useless.

Another question is how easy it is to describe or name what is perceived. For example, Gergle et al. (2003) found that referring was more efficient when the stimulus was less complex. This is not very surprising, but it implies an area in which designers may positively impact referring.

The studies we have described so far all involve distributed collaboration, that is, collaborative tools where speaker and listener use separate screens. There are also studies that show how referring may vary using a common screen. Bangerter (2004) showed that when the distance to the screen increased, the collaborators used pointing less frequently, and descriptions more frequently. They also found variation in how people used information about location and visual features to refer to elements.

Bangerter's study is an example of how people opportunistically use various referring strategies. It also indicates that not only deixis, but also the other strategies for referring can be influenced through design. Finally, it reminds us that referring is relevant to collaborative tools used in both distributed and co-located settings.

Methodology

In the next section, we will present a framework that should make it easier for designers to see what factors may influence referring. To accomplish this, we have made several simplifications. Further research and practice will reveal if these simplifications are justifiable. Although we do not provide any empirical evidence, it should be noted that the principle was developed concurrently with our own design of collaborative tools in complex work environments (e.g. Veland and Andresen, 2010).

Results

The framework consists of a) the vocabularies speakers use for referring, b) the elements of the tool they refer to, and c) the methods they employ in the act of referring.

Vocabularies: surface, artefact, and domain

The words a speaker uses to refer to elements of a collaborative tool can belong to three vocabularies: the surface, artefact or domain vocabulary. The surface vocabulary contains words that concern the look of visible elements. Except for visual appearance, these words convey no meaning. The surface vocabulary is understood by virtually everyone.

The artefact vocabulary contains words used for referring to elements commonly found in collaborative tools. This vocabulary is also understood by many people, but because of the rapid technological development, it undergoes more frequent changes.

The domain vocabulary contains word that are pertinent to a particular work domain. This vocabulary is only mastered by people with experience from the respective domain. Even more so than for the artefact vocabulary, the referring is closely tied to the particular tool and its use.

Although it might be possible to find situations where a speaker is primarily relying on one of these vocabularies, it is probably most common for speakers to opportunistically use all three. It should therefore be noted that the reason for introducing the three categories is not to argue that referring happens within certain vocabularies, but rather to expose the range of possible ways a speaker may refer to elements of a tool.

Let us now take a closer look at what elements of a collaborative tool the speaker may want the listener to attend to, and what words the speaker may use to refer to those elements.

Elements: frame, object and attribute

As will become clear later, to be able distinguish between different referring methods, it makes sense to separate elements into frames, objects and attributes.

A frame is an element used for organising other elements. The organisation may reflect a simple principle (e.g., a list ordered alphabetically), or it may convey an elaborate set of principles (e.g., a map). It is no coincidence that we have chosen to call this element frame; it literary builds on the notion "frame of reference".

In the surface vocabulary, a frame can be referred to as an 'area' or a 'field' or any other word people use to refer to a certain region of a display. In the artefact vocabulary, 'page', 'window', and 'panel' are examples of frequently used frames. In the work domain vocabulary, referring will vary from domain to domain and from tool to tool, but we might, for instance, find frames referred to as 'list', 'log', or 'plan'.

An object is the element the speaker typically would like the listener to attend to. In the surface vocabulary, objects can be associated with shapes with object-like properties such as 'square' or 'circle' or the speaker may rely on metaphors or similes ("it looks like a light switch"). In the artefact vocabulary, common objects are 'button' and 'icon'. In the work domain, objects can for instance be work orders or process components.

In principle, a speaker may refer to an attribute of a collaborative tool without having any particular frame or object in mind ("Look at the red!"). On most occasions, however, attributes will be associated with frames or objects. An attribute in the surface vocabulary can for instance be the colour of a region of the screen; in the artefact vocabulary it can be the status of a button; and in the work domain it can be the value of a process component. For frames, the most interesting attributes are the organising principles.

Methods: look-for-element and move-to-location

In the introduction we learned that there are three strategies for referring: deixis, naming and describing. In this paper we focus on referring that involves words only. This excludes deixis based on actions and gestures (e.g., pointing), but with that exception, all three strategies are relevant. We have chosen a slightly different categorisation, however. Instead of the three strategies we distinguish between two methods: look-for-element and move-to-location.

With a look-for-element utterance the speaker wants the listener to look for a certain element of the tool. In response, the listener performs a visual scan of the tool until a correct identification is made (or asks for more clues). The speaker assumes that simply by using words that directly refers to the target element, the listener will be able to identify it. A look-for-element utterance is achieved through naming or describing.

In contrast, a move-to-location utterance is deictic, or more correctly, it has a deictic function. That is, it "points" the listener in the direction of the target or "points" at the location of the target. The deictic function of a move-to-location utterance is most obvious when the speaker uses the surface vocabulary, "look at the top of the screen".

Implications for design

The interesting question is then how the framework can be used by designers. It is beyond the scope of this paper to conduct a comprehensive analysis, but let us try to extrapolate some implications based on the two referring methods, and see how they can manifest themselves in the design of a collaborative tool.

Look-for-element referring is basically a matter of naming or describing objects. Consequently, designers should try to create objects that are easy to "put words on". The most obvious solution is to assign the naming to the tool itself; i.e. to label objects with alphanumeric symbols.

Ideally, the labels should originate from a notation system belonging to the work domain, making the act of referring an integral part of talk-about-work. In reality though, introducing an entirely new notation system will be difficult due to its wide-ranging implications. A pragmatic solution can then be to include a dummy reference; i.e., a reference that is included for no other reason than to make referring easier in a particular collaborative setting.

Move-to-location referring is primarily a question of utilising frames. One obvious design strategy is to use frames to divide objects into groups and thus reduce the number of objects the speaker and listener must take into consideration during referring.

Another design strategy can be to make the organising principles of a frame easy to perceive and interpret. The frame can then serve as a coordination system, enabling the speaker to make precise and unambiguous referring to locations. Maps and calendars are good examples of frames that can facilitate efficient move-to-location referring.

Example: a collaborative planning tool

The collaborative tool we will use as an example is a research prototype implemented on a large touch-screen. It was designed to support an interdisciplinary team responsible for planning work on an offshore oil platform. The tool integrated data from various sources and used visualisation to show relations between variables and to highlight deviations. The tool was developed by the authors in parallel to the referability principle. Thus, the design was partly influenced by, and partly influencing our conception of the principle. Figure 1 shows a snapshot of the tool used during a user test.

Let us start with the implications derived from the look-for-element method that concerned use of alphanumeric symbols. We decided early to represent variables both through graphical and alphanumeric forms. The graphical representation should enable direct perception, while the alphanumeric values should support referring. An example of this can be seen in the upper part of the display in figure 1. Here there are both vertical bars and numeric values showing the number of available beds on the platform – an important constraint of the planning. Also, each

Figure 1.   The collaborative planning tool.

job, represented as horizontal bars on the timeline, was assigned numerical values. One of those served as a dummy reference. This was a number initially included for debugging purposes, but through user testing we found that it was an effective means of referring to jobs.

In respect to frames, the tool built on traditional timeline and Gantt formats and should thus be efficient for move-to-location referring. However, here we made a trade-off where referability was not prioritised. To increase the number of jobs that could be shown at the same time, we abandoned the convention of showing only one job per row. In addition, to reduce clutter, the vertical lines representing days on the timeline were barely visible. In effect, there was no coordination system to leverage in the referring. It should be noted that an important reason why we were willing to make this trade-off was that the screen was touch-based and allowed gestural deixis. Still, we can speculate that this was part of the reason why the dummy reference became a popular means of referring.

Discussion

Although referability is based on a theory that has been investigated in numerous HCI/CSCW studies, it does not necessarily follow that the principle is significant to performance in actual work environments. Is there any evidence that supports its practical relevance? Gergle, Kraut and Fussell (2013) demonstrated that information visualisation may influence referring, common ground and situation awareness. Gergle's experiments were performed on relatively simple tasks, but we can also find examples where it has been used to model complex work domains (see Klein, Feltovich, Bradshaw and Woods, 2005; Carroll, Rosson, Convertino and Gano, 2006).

When it comes to existing design recommendations we can find both principles that may promote referability (e.g. standardisation and transparency) and undermine it (e.g. tailoring and aesthetics). However, from the perspective of visualisation, we are primarily concerned about a potential conflict between referability and direct perception. Direct perception tends to drive design towards relatively complex graphical representations and less use of alphanumeric representations. Of course, we are not arguing against visualisations that support direct perception; our concern is that it may reduce the referability of a display. The challenge is then to find a good compromise. One solution can be to use both graphically advanced elements and simple alphanumeric elements (e.g., Bennett, Posey and Shattuck, 2008).

This research is still at a very early stage. One major limitation is that the framework has not been tested empirically or evaluated through practical use. Another limitation is that the framework needs to be expanded to assist designers in addressing some very central questions. For example, the framework does not cover gestural deixis. In terms of its relation to other research, several other domains should be explored. For example, there are undoubtedly lessons to be learned from studies of communication intensive work such as air traffic control.

Conclusion

We propose a new design principle: referability. The principle says that to support collaboration, the contents of collaborative tools should be easy to refer to. There are several trends that poses a threat to referability: different form factors of devices; individual tailoring of displays; visualisations that are primarily designed so support direct perception. These trends can negatively influence performance because they make referring to elements of collaborative tools less efficient. We believe a principle like referability can help designers make better tradeoffs.

Acknowledgement

This research was sponsored by the Center for Integrated Operation in the Norwegian Petroleum industry.

References

Bangerter, A. 2004. "Using pointing and describing to achieve joint focus of attention in dialogue." *Psychological Science* 15(6): 415–419.

Bennett, K. B., Posey, S. M. and Shattuck, L. G. 2008. "Ecological interface design for military command and control." *Journal of Cognitive Engineering and Decision Making* 2(4): 349–385.

Carroll, J. M, Rosson, M. B., Convertino, G. and Ganoe, C. H. 2006. "Awareness and teamwork in computer-supported collaborations" *Interacting with Computers* 18: 21–46.

Clark, H. H. 1996. *Using language.* (Cambridge University Press, Cambridge, UK)

Clark, H. H. and Wilkes-Gibbs, D. 1986. "Referring as a collaborative process" *Cognition* 22(1): 1–39.

Cruse, A. 2010. *Meaning in language: an introduction to semantics and pragmatics.* (Oxford University Press, Oxford, UK).

Gergle, D., Kraut, E. R. and Fussell, R. S. 2013. "Using visual information for grounding and awareness in collaborative tasks." *Human-Computer Interaction* 28(1): 1–39.

Hindmarsh, J. and Heath, C. 2000. "Embodied reference: A study of deixis in workplace interaction" *Journal of Pragmatics* 32: 1855–1878.

Klein, G., Feltovich, P. J., Bradshaw, J. M. and Woods, D. D. 2005, Common ground and coordination in joint activity. In W. B. Rouse and K. R. Boff. (eds). *Organizational Simulation*, (Wiley & Sons, Inc., Hoboken, NJ, USA), Retrieved 21st September 2013, from http://onlinelibrary.wiley.com.

Tatar, D. G., Foster, G. and Bobrow, D. G. 1991. "Design for conversation: lessons from Cognoter" *International Journal of Man-Machine Studies* 34(2): 185–209.

Veland, Ø. and Andresen, G. 2010. *Focusing design and staging conversations in complex settings through "design documentary" filmmaking.* Create 2010, the interaction design conference, 30th June to 2nd July 2010, Edinburgh.

Whittaker, S. 2003. "Things to talk about when talking about things." *Human-Computer Interaction* 18(1): 149–170.

Yule, G. (2010). *The study of language. 4th edition.* (Cambridge University Press, Cambridge, UK).

ANALYSIS OF HUMAN FACTORS ISSUES RELATING TO THE DESIGN AND FUNCTION OF BICYCLES

Derek Covill, Eddy Elton, Mark Mile & Richard Morris

University of Brighton

The cycling industry is now the largest sporting goods market in the world, with global sales of £28 billion (€33 billion) amounting to 15% of all sporting goods revenue. There is a large body of research relating to cycling related activities and mostly these consider the common diamond "safety" framed bicycles. There are a number of areas that could benefit from direct input from ergonomics, including the redesign of braking grips for drop handlebars on road bicycles, and the verification and validation of products for specialist performance. The design of bicycles for human wellbeing and performance comes into question with the ubiquity of standard upright "safety" bicycles when compared with the performance and ergonomic benefits of recumbent bicycles.

Introduction

In 2010 the cycling industry became the largest sporting goods market in the world, with global sales of £28 billion (€33 billion) amounting to 15% of all sporting goods revenue and approximately 137 million bicycles (including electric bicycles) sold that year (NPD group, 2011). Recent victories for British athletes in high profile events such as the 2012 and 2013 Tour de France and the 2012 Olympics have ensured that cycling is one of the top sports in the UK. Cycling is the fourth most popular sport in the UK in terms of weekly participation with just under 1.9 million people cycling each week compared with swimming which is the most popular sport at almost 2.9 million participants (Sport England, 2013). Over £26 m was invested for the 2012 Olympic cycling team through UK sport, second only to rowing (UK Sport, 2013). There is considerable evidence of increasing participation and interest in cycling (Grouse, 2011), and the UK has seen the National Cycling Network grow from 5,000 miles in 2000 to over 12,000 miles in 2013 (Sustrans, 2013).

There is a large body of research relating to bicycles and cycling related activities. Mostly these studies relate to the common diamond "safety" framed bicycles, ranging from the biomechanics and physiology of cycling, design requirements and manufacturing technology of bicycles, injury prevention and safety of cyclists, and the anthropometric requirements of cyclists. From a human factors perspective, much of this research appears fragmented as it draws from various subject areas with pockets of large amounts of very specific research in some areas, and gaps elsewhere. The aim of this paper is to bring much of this research together, along with an analysis of the bicycle designs and their components as part of a literature and product review of the design and use of bicycles from a human factors perspective. Our research question is: "to what extent has the bicycle design been optimised for human wellbeing and performance, according to a standard defined goal of ergonomics (Dul and Weerdmeester, 2003). This review has been broken down into analyses of research and products relating to the user and the task according to the BS ISO EN standard for Ergonomics (BS EN ISO 26800:2011), using subheadings and key questions from the approach suggested in Dul and Weerdmeester (2003).

The user

Biomechanical, physiological and anthropometric factors

Men dominate cycling in the UK, setting out on 70% of all bicycle trips – for comparison in the Netherlands men undertake only 45% of all bicycle trips (Grous, 2011). The key reasons why women cycle more in northern Europe are safety and accessibility to appropriate cycling infrastructure (e.g. bike lanes and parking) (Grous, 2011). There is an "under-representation of women in cycling journeys, which is believed by some to have 'turned a corner' as more women look to cycling within their local communities" (Grous, 2011, p. 8). The dominance of men in the UK cycling scene is reflected in (perhaps exaggerated by) the proportion of male specific bicycles that are available on the market. We analysed four major online bicycle stores in the UK in terms of the number of bicycles on offer specifically for men and women (although some stores forgo the classification of bicycles for males and use "unisex" as the alternative option to bikes for females). Female specific bicycles currently available in stores make up only 21% of the total bicycles available. For adult bicycles, this figure drops to 19%, for adult hybrid/city bicycles this is 38%, and for adult road/MTB bicycles this is only 12%. Overall the proportion of female specific bicycles on the market (19%) is below the proportion of UK cyclists who are female (30%) (Grous, 2011), and this could be a result of the female market being under represented, or a response to female riders preferring to ride men's bicycles.

There are many basic requirements for users of bicycles ranging from the ability to store, ease of maintenance, ease of cleaning, ease of disassembly, comfort when riding, ease of mounting, specific gear ratios for climbing and descending, protection from road spray, visibility in traffic or at night, security and so on. Clearly these differ depending on whether the user is riding for sport, riding for pleasure or riding for transport, although this paper does not cover these specifically. Instead we focus on some of the fundamental requirements that relate most significantly to the human wellbeing and performance associated with riding a bicycle.

The process of fitting bicycles, including saddle position relative to the pedals, head and upper body positions relative to the stem have been covered in great detail in the literature (Kolin and de la Rosa, 1979; Burke, 1994; Burke, 2003). In the more recent literature, further guidelines are provided for setting handlebar heights,

Table 1.   **Analysis of male and female bicycle models available at four large UK stores.**

		Female	Male	Total	% Female
Chainreaction (female vs male)	Road	12	121	133	9%
	MTB	18	235	253	7%
	Hybrid/city	36	67	103	35%
	Kids	25	27	52	48%
Wiggle (female vs unisex)	Road	11	126	137	8%
	MTB	4	57	61	7%
	Hybrid/city	12	33	45	27%
	Kids	4	4	8	50%
Evans (female vs unisex)	Road	74	373	447	17%
	MTB	58	402	460	13%
	Hybrid/city	115	200	315	37%
	Kids	24	46	70	34%
Halfords (female vs male)	Road	2	76	78	3%
	MTB	29	70	99	29%
	Hybrid/city	47	49	96	49%
	Kids	32	38	70	46%
TOTALS (all)		503	1924	2427	21%
TOTALS (adults only)		418	1809	2227	19%
TOTALS (adults hybrid/city only)		210	349	559	38%
TOTALS (adults road/MTB only)		208	1460	1668	12%
TOTALS (kids only)		85	115	200	43%

adjustments for leg length discrepancies and setting crank lengths, cleat and foot position adjustments and handlebar adjustments (Burke, 2003; Silberman et al, 2005). Many of these are supported by research using wind tunnel tests and power tests to analyse the efficiency and effectiveness of (mostly experienced) riders in various positions on the bicycle (Burke, 2003). Alternative methods and details of tools, software and measuring techniques are provided for fitting bicycles to riders (Burke, 1994; Burke, 2003; Silberman et al, 2005), not to mention the myriad of websites with guidance and the many professional fitting services that are now available across the country.

The task

The basic functional task of riding the common diamond "safety" framed bicycles of today has developed little since the days of Starley's Rover safety bicycles in 1885 (Wilson, 2004). While the functionality of components, the quality of materials and manufacturing, the inclusion of pneumatic tyres and derailleur gears and the standardisation of components and operation have all vastly improved the usability of bicycles, the basic functionality of direct steering, variable-ratio transmission of chain driven bicycles remains intact.

Factors related to posture: sitting and general performance

While the diamond framed bicycle dominates the commercial world of bicycles (including track, road, touring, commuter, mountain, bmx bicycles), other models do exist which have been "relegated to the fringe" (Sorenson, 1998), while others (e.g. the folding bicycle) have carved out niches of their own. The recumbent bicycle (available in short/long wheel base and trike options) is considered to have advantages over the upright diamond framed bicycles in a number of categories including reduced frontal area (improving aerodynamics), safer head (i.e. behind the body) and body (i.e. lower to the ground) positions, improved visibility (with the head naturally facing forwards), comfort (with the head in a natural forward facing position, the seating position reducing the likelihood of localised saddle related sores and the lower back and wrists are in more anatomically neutral positions) (Ballantine, 2000; Wilson, 2004; Burrows, 2008). Figure 1 highlights the basic differences in design and posture associated with a standard diamond framed road bike (left) and a recumbent bike (right). While there are clear advantages in some areas, the recumbent position has the disadvantage in that it may not allow for peak power production and sustained aerobic performances as those obtained in the upright "safety" position (Kolin and De La Rosa, 1979; Reiser, 2000). Other more practical disadvantages are the higher baseline cost of recumbents and that they are generally longer and often more cumbersome to store (although generally less wide than most mountain bikes). A common objection to recumbents is the perception that they are unsafe in traffic. We use the term perceived, since we could find no specific evidence in a literature search on bicycle related injuries and cycling accident risks on the risk factors associated with recumbent bicycles, however some studies provided some implicit insights into the effect of speed (Peden et al, 2004; Kloss et al, 2006; Kwan and Mapstone, 2009), and visibility (Thompson and Rivara, 2001; Kloss et al, 2006; Thornley et al, 2008; Kwan and Mapstone, 2009; Washington et al, 2012) on the risks associated with cycling which are relevant to recumbent riding. In theory the lower visibility and increased speeds associated with recumbents makes them even more vulnerable in urban environments, although it

Figure 1.   The differences in design and posture associated with a standard diamond framed road bike (left) and a recumbent bike (right). Images from http://www.jimlangley.net (left), http://venturecyclist.blogspot.co.uk (right).

is disputed that this may be offset or even overcome on the basis of riding skill, experience and technique (Ballantine, 2000, Burrows 2008).

Regarding the use of saddles in the most common upright bicycles, clearly there is much evidence regarding the influence of saddle height on power output from the rider (Too, 1990; Burke, 2003). One trend that seems to be emerging are claims from saddle manufacturers relating to improved power output, impact absorption and other performance measures with no supporting evidence available on their websites or references to published work on the studies conducted. A recent study by Heneghan et al (2012) began to assess the evidence underpinning sports performance products with a rather damning conclusion such that: "The current evidence is not of sufficient quality to inform the public about the benefits and harms of sports products. There is a need to improve the quality and reporting of research". Whilst there is much evidence surrounding the general fitting, biomechanical and physiological aspects of cycling, including the problems associated with saddle pain, discomfort and other symptoms (Keytel et al, 2002; Bressel & Larson, 2003; Partin et al, 2012) there is often a lack of published data relating to the performance, comfort and behaviour of specific products even where benefits or advantages are explicitly claimed.

Factors related to posture: gripping

With more recent focus on women cyclists, the ability for females and smaller males to grip brake levers – which is a fundamental requirement of riding a bicycle, has been shown to be inadequate. As one rider (Jami) claims "I don't feel confident with my brakes because I have to strain to reach the brakes when I'm in the drop bars. Gripping the brakes on the hoods is also tiring because the levers are out so far" (McKee, 2010). Apart from a few smaller components that are now more suitable for women and men with smaller hands, typical solutions are simply adjustments that can be made to the component from the "male" default, e.g using a shim within the brake lever or make adjustments from the default setting, softening brakes to bring the actuation point of the brakes closer to the bar allowing the braking action to occur in a closed position, but as McKee puts it: "the only problem is that it doesn't really address the real root cause of the problem and that is that I can't reach my brake levers easily", followed by a disclaimer that "This is NOT a long term fix, for some people it may work. But there can be safety concerns if the brakes are not perfectly adjusted" (McKee, 2010). It seems the designs of braking components need a rethink, to allow for more inclusive solutions for riders with smaller hands and fingers, and to improve wrist and finger positioning on drop handlebars.

Factors related to information and operation

There has been much progress in terms of information afforded to the rider and in terms of operation of a bicycle. The moving of gear levers from the downtube to the

Figure 2.   Extended fingers on braking levers can cause strained wrist and fingers joints especially for longer rides. Image from http://cycleandstyle.com.

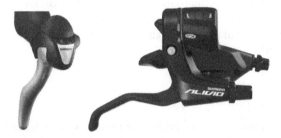

Figure 3.   Progress in gear shifters now allows more affordance in their design, with road (left) and mountain bike (right) shifters containing an optical gear display. Images from http://www.leisurelakesbikes.com.

handlebars (e.g. Shimano Total Integration (STI) shifters) has allowed users to operate without looking down at ill placed gear levers, which also ensures both hands can be kept on the handlebars. Perhaps the biggest impact on safety however is that the fingers and hands are kept away from the moving wheel during this operation. Furthermore, an important milestone for the affordance of gear shifters has been greatly improved by the inclusion of optical gear display systems to display current gear status rather than looking down at the drivetrain to assess the combination of gears (see Figure 3).

The standardisation of components and tools required to support the ease of maintenance and affordability of bicycles has come a long way in recent years. The use of Allen keys and standard components makes maintenance of most bicycles more straightforward and intuitive. There remain issues however with some awkward tasks, for example the alignment of brake pads which sit on spherical seatings for cantilever brake pads can be problematic and similarly the installation of a front derailleur installation, both requiring a skilful technician or even a "third hand" product to support the action.

There are also potential issues with such expensive high-end specialist components that there is a reluctance to adjust/handle because of fear of damage. In some cases components are so lightweight they are damaged easier, either during general maintenance or use. Similarly, the use of specialist performance components (e.g. deep rims), makes maintenance difficult or cumbersome (e.g. requires the tyre to be taken off in order to true the wheel).

Conclusion

Clearly there has been much progress in terms of the ergonomic interface of many bicycle components and the bicycle frames themselves. There is however a number of areas that could benefit from direct input from ergonomic design experts, including the redesign of braking grips for drop handlebars on road bicycles, and the verification and validation of products for specialist performance and comfort. Fundamentally the design of bicycles for human wellbeing and performance comes into question with the ubiquity of standard upright bicycles when compared with the performance and benefits of recumbent bicycles since recumbents offer clear ergonomic advantages in many respects.

References

Ballantine, R. 2000, "Richard's 21st Century Bicycle Book". (Pan Books, London, England).

Bressel, E. and Larson, B.J. 2003. "Bicycle seat designs and their effect on pelvic angle, trunk angle, and comfort". Med Sci Sports Exerc. 35(2):327–32.

British Standards. 2011. "Ergonomics — General approach, principles and concepts". (BS EN ISO 26800:2011).

Burke, E. 1994. "Proper fit of the bicycle". *Clinics in Sports Medicine*, 13(1), pp. 1–14.

Burke, E. 2003. "High tech cycling: the science of riding faster". 2nd Ed. (Human Kinetics, Leeds, England).

Burrows, M. 2008. "Bicycle Design: the search for the perfect machine". (Snowbooks Ltd. London, England).

Grous, A. 2011, *The British cycling economy*. Retrieved 20/9/2013, from: http://corporate.sky.com/documents/pdf/publications/the_british_cycling_economy

Dul, J. and Weerdmeester, B. 2003. *Ergonomics for Beginners: A Quick Reference Guide*. 2nd Edition. (Taylor & Francis, London, England).

Heneghan, C., Howick, J., O'Neill, B., et al. 2012. "The evidence underpinning sports performance products: a systematic assessment". BMJ Open 2012;2:e001702. doi:10.1136/bmjopen-2012-001702.

Keytel, L. and Noakes, T., 2002. "Effects of a novel bicycle saddle on symptoms and comfort in cyclists." S Afr Med J, 92(4):295–8.

Kloss, R., Tuli, T., Haechl, O. and Gassner, R. 2006. "Trauma injuries sustained by cyclists". (Trauma, 8, pp. 77–84).

Kolin, M. and De la Rosa, D. 1979. "The custom bicycle: buying, setting up and riding the quality bicycle". (Rodale Press, Emmaus, USA).

Kwan, I. and Mapstone, J. 2009. "Interventions for increasing pedestrian and cyclist visibility for the prevention of death and injuries (Review)". (Cochrane review, 2009, Issue 4).

McKee, T. 2010. "Women's Hands and Brake Levers". Retrieved 20/9/2013, from: http://cycleandstyle.com/2010/10/womens-hands-and-brake-levers/

NPD Group. 2011. Global Sport Market Estimate. Retrieved 20/9/2013 with subscription, from: https://www.npd.com/wps/portal/npd/us/news/press-releases/pr_110915a/

Partin, S., Connell, K., Schrader, S., et al. 2012. "The bar sinister: does handlebar level damage the pelvic floor in female cyclists?". J Sex Med. 9(5):1367–73.

Peden, M., Scurfield, R., Sleet, D., et al. 2004. "World report on road traffic injury prevention". (World Health Organization, Geneva).

Reiser, R. 2000. "Biomechanics of recumbent cycling: instrumentation, experimentation, and modelling". PhD thesis, Colorado State University.

Silberman, M., Webner, D., Collina, S. and Shiple, B., 2005. "Road bicycle fit". (Clin. J. Sport. Med. 15 (4), 271–276.

Too, D., 1990. "Biomechanics of cycling and factors affecting performance". (Sports Med. Nov;10(5):286–30).

Sorenson, E. 1998. *Rolling in from the fringe – the recumbent bicycle has benefits and may be ready to shed its 'weird' image.* Retrieved 20/9/2013: http://community.seattletimes.nwsource.com/archive/?date=19980616&slug=2756382.

Sport England. 2013. "Playing numbers by sport". Retrieved 20/9/2013, from: http://www.sportengland.org/research/who-plays-sport/by-sport/who-plays-sport/

Sustrans. 2013. *National Cycle Network – Sustrans.* Retrieved 20/9/2013, from: http://www.sustrans.org.uk/ncn/map/national-cycle-network

Thompson, M. and Rivara, F. 2001. "Bicycle related injuries", (Am Fam Physician, 63, 10, pp. 2007–2017).

Thornley, S., Woodward, A., et al. 2008. "Conspicuity and bicycle crashes: preliminary findings of the Taupo Bicycle Study". (Inj Prev, 14: 11–18).

UK Sport, 2013. *UK Sport – Sport by Sport – London 2012.* Retrieved 20/9/2013, from: http://www.uksport.gov.uk/sport/summer/

Washington, S., Haworth, N. and Schramm, A. 2012. "On the relationships between self-reported bicycle injuries and perceived risk among cyclists in Queensland, Australia". (In TRB 91st Annual Meeting, Washington, D.C.)

Wilson, 2004. *Bicycling Science*, 3rd Edition. (MIT Press, London, England).

HUMAN FACTORS AND PRODUCT DESIGN: UNCOMFORTABLE ALLIES

Eddy Elton

Brighton Product Lab, University of Brighton

The current state of Human Factors (HF) in Product Design (PD) is discussed within this paper. An analysis of consultancies, advertised jobs, the inclusive design area and product exemplars were conducted. Results showed that HF does not currently appear to play a significant role in the design of products. In order to address current and future societal changes e.g. the ageing population, HF needs to become the central emphasis of PD processes. A proposed way of addressing this is detailed in the form of a new educational pathway that combines HF and PD. It is envisaged that such a pathway would create individuals with necessary skill sets to address the societal challenges we face and overcome the barriers currently posed in this relationship.

Introduction

Ergonomics and/or Human Factors (HF) are two different terms that are used to refer to a discipline that focuses on 'designing for people/human use' (Institute of Ergonomics and Human Factors, 2013), and optimizing design for human well-being and overall system performance (British Standards Institute, 2011). In this paper, for consistency, HF will be used throughout.

Design is a process that is concerned with defining real world problems and creating appropriate solutions (Design Council, 2009). Product Design (PD), is specifically concerned with the creation of tangible solutions that will be commercially viable, i.e. appeal to mass and niche markets of people (Wright, 1998). A market consists of all the potential customers/users who share a particular need or want that might be willing to engage in exchange (of cash) to satisfy that need or want (Kotler, 1991). Thus, to summarise, PD is a process that is followed in order to create tangible objects that fulfil the wants and needs of various groups (markets) of people within society.

Evidently, there is significant overlap between HF and PD, i.e. both would appear to focus on designing for people. It would therefore be assumed that HF would be the central emphasis of all PD processes, resulting in a wealth of human-centred products available in today's market place. Unfortunately, this does not appear to currently be the case; the following section of this paper will discuss why this is so.

Consultancies

IEHF accredited consultancies

The Institute of Ergonomics and Human Factors (IEHF) has a total of 57 Accredited Consultancies, which are organisations that offer consultancy services in HF, with at least one employee being a Registered Member or Fellow of the IEHF. An analysis of their stated 'areas of work' was conducted from the details provided on the IEHF website. Results showed that out of the 57 consultancies, 18 (31%) stated that Product Design and Consumer Ergonomics was an area of work under practice.

Product design consultancies

The 'Design Directory' is an online directory of design consultancies (www.designdirectory.co.uk); the consultancies listed cover a very broad range of design areas, from Architectural through to Website Design. In the UK, there are currently 1,202 consultancies listed in the Design Directory under the area of Product and Industrial Design. A targeted search for the 'design skill' 'Ergonomics – Product Design' revealed 64 consultancies, 5.3% of the listed UK Product and Industrial Design consultancies who work/specialise in this area.

As part of a separate study, five of the Product and Industrial consultancies who listed the design skill 'Ergonomics in Product Design' were interviewed (Elton, 2012); only one of these consultancies employed HF specialists. At the other four consultancies, it was the product designers who were responsible for the application of HF. Interestingly, the owner/manager of one of the consultancies, when asked how they involved users in the design of their products, said *"we are users ourselves ... we as designers would sit down and fiddle with them, use them, give them to our friends, so it's an informal way of getting this information; this is quite common of how we work."* Pheasant (1990), identifies a very similar approach as being one of the five fundamental fallacies to HF i.e. the design is satisfactory for me – it will, therefore, be satisfactory for everybody else.

Consultancy findings

If these figures were taken at face value, it would suggest that out of the 1,257 consultancies registered on the IEHF and Design Directory websites (2 consultancies were registered on both sites), 82 (6.5%) are actively engaging in the application of HF to the design of products.

Initial insights from the Product and Industrial Design consultancies offering the 'design skill' 'Ergonomics in Product Design,' would suggest that it is the designers who are responsible for the application of HF to the design of products. In the one instance cited, it would appear as though HF is being applied incorrectly. It is important to note that this paper is by no means trying to imply that this type of user engagement/consideration of HF in PD is representative of the identified consultancies. However, it provides a hint towards how it might be considered in some design consultancies.

Advertised jobs in HF

A review of job vacancies advertised in 'The Ergonomist' (a monthly membership magazine of the IEHF) was conducted. The review focused on a 12-month period, between June 2012 and May 2013. During this period, a total of 31 jobs were advertised (including both commercial and academic roles, but not PhD studentships); posts ranged from HF specialists working in Nuclear and Defense, to training roles in Musculoskeletal Disorders (MSDs). Out of the 31 advertised vacancies, just two (6.5%) were advertised at PD consultancies. Only one of these vacancies specifically stated that the role involves applying HF understanding to the design of a broad range of products. The other role was more specific, (focused towards medical device design and usability testing, and stated design skills in wireframe software and mock-up tools as desirable.

Societal developments: are things beginning to change?

Societal developments and changes have meant new demands are being made on products and thus design practice. One of these changes is the dramatic increase in older adults (+65 years). In 1950, there were 200 million over 65s worldwide, in 2006 there were 487 million, and by 2050 it is predicted there will be 1.55 billion (U.S. Census Bureau, 2010). New design approaches have begun to emerge in order to address such issues; including one that has been the focus of much research over the past decade – inclusive design.

The inclusive design approach aims to deliver mainstream products and services that are accessible to, and usable by, as many people as reasonably possible, within the widest range of situations, without the need for special adaptation or design (British Standards Institute (BSI), 2005). In essence, inclusive design can be categorised as a human-centred approach to design, as it "... *encompasses all aspects of a product used by consumers ... its ultimate goal is to meet the needs of a diverse range of consumers by creating effective user-centred designs and better integrated product ranges through the application of HF principles* (BSI, 2005). HF is therefore deemed an essential component to the successful implementation of this approach and thus critical in ensuring products are suitably designed to address the aforementioned societal changes.

As previously mentioned, much research has been devoted to inclusive design over the past decade. In particular, gathering better data on users and creating better tools for designers to aid them in the development of inclusive products. Generally, tools have been developed to help designers either a) understand people, or b) access people. As part of this body of research, a number of studies (Cassim, 2005; Nickpour & Dong, 2010, 2011; Elton & Nicolle, 2011; Elton, 2012) have investigated how these tools can be developed so that they are *accepted* and *used* by the design community. This research suggests an underlying consensus that the only way to embed HF into the design of inclusive products is to provide product designers with HF (inclusivity) tools and information. This can be exemplified by citing recent publications within this area; for example Zitkus *et al* (2013) conducted a study that examined how to enable designers to make design decisions toward more accessible products; Elton & Nicolle (2011) investigated how to translate older adult capability data into a suitable form for designers; Nickpour & Dong (2010 & 2011) have conducted multiple studies into developing user data tools for designers.

The need for such research and thus design tools would imply the absence of HF specialists within the design teams/companies responsible for developing these types of products. In the case of inclusive design, it suggests the *designers* are applying HF to the design of products, not the HF specialists. Unfortunately, recent evidence suggests that these tools have not been widely used by industry (Zitkus *et al,* 2011) and such tools would only be accepted if they could be tailored to each design domain and the tools designers currently use (Zitkus *et al,* 2013). Also, Zitkus *et al* (2013), found that designers will only consider HF requirements if the clients require them to, but it is not common for clients to include HF requirements into their product requirements.

Technological developments

Aside from the aforementioned societal changes, there have also been rapid and unprecedented developments in technologies over the past few decades. These developments have led to the 'internet of things,' i.e. sensors, technology and networks all coming together to allow products, infrastructure and other resources to swap information. In fact, Pétrissans *et al* (2012) predicts that by 2020, 31 billion devices will be inter-connected; a large proportion being consumer products. Such products will require greater attention to their user interfaces, distant users, context of use, Standards, etc. Thus, HF will need to play a key role in developing these products in order to ensure they efficiently and effectively deliver their intended value to the end users.

Product exemplars

Another way of determining whether HF is considered accordingly in the design of products is to review products currently available in the marketplace/used as exemplars. The three key elements described in the 'hierarchy of user experience' (Cagan and Vogel, 2002) are often used to identify products that achieve good design from a HF perspective:

- **Functionality** – the extent to which the functionality of the product offers benefit to the user.
- **Usability** – the extent to which the user can achieve their specific goals with effectiveness, efficiency and satisfaction.
- **Desirability** (pleasurability) – the extent to which the product connects with the user on an emotional level e.g. social status, self-image, personal values, etc.

Product exemplars commonly used in business cases (Mieczakowski *et al*, 2013), teaching material and published literature (Jordan, 2000) include Apple, OXO Good Grips, BT (Big Button and Freestyle phone) and the Ford Focus car. The fact that 'a business case' proposition is indeed needed for HF in the first place is again evidence in itself to support the notion that Human Factors and Product Design are indeed uncomfortable allies. However, experts within the field of HF and PD have in fact used these very products to cite bad examples of the application of HF. For example, Hosking in his Include 2011 conference presentation (The inclusive design toolkit: on the practice frontline) highlighted the multi-touch gestures required to interact with Apple products as an area of concern; Clift (2013) highlighted aspects of the iPhone call function as bad design, along with the OXO good grips principle not actually solving the problem it apparently addresses. Elton and Nicolle (in press) cite an observation of an elderly participant who suffers from Parkinson's disease who removed the rubber grips from the OXO good grips cutlery before using them to eat with; the participant explained *". . . I don't need them . . . I'm not that disabled."*

It is therefore questionable whether the exemplars used to highlight good HF design actually achieve it. This raises the question as to whether the three factors identified as constituting good HF design (i.e. functionality, usability and desirability) are actually achievable within a single product. Clift (2013) raises an interesting point in reference to the product exemplars cited, in that; he states, *". . . these are product of designers thinking they understand Ergonomics."*

Discursive snapshot

The evidence presented suggests that:

- Only a small percentage HF and Industrial and Product Design consultancies focus on applying HF to the design of products. The very limited number of job vacancies that are advertised to fulfill such roles further supports this.
- Societal changes (such as the ageing population) result in a greater demand to integrate HF into the design of products; whilst this is an area that is rich in terms of research, evidence would suggest that the actual application of HF tools and techniques, by designers, to address such issues is minimal.
- Technology-driven consumer products of the future will require significant HF input in order to ensure they efficiently and effectively deliver their intended value to the end-users.
- There are a selection of products within the marketplace that are used as exemplars of good HF design; however, certain features of such products have been identified as providing bad examples of the application of HF.

Out of all of the factors discussed, the key pinch-point appears to exist with the relationship between HF and PD. For example, organisations not specifying it in project briefs, designers only considering it if specified, designers applying HF specialist knowledge and not HF specialists, and very few consultancies combining the two subjects in their areas of work. When considering the overlap between the two subject areas, this appears illogical. So, the key question to arise is how can these two subject areas become better allies in the future?

A way forward

Winograd and Woods (1997) have argued that the development of a discipline can be motivated either by the abstract logic of the discipline or by the needs of its areas of application. If we take the latter, and consider the societal changes occurring as a result of the ageing population, it is evident that HF needs to become better integrated in to the design of products, but more importantly, it needs to become the central emphasis when designing products for tomorrow.

Three potential ways of addressing this are:

1. Develop more profound business cases in order to better influence those who commission design work.
2. Better educate designers to work with HF specialists (Clift, 2013).
3. Create an educational pathway that will provide individuals with the necessary skill set and knowledge base to achieve this.

Case 3 (an educational pathway) is discussed further here. As noted previously, HF is a discipline focused on 'optimising design for human use/well being;' thus, a discipline that is centred around 'designing for people,' but does not actually design i.e. make ideas compelling and tangible through sketches, drawings, models and prototypes. Why? It would appear logical that individuals with such knowledge of how to optimise design would actually have the skills to design.

The 'hierarchy of user experience,' as detailed in the product exemplars section, provides the necessary HF framework that such an educational pathway should primarily be based around. There is also the consideration of the design skills that such an educational pathway should contain, i.e. the key skills an individual needs to make ideas tangible. The iD Cards produced by Evans and Pei (2010) provide a taxonomy of the key design representations used by product/industrial designers during product development. There are well-established design processes that place the user at the centre of design activities; such processes include the Human-Centred Design process (BSI, 2010) and the Inclusive Design Toolkit (www.inclusivedesigntoolkit.com). Thus, the necessary frameworks, knowledge and processes that lead to human-centred product solutions already exist, and synergising this knowledge into an appropriate educational pathway has the potential to address the issues raised in this paper.

Conclusion

This paper has given an insight into the current state of HF in PD. An analysis of consultancies, advertised jobs, societal and technological developments, and product exemplars were conducted. Results showed that HF does not currently appear to play a significant role in the design of products and would therefore suggests that HF and PD are indeed uncomfortable allies. An approach is proposed to address this in the form of a new educational pathway that combines HF and PD. Creating such an educational pathway has the potential to:

- Provide individuals with the necessary skill sets and knowledge to address current and future challenges posed by society;
- Ensure HF naturally becomes the central emphasis in the design of products, whether specified by the commissioning organisation or not;
- Reduce the need to develop HF tools for the design community – just the need to collect appropriate user data would remain;
- Ensure the HF discipline remains relevant and current in the design of products.

Maybe, offering a multidisciplinary educational pathway, such as this, is one way in which we can design a better tomorrow.

References

British Standards Institution. 2005. *BS 7000-6:2005 Design Management Systems. Managing Inclusive Design.* London: British Standards Institution.

British Standards Institution. 2010, *BS EN ISO 9241-210:2010 Ergonomics of human-system interaction, Part 210: Human-centred design for interactive systems.* London: British Standards Institution.

British Standards Institution. 2011, *BS EN ISO 26800:2011 Ergonomics – General approach, principles and concepts.* London: British Standards Institution.

Cagan, J., and Vogel, C. 2002, *Creating breakthrough products: Innovation from product planning to program approval.* Prentice Hall, New Jersey.

Cassim, J. 2005. Designers are Users too! Attitudinal and Information Barriers to Inclusive Design within the Design Community, In: *Proceedings of Include 2005*, London.

Clift, L. 2013. Dear Editor. *The Ergonomist.* No. 518, August 2013, P15 & P15.

Design Council. 2009, What is design. Retrieved 23rd August, 2013, from http://vimeo.com/5820010

Elton, E. 2012. Generating and translating context capability data to support the implementation of inclusive design within industry. Doctorial Thesis. Loughborough University.

Elton, E., and Nicolle, C. 2011, Translating inclusive capability data for designers. In: *Proceedings of Include 2011*, Helen Hamlyn Research Centre, London.

Elton, E., and Nicolle, C. in press, Inclusive Design and Design for Special Populations. In. J. Wilson, and S. Sharples (eds) Evaluation of Human Work, 4th Edition. London, Taylor & Francis.

Evans, M. A., and Peri, E. 2010. iD Cards: A taxonomy of design representations to support communicating and understanding during new product development. Loughborough Design School, England.

Institute of Ergonomics and Human Factors. 2013, What is Ergonomics? Retrieved 23rd August, 2013, from http://www.ergonomics.org.uk/learning/what-ergonomics/

Jordan, P. 2000, *Designing Pleasurable Products*. London, Taylor & Francis.

Kotler, P. 1991, *Marketing Management: Analysis, Planning, Implementation and Control. 7th Edition*. Prentice-Hall International, UK.

Mieczakowski, A., Hessey, S., and Clarkson, J. 2013. Inclusive Design and the Bottom Line: How can its value be proven to decision makers. *Universal Access in Human Computer Interaction*. Vol. 8009, 67–76.

Nickpour, F., and Dong, H. 2010, Developing user data tools: Challenges and opportunities. In: P. Langdon, J. Clarkson, and P. Robinson (eds.) *Designing Inclusive Interactions*. London, Springer: 79–88.

Nickpour, F. and Dong, H. 2011, Designing anthropometrics! Requirements capture for physical ergonomic data for designers. *The Design Journal* 14(1): 92–111.

Pétrissans, A., Krawczyk, S., Cattaneo, G., Feeney N, Veronsei, L. and Meunier, C. 2012. Design of future Embedded Systems toward system of systems – trends and challenges (SMART 2009/0063). Retrieved 29th November, 2013, from: http://cordis.europa.eu/fp7/ict/embedded-systems-engineering/documents/idc-study-final-report.pdf.

Pheasant, S., 1990, *Bodyspace: Anthropometry, Ergonomics and Design*. London, Taylor & Francis.

Winogard, T. and Woods, D. D. 1997, The challenge of human-centred design. In J. Flanagan, T. Huang, P. Jones and S. Kasif (eds) *Human-Centred Systems: Information, Interactivity, and Intelligence* (National science Foundation: Washington DC).

Wright, I. 1998, *Design methods in engineering and product design*. McGraw-Hill, England.

U.S. Census Bureau. 2010, Population division 1-866-758-1060. Retrieved 27th August, 2013, from http://www.census.gov/prod/cen2010/briefs/c2010br-09.pdf.

Zitkus, E., Langdon, P., and Clarkson, J. 2011, Accessibility evaluation: Assistive tools for design activity in product development. *SIM Conference Proceedings*, 1, 659–670.

Zitkus, E., Langdon, P., and Clarkson, J. 2013, Inclusive Design Advisor: Understanding the design practice before developing inclusivity tools. *Journal of Usability Studies* 8(4), 127–143.

PLAYING IN THE PARK: OBSERVATION AND CO-DESIGN METHODS APPROPRIATE TO CREATING LOCATION BASED GAMES FOR CHILDREN

Cathy Grundy, Lyn Pemberton & Richard Morris

University of Sussex/Brighton
University of Brighton

A series of co-design activities were carried out in response to a 'real world design problem' initiated by the Sussex Wildlife Trust. Investigations considered how technological interventions could encourage outdoor play for older children (aged 8–12). The focus for the work was on developing participatory techniques for this age group that improve communication with adult partners and encourage creative thinking. In particular the use of character and story design activities facilitated open discussions about preferences and opinions and avoided the constraints of a real design problem. The children helped to co-create a series of Location Based Games that provided a useful design template for further concept development.

Introduction

The following paper describes research carried out to address a real world problem posed by the Sussex Wildlife Trust. The question was "How can we use technology to motivate older children to play outdoors and appreciate nature." According to the Sustainable Development Commission, an appreciation of nature at a key age can lead to greater enthusiasm for environmental issues and encourage people to engage more fully in a sustainable society (Sustainable Development Commission, 2009). There is also evidence that visiting the countryside at an impressionable age leads to a lifetime habit of enjoyment (Natural England, 2009). Young pre-teenagers can present a challenge however as, by this stage, many become less interested in outdoor play than might be desired, and typical activities organized by the trust, e.g. den building or digging for bugs have been less popular with older age groups. Interest in digital media is relatively common, however, and can provide a key channel to access learning about the environment and biodiversity. Varied research projects have been carried out to consider how different technologies can influence location-based learning about the outdoors, particularly for scholastic activities (Rogers et al., 2010). However, frequently these studies are focused on functionality and how the technology could improve interaction and data storage (Druin, 1999a). There is less emphasis on why the young person might wish to cooperate in such activities in the first place, or how it might be made more enjoyable.

There has been considerable interest in the use of computer games for learning due to their powers of motivation; it is generally accepted that computer games have

Figure 1.   ARIS Game Menu.

great capacity to engage young people. Williamson found that 87% of 8–11s played games on a console at home in the UK (2009). Games can simply be fun, but by their nature usually include a system of rewards and recognition of success or effort, which can be highly motivating. For this context, because the outcome is outdoor activity, a 'Location Based Game' is likely to be the most appropriate model for the investigation. There are a series of subcategories of play in this area, for example digital Geo-caching, where the player tries to find artefacts in a particular place that have been left by fellow players; Treasure Hunts where players look for information or carry out tasks in a given place, and more social gaming versions where proximity to other players is automatically registered and interactions are a game feature. For the purposes of the project the activity will be based on specific locations due to the subject under study. Previous research had indicated the significance of characters and stories to children, so the game was also intended to use this vehicle to add a narrative to further improve the motivational properties of the game.

After a literature review of appropriate forms of technology, it was considered that a phone or I-pad based App was likely to be most accessible medium to further investigate the play requirements. From the possibilities studied, ARIS, a relatively user-friendly open source platform for creating and playing mobile games was selected. It provides a relatively simple web based, Flash system for making tours, treasure hunts or adventure games with stories that children should be able to use. Further use of an app called Morfo was introduced as a simple way for the children to create video to project the narrative and characters from their drawings or natural phenomena.

Rationale

The nature of a 'Real World Problem' usually calls for a flexible methodology and methods that are participatory and inclusive. Due to the complex nature of the

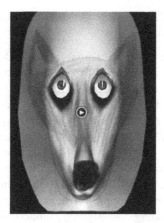

Figure 2.   Child's drawing animated with the Morfo App.

issues involved, with a range of stakeholders and questions that evolve in speci-
ficity over time, a variety of different methods will be required, including Action
Research, Ethnographic studies and a Participatory approach to development of
ideas. The approach here is analogous to Design Based Research, a methodology
frequently applied to research that relates to the development of learning tools,
however fun and motivation are more significant than imparting knowledge for the
project aims. Research for a design outcome can be considered a methodology,
according to Buchanan (1992), if it is properly validated during the process. An
aspect of the literature review towards this project has therefore been focused on
finding appropriate techniques to investigate the requirements of this group with
interrogations that are age appropriate and can be 'triangulated' to get valid data,
including Participatory and Co-Design methods.

Which participatory design method?

As an adult it is almost impossible to second-guess the preferences of an
8–12 year old child. (Naranjo-Bock, 2011) According to Mazzone et al. (2010)
"Children's ideas should be harvested appropriately when designing for them since
they observe and perceive things differently compared to adults". The work of Alison Druin (1999b) and many others also illustrates the significance of including
children in a design process at all stages. However, gathering *useful* information for
innovation can be more difficult than for adults. Not only do children have different
needs, behaviours and motivations, they also communicate in a different fashion. In
other words, the difference in perception and culture between children and adults
can lead to errors in interpretation during the process. Traditional participatory
methods therefore need to be modified in order to compensate for typical differences in cognitive development in order to gain useful results. However, Obrist
et al (2011) stated that there is still a profound lack of knowledge of how to involve

children in the different phases of a product development, in particular the early conceptualization and evaluation.

Elizabeth Sanders was one of the early pioneers of Co-Design techniques and part of her work while at Sonic Rim was to invent and apply the 'Say, Do, Make' model (Sanders, 2009). '**Say**' represents techniques that explore verbal information gathering, for example through questionnaires and interviews. Traditional forms of market research rely on verbal communication between people, i.e., on what people say. The "say methods" are useful for getting an idea of what people can and want to tell you in words, but they can be limited in predictive terms. '**Do**' represents observational techniques, like Video ethnography. These are useful but in many cases quite time consuming, often requiring extensive note taking, editing and analysis. **Make** describes more projective methods that involve creative techniques with the user, including co-designing. Sanders' work provides a semantic and practical categorization of the variety of methods available. An ideal situation would be a triangulation of information from all three **Say, Do, Make** categories. This project includes an attempt to investigate how we can apply the *Say, Do Make* tools for triangulating data.

Children as design partners

One of the more obvious reasons why 'Say' tools can be impractical for younger children is their relative inability to verbalize their responses. Methods that rely on description and memory are less likely to yield fruitful results, in either written or spoken form, than some alternatives. 'Do' tools are based largely on the observations of the design researcher and therefore a certain amount of interpretation is required. A child's world seen through adult eyes is open to misinterpretation and needs to be triangulated and tested against other results. Of the three approaches the 'Make' tools seem to fit with a child's ability and engagement most successfully. Despite skepticism in some quarters, Garzotto (2011) suggests that beyond the traditional roles of children as users, testers and informants, children are highly capable of being Design Partners and taking a more creative role. Alison Druin is also a proponent of including children as active designers through a process called 'Cooperative Inquiry' This also includes children as observers and researchers of others besides being arbiters in the design decisions (1999b). Creative activities rely less on language and memory, which in childhood are less developed skills: Secondly, creative activities appeal to a child's imagination and this is one of their main assets. In the words of Vaajakallio et al. (2009) 'Children were motivated or even enthusiastic with the Make tools.'.

Bridging the conceptual gap

Cognitive Psychologists have shown that children find it difficult to conceptualize ideas that are abstract in nature (Gelderblom, 2009). Design problems at the beginning of a development process require 'blue skies' thinking about new artefacts whose form cannot be pre-defined and therefore may appear intangible to a child. Activities that involve defining a final design outcome may not be tangible or understandable to a child. Methods that avoid the constraints of the design problem directly simplify the task. Storytelling and storyboarding have been used as co-design methods to enable both adults and children to communicate in an accessible form (Ryokai, 2012) However, the stories are usually about the user as the narrator and encourage them to tell their own story. Earlier experimentation has shown that a process involving the creation of stories and characters can provoke conversation and also presents useful insights into the world of the child (Grundy, 2012). Frequently their preferences and, in some cases, personality are expressed through the designed character and their adventures. Characters and stories both represent a familiar subject area for both adult and child. Characters are practically an everyday part of a child's life and they certainly have affection for them. They are also frequently used by counsellors to gain information from children in a sympathetic fashion. Thus the initial stage of the project involved creating characters and stories around the subject. These would later be relevant to the pupils designing the outdoor game, having been informed by the activity, however they also allow clear communication between the different members of the design team and provide information for later design efforts.

Methodology

A literature review had already been carried out to identify suitable technologies, games software and participatory design methods. Interviews with a series of wildlife experts and further study of environmental issues had also informed the experiments. The investigations with children were organized into two phases across two different schools. Phase 1: The project started with two school classroom visits, to St. Andrews school in Hove, in each case with a group of thirty pupils aged between 9 and 10. One group were asked to create characters and stories around their favourite natural subjects, the other were asked to create game scenes using their favourite natural subjects. (both relate to Sanders' 'Make' methods) This was done to compare results in terms of use of information for the game and how understandable the activity was. They were also asked to complete an online survey that related to their habits for outdoor play ('Say' methods). Phase 2: A group of twelve pupils from Fairlight primary school, aged between 10 and 11 were asked to design a Location Based Game for a local park over a five week period. The pupils worked in groups of three during the project. Week 1 they were taken to Stanmer Park and at specific locations, asked to take images of what they considered interesting phenomena. This was done from inside the ARIS game environment, so that the points were geo-tagged and ready to be used in the developing game. While one pupil had an I-pad, another took video footage of the experience and the third made notes. During week 2, the pupils were introduced to the concept of location-based games and worked with the ARIS interface. For week 3, Character Designs and Stories were developed for the game scenes. During this process the conversations were observed and recorded. In week 4, the games were further developed and videos created of natural objects in Morfo. Eventually in week 5 the pupils played the

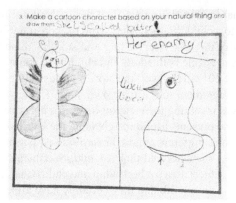

Figure 3.   Friend and Foe characters.

games outdoors in the park and again one member took video of the experience. Four successful games were eventually produced.

Results

For Phase 1, evidence for games characters and suitable game scenes was successfully gathered for both groups. The activity also provided a good discussion opportunity with group facilitators to find out more about the children and their preferences for the context.

Sometimes, pupils who were asked to create games resorted to copying other known games (one example is illustrated in fig 3) and in some cases were confused about what the required outcome should be. This yielded further evidence that the character design and story activity is preferable to a request to provide potential game designs. Nature and creature preferences were also identified, e.g. squirrels, rabbits etc. were preferred, slugs and bugs were considered 'the baddies'. More broadly the activities helped to identify key properties and themes of the game, for example social aspects like parties with friends were a common theme, typical power games like defeating evil and so on. The survey helped to put the pupil's interests in a broader context.

During Phase 2: The pupils all managed to successfully create a playable game, however evidence indicated that they were more interested in playing their own games than each other's. It was noted by all of the adult participants that it was difficult in the timescales for the children to produce a game that coherently communicated its goal and the tasks in hand. This suggested that either more time was necessary, or that the results should be interpreted in a more participatory fashion, to provide information for further development by the designer at a later stage for testing with the children. The character design and story co-design activities for

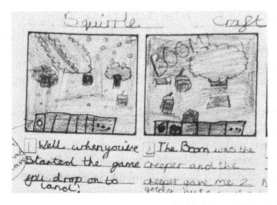

Figure 4.   Game design similar to existing examples.

this phase also provided a useful basis for deciding the content of the game characters and their activities. The videos and observations by group members helped to identify game characteristics, e.g. The distance between points should not be more than 30 metres or children lose focus and get distracted by other features, directional clues and illustrations of the next point should be provided as map reading is difficult for this age and sometimes frustrating. All design conclusions are too many to detail here.

Conclusions

The character and story design method, provided an excellent opportunity to discuss ideal preferences with children without the technical constraints of the design problem. The game creation co-design activity itself generated useful data but would need further development from a design expert to create communicable games to others. The methods chosen, when added to the ethnographic observations and survey results were effectively triangulated using the Say Do Make method.

References

Buchanan, R. 1992. Wicked Problems in Design Thinking. *Design Issues*, 8, 5–21.
Druin, A. 1999a. Children as our Technology Design Partners. *In:* Druin, A. (ed.) *The Design of Children's Technology.* San Francisco: Morgan Kaufmann.
Druin, A. 1999b. Cooperative Enquiry: Developing New Technologies for Children with Children. *In:* 99–14, H. N. (ed.) *CHI 99.* PIttsburg, USA.
England, N. 2009. *No charge? Valuing the natural environment.* Natural England.
Garzotto, F.A.G.R. 2011. Children's Co-Design and Inclusive Education. Opportunities and Challenges when Designing and Developing with Kids @ School Workshop. *IDC 2011.*

Gelderblom, H., Kotze, P. 2009. Ten Design Lessons from the Literature on Child Development and Children's Use of Technology. *Interaction Dsign for Children.* Como, Italy: ACM.

Grundy, C., Morris, R. & Pemberton, L. 2012. Using Character Designs To Represent Emotional Needs. *Participatory Innovation Conference 2013.* Lahti, Finland.

Mazzone, E., Livari, N., Tikkanen, Read, J., Beale, R. 2010 Considering Context, Content, Management and Engagement in Design activities with Children. IDC, 2010 Barcelona, Spain. ACM.

Naranjo-Bock, C. March 7, 2011. *Approaches to User Research When Designing for Children* [Online]. UX Matters. Available: http://uxmatters.com/mt/archives/2011/03/approaches-to-user-research-when-designing-for-children.php – top [Accessed 19 March 2011 2011].

Obrist, M., Markopoulos, P., Hofstatter, J. 2011. Opportunities and Challenges when Designing and Developing with Kids at School. *In:* ACM (ed.) *Interaction Design for Children.* Ann Arbor, USA: ACM

Rogers, Y., Connelly, K., Hazlewood, W. Tedesco, K.C. 2010. Enhancing Learning: A Study of How Mobile Devices Can Facilitate Sensemaking, *Personal and Ubiquitous Computing Journal,* 111–124.

Ryokai, K., Raffle, H. & Kowalski, R. 2012. StoryFaces: Pretend Play with Ebooks to Support Social-Emotional Storytelling. *Interaction Design for Children* Bremen, Germany: ACM

Sanders, L. & Simons, G. 2009. A Social Vision for Value Co-creation in Design. *Open Source Business Resource,* 0.

Sanders, L. & William, T. 2001. Harnessing People's Creativity: Ideation and Expression through Visual Communication. *In:* (EDS.), L. J. A. M.-P. D. (ed.) *Focus Groups: Supporting Effective Product Development.* Taylor and Francis.

Sustainable Development Commission 2009. Breakthroughs for the twenty-first century. *In:* www.sd-commission.org.uk/file_download.php?target=/publications/downloads/SDC_Breakthroughs.pdf.

Vaajakallio, K., Lee, J-J. & MattelmäKi, T. 2009. "It has to be a group work!" – Co-design with Children. *Interaction Design for Children* Como, Italy: ACM.

Williamson, B. 2009. *Computer games, schools, and young people A report for educators on using games for learning.* Futurelab report, March 2009.

BEYOND BOX TICKING – THE ROLE OF HUMAN FACTORS IN DESIGN

Daniel P. Jenkins

DCA Design International, Warwick, UK

In many industries, Human Factors (HF) is now a mandatory part of the product development process. Regardless of whether this stems from a regulatory or contractual requirement, it has played a significant role in the recruitment of human factors specialists into project teams, and the adoption of their tools and techniques. The integration of HF in a project team, should lead to safer and more productive systems. However, this 'free pass' onto the development team, has the potential to reduce the role of the human factors specialist to one of a 'box-ticking' exercise. This paper explores the importance of communicating the value of HF, and discusses how the unique skills and tools held by HF specialists can allow them to assume a pivotal role.

Introduction

There is a commonly held perception that the role of ergonomics and human factors specialists (for convenience, human factors specialists hereafter), is to supplement a design team and provide advice. For many, the role of human factors specialists working in consultancies and in-house development teams is often perceived as providing upfront input into the development of design specification, and the subsequent acceptance sign-off. In this perceived view, there are clear points in the design process where the human factors specialist needs to be consulted. They support the development of a product specification by providing key information, such as acceptable pushing or turning forces, and optimal handle heights and sizes (typically relying on standards and key texts, such as Pheasant and Haselgrave; 2006). While, in the later stages of the design process, the human factors specialist is called upon once more to assess the compliance of a concept, or range of concepts, against this specification. This often involves testing the push forces required, compliance against usability checklists (e.g. Neilsen & Milich, 1990), or acceptance testing with end-users.

There are, undoubtedly, instances where the relationship just described is largely accurate. Indeed, many international standards (e.g. BS EN 62366:2008) and guidelines provide graphical examples of a classic design cycle that is annotated to show: what information is required from human factors specialists, at which stages of the project they should perform evaluations, and the kind of documentation that should be created. However, somewhat reassuringly, for many who work as practitioners in design consultancies, or in-house design teams, this way of working will seem at least over-simplified, but more likely antiquated or simply unfamiliar.

One possible reason why the role of the human factors specialist is often perceived in this manner lies in the regulatory requirements for human factors integration. This, often legal, or sometimes contractual, requirement is captured and explained in human factors integration plans (HFIPs). These plans provide a description to the project team of the human factors specialist's role. They also form an important part of documenting the process for regulators and auditors. As such, human factors specialists are required to write these plans, conduct assessments, and produce reports.

Clearly, the mandatory requirement for human factors involvement is welcome, and has had a significant influence on both safety and productivity, not to mention the growth of the discipline. However, not having to justify explicitly one's role, or the value that the consideration of human factors can bring to a project, can also be considered as a hindrance. Particularly in gaining acceptance and influencing key project decisions. When considered in this way, there is the very real risk that human factors integration becomes a tick-box exercise. While mandated human factors involvement may lead to positive impacts on the design, these are often viewed as a convenient side effect of the regulatory process, rather than an objective.

Regardless of whether the industry in question is regulated or not, most would agree that good human factors involves going beyond 'box-ticking'. As a discipline, we have a wide range of skills and tools that are either designed to support design and innovation or can be readily repurposed to do so.

Skills

Those trained in human factors typically come armed with a number of skills that are extremely valuable in the process of designing products. These include capturing and filtering salient information from real-world situations (e.g. ethnography), eliciting stakeholders needs, values and mental models (e.g. interviewing), and building and testing hypotheses (e.g. experimental design). These kinds of skills are required throughout the design process. However, they are particularly useful at the start, in helping to define the purpose of a product, the constraints imposed by multiple stakeholders and the required context of use.

When designing consumer goods such as children's toothbrushes, this may involve visiting users in their home and observing not only the act of tooth brushing, but also how tooth brushing fits into their routine. Interviews may be conducted with experts such as dentists, as well users and stakeholders (parents). Experimental design may include the development and testing of a range of concepts.

Also, as systems thinkers, human factors specialists often hold a unique view of the systems that they support. While many of the engineers and designers in a project team will be required to specialise in one aspect of a product, human factors experts are, more often than not, required to think of the system in its entirety. Because human factors specialists hold this unique macro view, it is not uncommon for them to be called upon to provide high-level explanations of larger projects to visitors. The ability to think of the wider system that a product inhabits is of clear value. The growth of the service design industry is testament to this. This systems-level consideration often allows a human factors specialist to define the purpose of a system, as well as important metrics for assessing its performance.

Tools

The choice of tools will be largely dependent on the type of project. Where incremental improvement of a product is required, for example developing the next generation of product using the same manufacturing processes, descriptive and prescriptive tools are normally appropriate. Tools such as task analysis (Annett et al, 1971; Stanton, 2006) can be used to explore current practice, either idealised or observed. These simple diagrams often prove to be valuable resources for project teams, forming a common language. Moreover, they can encourage the team to think beyond the primary task to the steps either side of a traditional model. For example, models may be extended to consider the purchase journey associated with a product or its disposal.

There are many compatible tools that can be used to identify opportunities for incremental improvement. Typically, a different tool is used for each metric, for example, safety (e.g. TRACEr Shorrock & Kirwan, 2002; HEART, Williams, 1986; CREAM, Hollnagel, 1998), efficiency (e.g. critical path analysis), efficacy, intuitiveness, manual handling (e.g. REBA, Hignett & McAtamney, 2000; MAC, HSE, 2004), and resilience (e.g. FRAM, Hollnagel, 2012). A task analysis model can be used as a common reference point to tie these analyses together. To the initiated, this impact of these models is often under-estimated. However, engineers and project managers are often looking to inform evidence-based decision-making. As such, they are, more often than not, very receptive to tools that link performance changes of physical components to metrics that are more tangible to users and stakeholders.

Where the objective of a product is less well-defined, or there is scope to revisit the current proposition, more formative tools such as cognitive work analysis (Rasmussen et al, 1994; Vicente, 1999; Jenkins et al, 2009) can be used to explore the constraints that shape behaviour. Alongside design strategists, human factors specialists can use these tools to form the basis of defining what a product should be, and defining how they should be designed to support known and emergent user needs. These tools are extremely useful in markets where known needs (those that can be captured by simply asking a user what they would like from a product) have been met, but there remains scope to improve performance.

The exact choice of tools and techniques will be heavily influenced by the product, and the domain that is being designed for. Notable constraints that shape an approach include restrictions on manufacture, as well as the availability of time and resources.

The tools used as examples above, each have their relative strengths and weaknesses, and there are many that could be used in their place (the latest human factors methods book covers 107 methods; Stanton et al, 2013). The purpose of this paper is not to advocate specific tools, nor is it to prescribe a new technique or framework, but rather to encourage a philosophy that, to many, will be simply considered best practice. The examples above are intended to highlight how a wide range of tools and techniques can be combined to not only describe work situations, but also to quantify their performance and identify opportunities for improvement. This last point is of critical importance in product design, as analyses that do not identify ways of improving a product are of limited value.

As researchers, we can often become overly concerned with the reliability and validity of the approaches we use. However, in product design, absolute values of human performance are typically far less interesting than the relative differences between concepts or between new and legacy products. Clearly, there are times when validity is imperative, such as when marketing claims are made. However, for supporting evidence-based decision-making, tools that can identify gross differences between concepts are adequate.

Role

Human factors specialists are using some of the skills and tools discussed to gain valuable insights into products and contribute in the innovation process. This up-front emphasis on defining the product purpose and values, as well as direct input into the ideation process, represents a clear step-change in role. As such, these refined roles draw a stark contrast with the perceptions described at the start of this paper of individuals who simply set specification points and test them.

As a direct result of these insights and inputs, human factors specialists are now commonly taking fully integrated roles in the design team and are becoming involved in all key decisions throughout the design process, regardless of their pay grade. Furthermore, it is not uncommon for them to take senior roles in design teams. To put it simply, in many organisations, the idea of human factors experts working as a bolt-on resource is a thing of the past.

Conclusions

In summary, regulatory requirements for human factors' integrations have increased awareness of the discipline in many industries. Even in non-regulated domains, clients and stakeholders now come with preconceived views of what human factors specialists do. The importance of ergonomics is often largely understood from a theoretical perspective. However, the direct value to the project is, typically, less clear.

In many organisations, the role of the human factors specialist has evolved to leverage their unique skills and tools. They are now assuming more strategic roles, helping to define the purpose of products, and providing an important part of the evaluative process.

Looking to the future, it is important that as a discipline, human factors specialists continue to innovate and keep pace with the changes in product development and market requirements. The existing core skill and tool set is largely fit for purpose. However, continual innovation is needed to ensure that these tools are used effectively and efficiently to provide demonstrable value to the design process. The exact mix of tools and techniques, along with the fidelity of the analysis, will need to be specific to the project at hand. This will be influenced by the size and scale of the project, its place in the design cycle, and the size and experience of the human factors team.

Fortuitously, human factors specialists come armed with a suite of tools that allow them to observe the environments that they work in and identify how the organisations, as well as the products and systems they are developing, can be improved. Through an evidence-based approach to design, based on quantifying change, we can move from the rhetoric of evangelising the philosophy of user centred design to letting the evidence sell the value of the proposed change and, in turn, the value of human factors.

Furthermore, an evidence based approach often acts as a useful leveller; the data often speaks for itself giving junior members of a project team a powerful voice. Ultimately, though, if human factors specialists are to have a positive impact on product performance through design, they not only need a seat at the decision-making table, but also need influence. In order to assume key roles in the design team, it is imperative that human factors specialists communicate their value and influence on improving the quality of design – irrespective of whether or not they are mandated to be involved.

References

Annett, J., Duncan, K.D., Stammers, R.B. & Gray, M. (1971). *Task Analysis*. London: HMSO.

Hignett, S. & McAtamney, L. (2000) Rapid Entire Body Assessment: REBA, *Applied Ergonomics*, 31, 201–5.

Hollnagel, E. (1998). *Cognitive Reliability and Error Analysis Method – CREAM*. Oxford: Elsevier Science.

Hollnagel, E. (2012). *FRAM, the Functional Resonanace Analysis Method*. Aldershot, UK: Ashgate.

HSE (2004). *Manual handling. Manual Handling Operations Regulations* 1992 (as amended). Guidance on Regulations L23 (Third edition) HSE Books

Jenkins, D. P., Stanton, N. A., Salmon, P. M. & Walker, G. H. (2009). *Cognitive work analysis: coping with complexity*. Ashgate, Aldershot, UK.

Nielsen, J. & Molich, R. (1990). Heuristic evaluation of user interfaces. Proceedings of ACM CHI'90, (pp. 249–56).

Pheasant, S. & Haselgrave, C. (2006). *Bodyspace: Anthropometry, Ergonomics and the Design of Work*, Third Edition. Taylor and Francis.

Rasmussen, J., Pejtersen, A. & Goodstein, L. P. (1994). *Cognitive systems engineering*. New York: Wiley.

Shorrock, S. T. & Kirwan, B. (2002). Development and application of a human error identification tool for air traffic control, *Applied Ergonomics*. 33(4): 319–36.

Stanton, N. A. (2006). Hierarchical task analysis: Developments, applications, and extensions. Applied Ergonomics, 37, 55–79.

Stanton, N. A., Salmon, P. M., Rafferty, L., Walker, G. H., Baber, C. & Jenkins, D. P. (2013). *Human Factors Methods: A Practical Guide for Engineering and Design*. Second edition. Ashgate, Aldershot.

Vicente, K.J. (1999). *Cognitive work analysis: Towards safe, productive, and healthy computer-based work*. Mahwah, NJ: Lawrence Erlbaum Associates, Inc.

Williams, J. C. (1986). HEART – a proposed method for assessing and reducing human error. *In 9th Advances in Reliability Technology Symposium*, University of Bradford.

HUMAN FACTORS CONTRIBUTIONS
TO CONSUMER PRODUCT SAFETY

Bonnie B. Novak

Director, Division of Human Factors, U.S. Consumer Product Safety Commission, Office of Hazard Identification and Reduction, Engineering Sciences Directorate

The mission of the Office of Hazard Identification and Reduction at the U.S. Consumer Product Safety Commission (CPSC) drives the Division of Human Factors to evaluate corrective action plans and proposed product redesigns of consumer products. Human factors studies and research focuses on consumer product-related injuries, including assessing system design hazards for the general US consumer population, ranging from newborns to seniors. These assessments include usability, safety features, choking hazards, and evaluation of potential misuse of a product that could lead to injury or death. This paper will provide an overview of human factors work conducted at the CPSC and how this specialized work contributes to reducing risk to the American public.

Introduction

Human Factors (HF) refers to designing products and systems for human use. As Don Norman, leader in human-centered design applications, (1988) has stated, human interaction with any product is heavily influenced by the users' goals and interpretations of the product. At the CPSC, HF involves studying the user, product characteristics and use, and environment separately, and in combination, determining the factors that contributed to an injury-producing situation. HF staff evaluates the consumer and the product as a system – in other words, how human behavior affects the safety of the product, *and* how the design of the product affects human behavior. Human behavior includes all the aspects of the person – from their physical attributes and how that affects their interaction with the product – to information processing, reaction times, and expectations.

CPSC history and background

Congress established the CPSC in 1972, as an independent federal health and safety regulatory agency, charged with protecting the public from unreasonable risks of injury or death associated with the use of the thousands of consumer products under the agency's jurisdiction. These products are used in the home, schools, in recreation, other public places, or otherwise. Based on estimates from the CPSC's

Directorate for Economic Analysis, deaths, injuries and property damage from consumer product incidents cost the nation more than $1 trillion annually. The CPSC is committed to protecting consumers and families from product-related hazards that pose an unreasonable risk of injury. The CPSC achieves that goal through education, safety standards activities, regulation, and enforcement of the statutes and implementing regulations.

As a result, the CPSC is charged, specifically, with the following:

1. protecting the public against unreasonable risks of injury associated with consumer products;
2. assisting consumers in evaluating the comparative safety of consumer products;
3. developing uniform safety standards for consumer products and to minimize conflicting state and local regulations; and
4. promoting research and investigation into the causes and prevention of product-related deaths, illnesses, and injuries.

The CPSC's statutory authority generally requires the agency to rely on voluntary standards rather than promulgate mandatory standards, if compliance with a voluntary standard would eliminate or adequately reduce the risk of injury identified, and if it is likely there will be substantial compliance with the voluntary standard. CPSC staff works with both domestic and international standards development organizations that coordinate the development of voluntary standards.

Mandatory standards, on the other hand, are federal rules set by statute or regulation that impose requirements for consumer products. Mandatory standards under the Consumer Product Safety Act (CPSA) typically take the form of performance requirements that consumer products must meet or warnings they must display to be imported, distributed, or sold in the United States.

Under the CPSA, the CPSC may set a mandatory standard when it determines that compliance with a voluntary standard would not eliminate or adequately reduce the risk of injury or finds that it is unlikely that there will be substantial compliance with a voluntary standard. Under the CPSA, the Commission may also promulgate a mandatory ban of a hazardous product when the Commission determines that no feasible mandatory standard would adequately protect the public from an unreasonable risk of injury. In some cases, Congress directs and authorizes the Commission to promulgate a mandatory standard.

Human factors analysis in consumer product safety

HF staff works together with our partner divisions in Engineering Sciences, Health Sciences, and Laboratory Sciences to develop suggestions for product redesign after a product has been manufactured. Human factors studies and research applies the psychology of human behavior, learning theory, engineering principles, anthropometry, biomechanics, and industrial design to consumer product evaluation.

This total system evaluation includes the person + the product + the environment. The analysis and testing conducted by HF staff directly contributes to consumer product safety and risk reduction through design verification and validation efforts. Specifically, through technical analyses and reports, HF staff addresses:

- what age range of children is most likely to use the product if the product is intended for use primarily by children;
- how consumers interact with and use the product, such as foreseeable use and misuse;
- the likely effectiveness of alternative designs, guarding systems, warnings, labels, and other hazard-mitigation strategies;
- the likelihood of consumers encountering a given hazardous scenario;
- anthropometric and strength evaluation (*i.e.*, the size of the person and their strength capabilities);
- the ability of consumers to perform a repair or carry out a desired corrective action; and
- how consumer behavior and product design interact.

At CPSC, human factors involves studying the interaction of humans, products, and the environment, the parameters of human performance, the relevant design features of consumer products, and how all of these elements integrate and contribute to a potentially hazardous situation.

Origin of human factors work

Human factors work comes to the CPSC through various avenues. This includes, consumers reporting incidents through the agency's hotline or website; field investigators conducting interviews with victims and their families and compiling photographs and details from police reports; port surveillance personnel who send potentially violative products that are intended for import into the United States from various countries; news reports of consumer product-related incidents; death certificates and medical examiner reports; and from manufacturers themselves who have a legal obligation to report to the CPSC. This legal obligation is specified under the Consumer Product Safety Act (CPSA), Section 15 and requires manufacturers to report on a product that: (1) is defective, (2) does not comply with standards and regulations, (3) was choked on, (4) has been specified as a substantial product hazard by CPSC, or (5) was subject to certain lawsuits. Human factors analysis is also part of all project work and petitions that are examined at the agency.

Age determinations

To enforce CPSC regulations, it is often necessary to determine the intended ages of the users of toys, products, and jewelry. These age determinations are needed to assess products for defects, evaluate a corrective action plan, and conduct enforcement activities. Currently, determinations are made on a toy-by-toy basis through consultation with child development resources, such as the publicly available *Age*

Determination Guidelines: Relating Children's Ages to Toy Characteristics and Play Behavior document (Smith 2002). Attention is given to the developmental abilities of typical children, with an understanding that a wide variation exists within the rates at which children achieve major developmental milestones.

Toy evaluations

Toy evaluations are conducted by laboratory engineers via tests that simulate the use and foreseeable abuse of toys by children. A toy must pass all of the applicable standards tests and are required to be certified by third party testing laboratories as complying with those standards. The CPSC National Product Testing and Evaluation Center (NPTEC) conducts either spot-checks of the compliance or evaluations based on an incident to determine if the product can continue to be sold to the public. Some examples include: drop tests, torque tests, pull tests, small-parts tests, and sharp-edge tests. Specific forces, sizes of clamps, and distances of drops have been determined to mimic as closely as possible those produced by a child during normal play situations and foreseeable misuse.

Product Safety Assessments (PSAs)

The CPSC's Office of Compliance negotiates recalls, liaises with companies, and encourages manufacturers to comply with CPSC regulations. A PSA, Product Safety Assessment, is the CPSC-generated document that provides the scientific and technical analysis conducted by technical staff on that specific product. Most PSAs are associated with particular hazards that human factors will focus on; and in general, only hazards that have been associated with known incidents are addressed. For example, if incidents resulted from a fall, HF staff evaluates the potential fall hazards associated with that product sample.

Often, if HF staff is asked to assess a corrective action plan that involves a product repair, the usability analysis involves the question: "Could the average consumer successfully complete the repair?" In these cases, Human Factors staff may conduct several different kinds of analyses. For example, staff would provide a usability review and make recommendations regarding manufacturer-provided instructions and also conduct a walkthrough of the recommended repair to determine whether the repair can be completed successfully based on the materials provided, the instructions, and the likely abilities of the intended user.

Often, the CPSC HF staff is asked to perform usability analyses on instructions, which might include evaluating warnings and labels. Generally, the CPSC HF staff recommends warnings adhere to the ANSI Z535 series of standards, which are the primary U.S. voluntary consensus standards for safety signs and colors. Figure 1 presents an example of a label that staff evaluated, which was determined not to have followed good warning practices. Although the label adheres to ANSI standard warning color use, the average U.S. consumer cannot understand the last bullet because "oversely" is not a word. In addition, Hilts and Krilyk (1991) reported that adults read at least one or two grade levels below their last school grade completed,

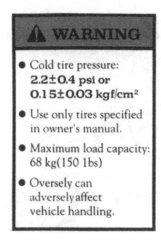

Figure 1.   Example warning label evaluated by Human Factors Staff at the CPSC as part of a PSA.

and the Flesch-Kincaid Readability Calculator estimates the language in this label at the 14th grade level.

Projects and petitions

PSAs examine hazards associated with specific models of a product, manufactured by a particular firm; whereas, projects and petitions examine hazards and hazard patterns associated with whole classes or categories of products. So while a PSA request might ask staff to analyze model X of a particular firm's line of clothes dryers, a project on clothes dryers may ask for an analysis of hazards that are common to most or all clothes dryers currently on the market. Projects and petitions generally take place over a timeframe of several months or years. The results of these analyses may be used to develop or modify consumer product safety standards, which would affect all future production of that product.

HF staff does a large amount of project work related to children's products because of the requirements in the Consumer Product Safety Improvement Act (CPSIA) of 2008. Section 104 of the Act requires the CPSC to study and develop safety standards for infant and toddler products. The CPSC's standard must be substantially the same as the relevant voluntary safety standard or more stringent than that voluntary safety standard. One example is the Children's Portable Bed Rails project. Portable bed rails are products with a vertical panel that rests against the side of a bed to prevent children from falling out of adult-sized beds. Portable bed rails typically have support legs that slide between the mattress and box spring. However, earlier versions of these bed rails can sometimes slip out of place and allow a child to slide between the rails and the bed. The child who becomes entrapped in this space can suffocate. HF staff was asked to design a probe to be incorporated into a standard and used during performance testing of bed rails. This probe was designed based on the anthropometric dimensions of the children at risk, and all portable bed rails are now required to prevent passage of the probe when a certain force is applied.

While projects and petitions are similar in scope, there are some differences between the two. Petitions are essentially requests made by the public for the CPSC to issue, amend, or revoke a rule or regulation associated with a particular product or product hazard. Typically, work performed in response to a petition typically is completed within 180 days; accordingly, technical assessments must be made using existing or "off-the-shelf" data. New studies or data collection are generally not considered.

In human factors, analysis might include incident analysis, and specifically, analysis of behavioural and anthropometric contributing factors. We could also evaluate provided instructions and warnings and recommend design modifications that increase safety and enhance human performance.

Methodology

Human Factors staff at the CPSC uses a wide variety of methodologies and approaches to analyze product safety and design, from heuristics to task analysis. The human factors engineers and psychologists have the autonomy to determine which approach is right for the product on hand and will work best to aid the mechanical engineers, health scientists, and Office of Compliance personnel to make a determination on how to proceed in interacting with the manufacturer.

Heuristics

Heuristics require the Human Factors analysts to use professional judgement, experience, and best practice education to guide product evaluation (Nielsen 1992). Heuristic analysis at the CPSC is conducted in both table-top and simulated settings at the CPSC National Product Testing and Evaluation Center (NPTEC). While this is a subjective analysis, conducted by subject matter experts, the approach is often a preferred method for completing analysis, due to the ease and speed which work at the agency often demands.

Surveys

Standardized questionnaires and surveys have been used by staff to assist in making informed and accurate expert assessments of consumer behavior. The user population that the CPSC supports ranges quite broadly from infants to senior citizens. Therefore, it is often advantageous to conduct a targeted survey to determine how the general public interacts with a specific product design. This methodology is often time-consuming and extensive, requiring funding and approval by the U.S. Office of Management and Budget. Thus, the methodology is used primarily to support long-term projects.

Task analysis

Task analysis refers to the study of what a user is required to do, in terms of actions and cognitive processes, to achieve a goal presented by a product or system (Kirwan and Ainsworth 1992). Staff employs a variety of task analysis techniques to identify how incidents and accidents occur when there is interaction between a product and user. These techniques include Observational Analysis, Task Decomposition, Walk-throughs, Fault Tree Analysis, and on occasion, Computer Modelling and Simulation.

Human factors task analyses at the CPSC are used to:

1. identify hazards to the user;
2. analyze product safety through good design techniques;
3. analyze human error when interacting with the product and environment; and
4. identify remedial measures.

Conclusion

The Division of Human Factors at the CPSC provides support for enforcement and safety initiatives, by evaluating the behavioral interactions between consumers and the products they use. Human Factors analysis involves studying the human interaction, product characteristics and use, and the environment, separately, and in combination, to determine the factors that contributed to a hazardous situation. This technical discipline is an important contributor to the analysis that the CPSC uses to understand and determine the interaction of people with consumer products to reduce the risk of injury and death.

References

Hilts, L. and Krilyk, B. J. 1991, *Write Readable Information to Educate,* (Hamilton, Ontario).

Kirwan, B. and Ainsworth, L. K. 1992, *A Guide to Task Analysis,* (Taylor & Francis, London).

Nielsen, J. 1992. *Finding usability problems through heuristic evaluation,* (ACM Press, Monterey, CA).

Norman, D. A. 1988, *The Psychology of Everyday Things,* (Basic Books, New York).

Smith, T. P. 2002, *Age Determination Guidelines: Relating Children's Ages to Toy Characteristics and Play Behavior,* (CPSC, Bethesda).

Therrell, J. A., Brown, P. S., Sutterby, J. A., & Thornton, C. D. (2002). *Age determination guidelines: Relating children's ages to toy characteristics and play behavior* (T. P. Smith, Ed.). Washington, DC: U.S. Consumer Product Safety Commission.

LIFELONG HEALTH AND WELLBEING

EVALUATING THE UNIVERSAL NAVIGATOR WITH CONSUMERS WITH REDUCED MOBILITY, DEXTERITY & VISUAL ACUITY

P.N. Day[1], J.P. Johnson[2], M. Carlisle[1] & G. Ferguson[3]

[1] *Consumer Experience, NCR Corporation, Dundee*
[2] *NCR Corporation, Duluth, Georgia*
[3] *UX Consultant, South Falfield, Leven*

An alternative input device for navigating on-screen content is presented, along with a repeated measures evaluation of this device with 20 participants with varying physical and sensory impairments including reduced mobility, dexterity and visual acuity. All participants were able to successfully complete the travel check in task using the universal navigator.

Introduction

The universal navigator (or uNav) is a physical input device concept which can be attached to a touchscreen such as that used on a self-service kiosk, and enables users who cannot see or reach the screen to activate the onscreen options. This is of increasing importance as there is a trend to use touch screens in self-service products (Penn et al., 2004, Digital Trends, 2011), along with ageing populations in many countries (UKONS, 2010, Zaidi, 2008). It has the potential to provide significant accessibility advantages to those with low visual acuity, reduced upper body mobility, and also low manual dexterity as it enables consumers to navigate any onscreen option using a tactually discernible, physically compact device with associated private audio feedback (i.e. through earphones rather than speakers). It consists of four direction keys laid out in a diamond shape, arranged around a central select button (Figure 1). The left and right buttons cycle through each on-screen option in turn, with the up/down buttons jumping to the next block of on-screen options (e.g. next row of an on-screen keyboard). An audio jack and volume button are also provided, as auditory feedback is essential for people who are blind or partially sighted. The uNav is a concept device and is currently not in development.

This paper follows on from a prior publication (Day et al., 2012) that described the initial concept & testing of the uNav, and a subsequent publication (Day et al., 2013) in which evaluations with blind and partially sighted consumers in the UK were presented. Although this previous evaluation was very useful, there was a desire to also validate with different types of impairment and in a different cultural context. In this paper we therefore present further studies evaluating the use of this device

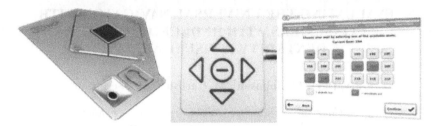

Figure 1.   The physical & overlay uNav and sample screen from evaluation.

by consumers in the USA including some with mobility impairments, some with reduced dexterity, and some with reduced visual acuity (blind or partially sighted).

Usability testing of the Universal Navigator with disabled users was conducted in Atlanta, Georgia, USA. The purpose of the additional testing in Atlanta was to validate the findings from the previous UK evaluation, and also to expand the participants to include wheelchair users and people with manual dexterity limitations.

Methodology

This evaluation used a modified version of an NCR travel check-in application previously used (Day et al., 2012, 2013), but with the addition of high-quality speech output. This task (checking in for a flight) was chosen as it included a number of complex options including using an on-screen alphanumeric keyboard and selecting a seat on the aeroplane.

The evaluation was a repeated measures design, with participants completing the travel check-in task twice using the uNav in the vertical and horizontal orientations, and those that felt able to use a touch screen completed an additional task with this touchscreen (thus enabling comparison with the current interaction style for sighted participants). The vertical orientation (in line with the display, 65° from horizontal) was expected to be easier to integrate into products, but there was a concern over comfort compared with a more typical horizontal orientation (13° from horizontal tilted towards user to aid comfort) as is commonly found on keyboards. NCR designed a fixture to hold the uNav and allow for easily changing between the two angles, shown in Figure 2. In addition participants were asked for their opinion of a (non-functioning) overlay designed to stick on the touchscreen giving uNav like functionality without the need for a separate physical device. Task orders were balanced to alleviate learning effects. However, participants who were blind were not expected to attempt using the touchscreen, as this had no additional accessibility features such as voice over or talking fingertips.

Testing occurred at disABILITY Link in Decatur, Georgia. disABILITY LINK is a Center for Independent Living (CIL), committed to serving the disability

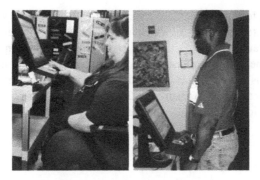

Figure 2.   Photos of participants.

community through advocacy and education, and serving the metro Atlanta area (DisabilityLink, 2014). Additional testing with visually impaired users occurred at CVI, Center for Visually Impaired in Atlanta. The Center for the Visually Impaired (CVIGA, 2014) is Georgia's largest comprehensive, fully accredited, private facility providing rehabilitation services for individuals of all ages who are blind or visually impaired. Since 1962, the Center has grown to serve as a model of innovative services for people who have a wide range of vision impairments from low vision to total blindness.

Participants

There were 20 participants (5 male, 15 female) with a good representation of ages from 25–64. 6 participants were in a wheelchair. 5 participants reported having upper body impairment; with 2 mild, 2 moderate, and 1 severe level of self-reported impairment. These levels (derived from Ofcom, 2009) were defined as follows. Mild: *"the condition does not have a significant impact on my normal day-today activities; my upper body mobility is generally good and any pain or discomfort wears off quickly"*. Moderate: *"the condition has a significant impact on my normal day-to-day activities; my upper-body mobility is regularly restricted"*. Severe: *"the condition has a significant and detrimental impact on my normal day-to-day activities; I have very limited upper-body mobility"*.

4 participants reported a mild impairment in their level of manual dexterity, with 3 reporting moderate impairment and 1 a more severe level of impairment.

2 participants considered their level of sight as blind without any useful residual vision, 5 stating that they were blind with some useful residual vision, 4 were partially sighted (including 1 with peripheral vision only) and 9 were sighted.

Attitude towards technology

Participants were asked to rate their response to six individual statements about their attitude to technology, using a five-point scale. These questions were derived from a nine-item technology anxiety scale specifically concerned with self-service technology (Meuter et al., 2003), which in turn was a development of a computer anxiety scale (Raub, 1981). As this was a peripheral activity in the study, only six out of the nine scales were used, with the six being selected that Meuter et al. (2003) reported as scoring highest for predicting technology anxiety when assessed by varimax factor analysis (to ensure that this was distinct from other related factors such as self-confidence or venturesomeness).

Results & discussion

Overall the uNav was very well received by participants. When asked whether the device was acceptable for use, the majority (15) stated that it was acceptable as is, with the remainder (5) stating that it was acceptable with minor changes. This is in contrast to the finding from the London test (Day, 2013) where the majority (72%) of those users wanted minor changes.

Overall ratings

There were a number of ease of use ratings recorded on a 5 point scale where 1 was 'very bad' and 5 was 'very good'. For moving between options using the uNav participants tended to rate it highly with 15 rating 'good' or 'very good', 3 as 'neutral' and 2 as 'bad' (mean rating 3.9). Of those who rated it badly they commented on some initial confusion over the transaction, and trying to learn the system.

Comparison between devices

Participants were invited to state which device they preferred to use, and were also asked to rank in order of preference. Of the 20 participants there was a preference for the touchscreen (11 participants) (with sighted participants in particular tending to prefer it), then the physical uNav was next preferred (7) with an overlay uNav (stuck on screen) that was also shown being least preferred (2). Interestingly when asked to rank them a different order was given as follows: 1 – Universal Navigator (with physical buttons), 2 – Touchscreen, 3 – Overlay Universal Navigator.

Respondents reported a clear preference for the horizontal orientation (17 participants) which was also supported by individual ratings throughout the transaction (Table 1). There were significant differences between ratings for comfort ($t(19) = 2.78$, $p = 0.01$) and ease of use ($t(19) = 2.24$, $p = 0.04$) in the two orientations. However, all participants could complete the task with the uNav in either orientation.

Order effects

The order in which a device was used appeared to indicate a trend to higher ratings on the 2nd and 3rd time that participants completed the task. However, these differences

Table 1.   Mean ratings for all devices.

Input device	Measure	Trial 1	Trial 2	Trial 3	Mean
Horizontal uNav	Ease of use	4.3	4.0	4.5	4.3
Horizontal uNav	Comfort	3.4	4.0	4.5	4.0
Touchscreen	Ease of use	4.2	4.5	4.6	4.4
Touchscreen	Comfort	4.0	4.8	4.2	4.3
Vertical uNav	Ease of use	3.8	3.4	4.0	3.8
Vertical uNav	Comfort	3.2	2.9	2.8	3.0

were not statistically significant and ordering was balanced, so ratings for each device can be combined irrespective of when they were used to compare between the different devices.

Ratings for horizontal uNav

The horizontal uNav device was well-received by the participants, especially those who used the device third. There seemed to somewhat of trend of rating the devices more favourably after the second and third trials.

Ratings for vertical uNav

The vertical orientation of the uNav received the lowest ratings, especially for comfort. The 3.0 average comfort rating was due to participants not having a place to rest or steady their hand to operate the device.

Ratings for touchscreen

The touchscreen received the highest ratings, due to the fact that sighted and partially sighted users perceived the touchscreen to be the fastest and most accurate input device.

Effects of other factors

Although the sample size was quite small (20), a Shapiro-Wilk test revealed that the ratings for ease of use & comfort did follow a normal distribution, and therefore parametric statistics could be used.

The following factors were analysed to determine whether they had an effect on the results (ratings of ease of use, comfort, voice output, acceptability, and preference for device and angle): visual acuity, mobility, dexterity, height, age, gender, experience with ATMs, travel kiosk and self-checkout in retail, experience of the EZ Access device, and attitude to technology. Only significant effects are reported.

There was a significant difference between ratings of the voice output between males and females ($F(1,18) = 6.62, p = 0.02$), with males tending to rate the voice higher than females. This may be related to the fact that the voice output used was a female voice.

There was a significant effect of whether participants used ATMs independently on rating of comfort for the horizontal uNav ($F(1,18) = 5.74, p = 0.03$), and rating of the raised features on the uNav ($F(1,18) = 6.32, p = 0.02$). Those who did use ATM independently tending to rate horizontal uNav comfort higher and usefulness of raised features lower than those who did not use an ATM independently. This may well be as the majority who did not use an ATM independently were blind and therefore this shows the increased reliance on tactile features for such a population. However, as there were only six participants who did not use an ATM independently a stronger conclusion could not be reached.

In a similar manner there was a significant effect of previous usage of an ATM with voice guidance on rating of the usefulness of the raised features on the uNav ($F(1,18) = 10.05, p = 0.005$), with those who had not used an ATM with voice guidance rating the features lower. Again, this may be due to those who had used an ATM with voice tending to be blind or partially sighted.

There was a significant effect between previous usage of self-checkout on ease of use ratings for the horizontal uNav ($F(4,15) = 3.63, p = 0.03$), and ease of use for touchscreen, $F(3,10) = 5.36, p = 0.02$, with those who had used it once and never again rating both the uNav and touchscreen lower than others.

There appeared to be significant effects between familiarity with the EZ Access device and ratings of ease of use for the horizontal uNav ($F(2,17) = 4.89, p = 0.02$), and ease of use for touchscreen ($F(2,11) = 7.28, p = 0.01$), with those who both knew of it and had used the EZ Access device giving higher ratings for ease of use of the horizontal uNav and touchscreen than the other groups. This may be due to the small sample sizes involved though.

There appeared to be a effect of level of upper body mobility on preference for uNav angle, but sample sizes were too small to make any meaningful judgment on whether this was significant with those with severe and moderate impairment preferring vertical, all others preferring horizontal.

Significant effect of level of manual dexterity on rating of comfort of the vertical uNav ($F(3,16) = 3.58, p = 0.037$), with decreased dexterity tending to give a lower rating for the vertical uNav, and on rating of comfort for the horizontal uNav, ($F(3,16) = 3.80, p = 0.03$), with decreased dexterity tending to give a higher rating for comfort of the horizontal uNav.

Correlations between ratings

Significant correlations were found between ratings of comfort and ease of use for the horizontal uNav ($r(18) = 0.67, p = 0.001$), and touchscreen ($r(12) = 0.87$,

$p < .0001$), size of uNav keys and ease of use for the vertical uNav ($r(18) = 0.57$, $p = 0.008$), uNav ranking and rating for comfort comfort of the vertical uNav ($r(18) = 0.50$, $p = 0.02$), ranking for touchscreen with ease of use ($r(12) = -0.90$, $p < .0001$), and comfort ratings ($r(12) = -0.84$, $p = 0.0002$), for touschscreen, ranking for overlay uNav with ease of use for touschscreen ($r(12) = 0.75$, $p = 0.002$), and comfort for touchscreen ($r(12) = 0.62$, $p = 0.02$), and also between ranking for overlay uNav with ranking for touchscreen ($r(12) = -0.81$, $p = 0.0005$).

Prior experience of self-service

ATM usage varied from never (1), used it once (2), once a month or less (5), 2–3 times a month (4), once a week (6), at least 3 times a week (2). Reasons for not using an ATM very often included worry about fees, lack of talking ATMs in their area, safety, and convenience with some stating that their spouse or relative does it for them. Of those who used an ATM 14 used it independently, with 6 not using it without help. Only 8 participants had ever used an ATM with voice guidance, with 14 never having used one. When asked whether they would use voice guidance at an ATM if it was available, 15 replied that they would, 5 said no.

Travel check-in kiosk usage varied from never (7), only use with others help (2), used it once (3), occasionally (4), most of the time I travel (2), every time I travel (2). Use of self-checkout in retail stores also varied from never (3), only use with others help (6), used it once (1), occasionally (7), every time I shop (3).

18 (90%) participants had never heard of the EZ Access device. 1 participant had used it before, and 1 had heard of it but never used it.

The raised features on the device were felt to be useful by the majority (9), or very useful (6), with 3 being neutral and 2 rating as not very useful. Sighted participants tended to rate them as less useful than blind and partially sighted participants.

Suggested improvements and changes

Participants were invited to comment on what could make the device better for them. A number of sighted and partially sighted participants expressed a desire to use touchscreen wherever possible. Those who wanted uNav preferred the horizontal layout. The ability to zoom in and out of the screen using touch screen gestures was raised as was the addition of audio to the touchscreen.

Conclusions

This evaluation not only validated the findings of the previous study (Day, 2013), it also demonstrated that this device is of benefit not only to blind and partially sighted consumers but also to those with manual dexterity or reach issues All participants were able to complete a travel check-in kiosk without prior training or experience of such an application thus affirming the fact that a device like this could readily be used in a self-service environment without the need to significantly change the application or flow. The results demonstrate a preference for having the device in an orientation close to horizontal, and also demonstrate the importance of offering those with low visual acuity the choice to use touch screen or an alternative input device as those with some vision tended to prefer the touch screen.

Acknowledgements

The team acknowledges the assistance from disABILITY LINK and the Centre for the Visually Impaired and thanks all participants for their time and input.

References

CVIGA, 2014, The Centre for the Visually Impaired, Atlanta, Georgia, http://cviga.org

Day, P.N., Chandler, E., Colley, A., Carlisle, M., Riley, C., Rohan, C. & Tyler, S. 2012, The universal navigator: a proposed accessible alternative to touchscreens for self-service. In: M. Anderson (ed.) *Contemporary Ergonomics and Human Factors 2012*, (Taylor and Francis, London), 31–38

Day, P.N., Carlisle, M., Chandler, E. & Ferguson, G. 2013, Evaluating the Universal Navigator with blind and partially sighted consumers. In: M. Anderson (ed.) *Contemporary Ergonomics and Human Factors 2013: Proceedings of the international conference on Ergonomics & Human Factors 2013*, (Taylor and Francis, London), 355–362

DisabilityLink, 2014, http://disabilitylink.org

Meuter, M., Ostrom, A., Bitner, M. & Roundtree, R. 2003, The Influence of Technology Anxiety on Consumer Use and Experiences with Self-Service Technologeies. *Journal of Business Research*, 56(11), 899–906

Penn, Schoen and Berland, 2004, *Elo TouchSystems In Touch Survey*, from http://www.elotouch.co.uk/AboutElo/PressReleases/040617.asp

RNIB, 2009. RNIB Group current five year plan 2009 to 2014. http://www.rnib.org.uk/aboutus/organisation/thefuture/Pages/rnib_current_plan.aspx. Last updated 16th July 2012

Ofcom, 2009, Experience of people with upper-body mobility and dexterity impairments in the communications market, Research report prepared by GfK NOP, http://stakeholders.ofcom.org.uk/binaries/research/consumer-experience/GfKNOP.pdf

UKONS, 2010, *Older People's Day 2010 Statistical Bulletin*, UK Office for National Statistics, from http://www.ons.gov.uk/ons/rel/mortality-ageing/focus-on-older-people/older-people-s-day-2010/focus-on-older-people.pdf

Zaidi, 2008, *Features and challenges of population ageing*, from http://www.euro.centre.org/data/1204800003_27721.pdf

HEALTHY AGEING IN THE CONSTRUCTION INDUSTRY; BACKGROUND AND PRELIMINARY FINDINGS

S. Eaves, D.E. Gyi & A.G.F. Gibb

Loughborough Design School, Loughborough University
School of Civil and Building Engineering, Loughborough University

With the growth of an ageing population and recent extension of retirement age we are beginning to see an increasingly ageing workforce. Older workers have expressed an interest in remaining in work. However, this is more difficult in certain occupations, such as construction. This research explores how the experience and knowledge of older workers can be harnessed in order to facilitate healthy working behaviours, leading to a better quality of working life for longer. Using a mixed methods approach including in-depth semi-structured interviews and surveys, preliminary observations (n = 10) are presented, which have shown that construction workers are concerned about their health and changes can be made to improve their quality of life at work.

Introduction

We are living in an ageing population and subsequently seeing an ageing workforce. Retirement can be a stressful life transition encouraging development of behaviours such as smoking, drinking and social isolation whilst remaining in work provides people with a sense of purpose and maintains good social networks (Holcomb, 2010). With the extension of retirement age, remaining in work during later life is common, however large individual differences in ageing mean that it can be difficult for older workers to remain in certain occupations.

Physical and mental abilities decline with age. The lens of the eye hardens which decreases visual ability meaning that older workers may be less aware of obstacles in the workplace. Hearing ability also decreases, presenting issues with conversations instructions in the workplace. Older workers also have reduced stamina and muscle strength, something which may become particularly apparent in construction, where heavy lifting and constant movement are common.

Construction is renowned for hard manual labour in sometimes dark, noisy and dusty environments. Workers spend years learning their trade but are often forced to leave sooner than expected due to illness, injury or disability with heavy lifting and bending of the back being predictors for early retirement (Arndt et al., 2005; Hengel et al., 2012). Workers are in awkward, cramped positions for long periods of time, perform repetitive movements and lift and carry heavy loads throughout the day. Bricklayers lay up to 800 bricks a day (de Looze et al., 2001) and electricians have been reported to spend substantial time working above shoulder level (Rwamamara et al., 2010).

Musculoskeletal disorders are a common issue in construction. Many can be attributed to particular trades such as shoulder problems in scaffolders and lower back pain in floor fixers (Holmström and Engholm, 2003). Other long term health issues include prolonged exposure to cement resulting in dermatitis and dusty conditions contributing to occupational asthma or Chronic Obstructive Pulmonary Disorder causing shortness of breath and chest pains.

The natural decline in physiological functioning can be exacerbated by working in these tough conditions therefore it is important for workers to be aware of their health at work, which can be influenced by workplace design and ergonomics. This research is part of a larger study exploring whether the experience and knowledge of older workers can be harnessed in order to facilitate healthy working behaviours in construction.

Methodology

A mixed methods approach was used; site managers were contacted through professional and personal contacts. Participants were English speaking construction workers, working in awkward, cramped positions for long periods of time, performing repetitive movements or heavy lifting. Semi-structured interviews with workers lasted approximately 30 minutes and explored participants' trades, job, feelings about their health at work and if they make any changes in their environment to protect themselves, using the Stage of Change Questionnaire (Whysall et al., 2007). The Nordic Musculoskeletal Questionnaire (Descatha et al., 2007) was used to measure point (7 day) and period (12 month) prevalence of musculoskeletal symptoms and the Work Ability Index (Zwart et al., 2002) was used to investigate how the participants felt about their ability to work. Observations of workers carrying out particularly tough tasks or using materials in their environment to protect their health were recorded.

Preliminary findings

This paper presents findings from a pilot study of 10 interviews between May and June 2013 on a large construction site. Participants were aged under 25 ($n = 1$), 25–34 ($n = 2$), 35–49 ($n = 4$) and over 50 ($n = 3$). Trades included scaffolders, bricklayers, and carpenters with the most common being electricians ($n = 4$).

The over-riding opinion was that pain is a natural part of working in construction. Workers expected to be uncomfortable at work for long periods of time, but maintained that this was not a problem as their bodies got used to this;

"this shoulder don't really feel any pain, it's used to it" (Scaffolder, 35–49).

"at the end of the day that's just what happens" (Electrician, 35–49).

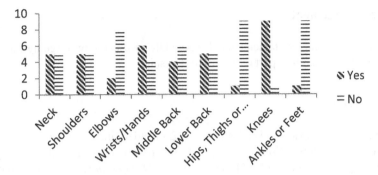

Figure 1. Musculoskeletal symptoms experienced in areas of the body in the past 12 months (*n* = 10).

A small number of participants protected their health at work by using materials around site such as pieces of foam or cardboard as temporary knee pads. Interestingly, it was the youngest (and only female) participant who had thought about the need to support her back and how she might achieve this on site;

"Sometimes I get a cable drum to sit on so that I can sit up straight . . . 'cus I get back ache . . . I might put my bag behind it as well" (Electrician, under 25).

Workers use their personal protective equipment (PPE) to protect their health at work; dust masks, ear defenders and gloves were cited as protecting workers, although the issue of PPE not always being fit for purpose was raised. For example, wearing a dust mask and eye goggles simultaneously leads to goggles steaming up, compromising vision and this has implications for design.

The majority of workers (*n* = 9) had experienced musculoskeletal symptoms in at least one area of their body in the previous 12 months (Fig. 1). They believed most of these symptoms were directly related to their work. Those who did not attribute their symptoms to work believed they were due to activities such as running and playing football outside of work.

Conclusion

In conclusion, preliminary observations suggest that workers are aware of the tough environment they work in and are concerned about their health, with some proactively making changes in their workplace to protect their health and reduce physical stress on the body. They have detailed knowledge about the risks of their job and ideas about healthy working which can be harnessed and contribute to encouraging healthy behaviours at work to improve the quality of working life. The next stage of the research will be to discuss the findings with stakeholders in construction and explore the barriers and opportunities for improving healthy working behaviours in the industry.

Acknowledgement

The authors would like to acknowledge Age UK's Research into Ageing Fund for funding this research as part of a PhD studentship.

References

Arndt, V., Rothenbacher, D., Daniel, U., Zschenderlein, B., Schuberth, S. and Brenner, H. 2005, Construction work and risk of occupational disability: a ten year follow up of 14 474 male workers. *Occupational and Environmental Medicine,* 62, 559–566.

De Looze, M. P., Urlings, I. J., Vink, P., van Rhijn, J. W., Miedema, M. C., Bronkhorst, R. E. and van der Grinten, M. P. 2001, Towards successful physical stress reducing products: an evaluation of seven cases. *Applied Ergonomics,* 32(5), 525–534.

Descatha, A., Roquelaure, Y., Chastang, J., Evanoff, B., Melchior, M., Mariot, C., Ha, C., Imbernon, E., Goldberg, M. and Leclerc, A. 2010, Validity of Nordic-style questionnaires in the surveillance of upper-limb work-related musculoskeletal disorders. *Scandinavian Journal of Work and Environmental Health,* 33(1), 58–65.

Hengel, K., Blatter, B., Geuskens, G., Koppes, L. and Bongers, P. 2012, Factors associated with the ability and willingness to continue working until the age of 65 in construction workers. *International Archive of Occupational and Environmental Health,* 85, 783–790.

Holcomb, T. F. 2010, Transitioning into Retirement as a Stressful Life Event. In T. W. Miller (ed.) *Handbook of Stressful Transitions Across the Lifespan,* (Springer, London), 133–145.

Holmström, E. and Engholm, G. 2003, Musculoskeletal Disorders in Relation to Age and Occupation in Swedish Construction Workers. *American Journal of Industrial Medicine,* 44, 377–384.

Rwamamara, R. A., Lagerqvist, O., Olofsson, T., Johansson, B. M. and Kaminskas, K. A. 2010, Evidence-based prevention of work-related musculoskeletal injuries in construction industry. *Journal of Civil Engineering and Management,* 16(4), 499–509.

Whysall, Z., Haslam, C. and Haslam, R. 2007, Developing the stage of change approach for the reduction of work-related musculoskeletal disorders. *Journal of Health Psychology,* 12, 184–197.

Zwart, B., Frings-Dresen, F.M.H. W. and van Duivenbooden, J. C. 2002, Test-retest reliability of the Work Ability Index questionnaire. *Occupational Medicine,* 52(4), 177–181.

EFFICACY EVALUATION OF RECREATIONAL PICTURE COLOURING IN ELDERLY NURSING HOME PATIENTS

Shinichiro Kawabata[1], Maki Nasu[2], Akiyoshi Yamamoto[3] & Noriaki Kuwahara[1]

[1] Graduate School of Kyoto Institute of Technology
[2] Soliton corporation CO. LTD.
[3] SUPER COURT Co., Ltd.

Aging poses a problem all over the world, and the number of dementia patients is increasing. The number of elderly nursing homes is also increasing. Accordingly, the health care sector is facing a serious labour shortage. There are therefore needs for a recreational activity that can prevent or reduce elderly people's dementia and ease a care worker's burden. In this study an evaluation of picture colouring was carried out; as a result behaviour problems of elderly people decreased dramatically and burden of care worker has also decreased.

Introduction

The population of the world is aging at an accelerated rate and it has become a serious problem in wide areas of the world. The number of elderly people increased more than threefold since 1950, from approximately 130 million to 419 million in the year 2000. The number of elderly people is now increasing by 8 million per year, and by 2030, this increase will reach to 24 million per year. The most rapid acceleration in aging will occur after 2010, when the large post World War II baby boom cohorts begin to reach age of 65. Declining fertility rates combined with steady improvements in life expectancy over the latter half of the 20th century have produced dramatic growth in the world's elderly population. People aged 65 and over now comprise a greater share of the world's population than ever before, and this proportion will increase during the 21st century. This trend has immense implications for many countries around the globe because of its potential to overburden existing social institutions for the elderly. As of October 1, 2010, the elderly population aged 65 and over became 29.6 million people to be the highest ever in Japan. Moreover the proportion of the population of the total population over the age of 65 was also recorded the highest of 23.1%. When this tendency continues, one person in four people will be the age of a senior citizen in 2015. Accordingly, health care sector is facing a serious labour shortage. That is largely due to the fact that health care positions do not tend to pay much more than minimum wage, even though they are incredibly demanding jobs. Furthermore the numbers of dementia patients are also increasing and various measures for dementia prevention are taken place. Recreational activities which have the possibility to ease a care

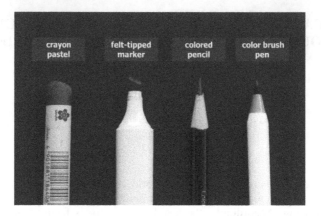

Figure 1.   Four kinds of writing instruments for colouring.

worker's burden and can also potentially prevent onset or decline of dementia are therefore beneficial. In this study, picture colouring, which is easily carried out for both elderly and care givers was examined. When starting colouring, people need to observe the original picture carefully. At this time, lobus occipitalis that take charge of the sight work. Moreover, to understand the original picture accurately, the temporal lobe that takes charge of the memory works to refer from the memory the shape and the colour seen in the past. The parietal lobe cooperates when the balance of the entire picture is gripped. As written above picture colouring has the effect to activate a widespread area of the brain. In this study picture colouring was taking place at the elderly nursing home for aged residents as part of a recreation activity, and the influence given to the resident was evaluated.

Methodology

Selection of writing instrument for picture colouring

For the first stage of this study, an experiment to select the optimal writing instrument for picture colouring was carried out. The colouring experiment was carried out with four kinds of writing instruments: crayon pastel (SAKURA COLOUR PRODUCTS CORP); colour pencil(MITUBISHI PENCIL CO., LTD.); felt-tipped pen(Too Corporation.); and colour brush pen (soliton corporation CO. LTD.). Four writing instruments are shown in Figure 1. These four writing instruments are commonly used instruments for picture colouring at elderly nursing home. The brain activity in each case was measured.

Experimental method

Electroencephalograph (EEG) of Digital Medic co., Ltd, was used to measure the brain activity. Five postgraduates aged 21–31 cooperated in this experiment as test subjects. To make experimental conditions impartial, each test with different writing instrument was conducted in the same time zone of a different day using the same laboratory with a tranquil environment. An interval of 7 days was given between each experiment. The sequential order for colouring the grain of the grape was determined to make equal condition between the subjects. After having installed the EEG, the test subject closed their eyes for 1 minute to record the brain wave at the rest situation, subsequently after that picture colouring was taken place for three minutes to record the brain wave during colouring. Assuming the brain waves at the time of eye closure as 100%, the brain waves of alpha wave and beta wave under colouring work in progress was compared.

Colouring recreation activity at elderly nursing home

Colouring recreation activity was taken place at the elderly nursing home to 56 residents using colour brush pen. The frequency of the recreation carried out was 2–3 times a week, and each recreation was about 1.5 hours. The frequency of the fall accident and the number of the nurse call (sensor mat type) before and after the recreation was recorded.

Five residents who have more behaviour problems comparatively with other resident, such as wandering around, petition of excretion and unnecessary nurse calls were chosen as a test subject. Each colouring recreation was operated for approximately thirty minutes and the frequency of the colouring recreation was two or three times a week, changing according to physical condition of the residents. The experiment was conducted for three months and the frequency of wandering around and petition of excretion was recorded.

An additional experiment was carried out with one other test subject to find out the relationship between number of nurse calls and sleeping time of the resident after working on colouring. Checking of sleeping hour was operated once in every fifteen minutes. Furthermore, in order to verify the influence of stopping the colouring recreation, as for this test subject, picture colouring experiment was stopped after two month and follow up observations were carried out.

Results and discussion

Selection of writing instrument for picture colouring

The results of the beta wave brain activity of four different writing instruments are shown in Figure 2 and the result of the alpha wave brain activity are shown in Figure 3. Colour brush pen and colour pencil showed the high value of Beta wave, which of 139% compared to rest condition. Beta wave is related with active thinking and concentration, therefore by using colour brush pen and colour pencil the user can achieve more concentration of the brain. The lowest value for the alpha wave was colour brush pen by 93% compared to rest condition. It is suggested that

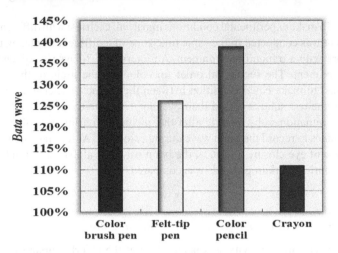

Figure 2.   Comparison of Beta wave between at rest and colouring using different writing instruments.

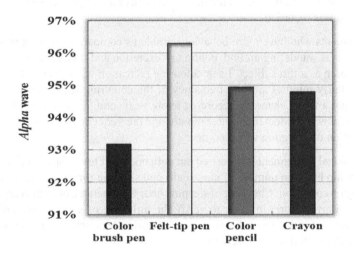

Figure 3.   Comparison of Alpha wave between at rest and colouring using different writing instruments.

colour brush pen has softest tip compared to other writing instruments. Thereby the test subject had to be vividly aware to hand movement of not only the XY-axis of left to right, but also to the Z-axis of up and down movement, leading alpha wave to decline consequently. From these results colour brush pen is demonstrated as the writing instruments which gives most stimulation to the brain. The reason for testing with postgraduates was because the physical and mental load was too heavy for the elderly people to wear EEG. Therefore in this study the experiment was conducted to the postgraduates.

Number of fall accident before and after recreation

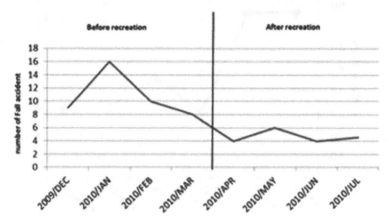

Figure 4. The number of fall accident before and after recreation.

Number of nurse call before and after recreation

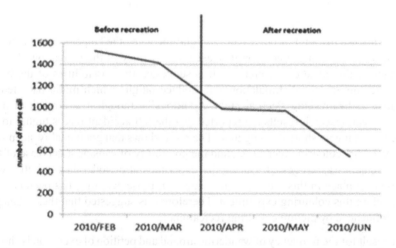

Figure 5. The number of nurse call accident before and after recreation.

3.2 Colouring recreation activity at elderly nursing home

The number of the fall accident per month before and after the recreation activity is shown in Figure 4 and the frequency of the nurse call is shown in Figure 5. The averages of fall accident per month at elderly nursing home decreased to 4.6 times from 10.7 times which in the percentage by 57% decrease. When the frequency of the recreation including colouring increased, the frequency of the nurse call decreased to average of 832 from 1469. The decreasing percentage was 35%.

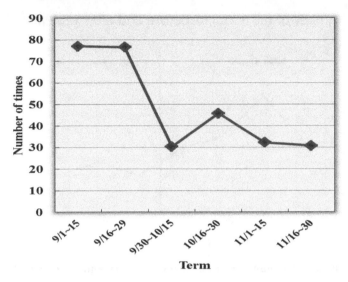

Figure 6.   Average frequency of wandering around and petition of excretion.

There are usually a lot of nurse calls at nighttime when the helper's round is fewer compare to day time. When you increase the frequency of the recreation, the frequency of the nurse call decreased. It is suggested the possibilities of the brain and the body received stimulation and produce fatigue which might have led to enough and refreshing sleep at night. Additionally, though it is a result of only the woman, research results are reported that the fall accident risk is higher to an aged woman with short sleeping time. The result shows that good quality sleep was urged by the colouring recreation, and the possibility of causing a decrease of the fall accident was suggested. During this experimental period, other activities was also taken place in this nursing home, however, those activities has been carried out before this colouring experiment. Therefore it is suggested that these changes are influenced by picture colouring recreation.

The result for the frequency of wandering around and petition of excretion is shown in Figure 6. The average number of wandering around and petition of excretion for five test subjects have decreased approximately 40% after three month of colouring recreation. According to the commentary of the physical therapist, the decrease of behaviour problems occur as a result to the plural domains such as cerebral cortex and basal nuclei, cerebellum, the brainstem were activated concurrently by feeling strain increase caused by the effect of colouring recreation. It is estimated that result occurred especially by the change in the function of frontal lobe participating in an accomplishment function. The workers at elderly nursing home have to bear a burden when behaviour problems occur. If the behaviour problem reduces, time required to deal with behaviour problems also reduces, that will make possible to use the time on the requisite care situation.

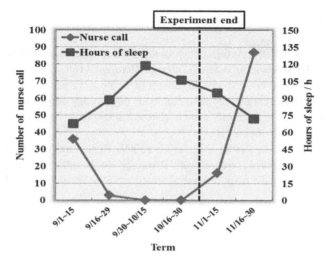

Figure 7.   Results for number of nurse calls and sleeping time.

The result for the relationship between the number of nurse calls and sleeping hours of the resident is shown in Figure 7. After starting the experiment, the number of nurse calls decreased dramatically. Due to this change, the hours of sleeping time has increased up to average of 4.5 hours per day to 7.9 hours per day in maximum. The number of the nurse calls increased up to 87 times and hours of sleeping time declined again to average of 4.8 hours per day after aborting the colouring experiment. Even though this experiment is a provisional examination, it was very interesting to see correlation between the number of nurse call and sleeping hour of the residents. We would like to conduct the experiment using more subjects. From this result, as for the increase of nurse calls and decrease of sleeping hours after stopping colouring experiment, it is suggested that colouring should be carried out continuously to gain effective influence.

Conclusions

In this study, the optimal writing instrument for colouring is suggested as the colour brush pen. The number of nurse calls, fall accident, wandering around frequency and petition excretion of the resident at nursing home decreased due to working on colouring recreation. The number of test subjects might be small as a representative sample, however in the case which the aging society are placed, this research has the possibilities to give a positive impact to the health care sector. Colouring is easily done compared to painting and descriptions, also regardless to the needing care degree. Moreover a care worker's burden can also be reduced because the time for handling unnecessary nurse calls, wandering around and petition of excretion can be reduced. As the result shows, colouring is suggested as an activity that should be taken as one of the recreations at an elderly nursing home.

References

Stone, K.L. et al. (2008), Actigraphy-Measured Sleep Characteristics and Risk of Falls in Older Women, *Archives of Internal Medicine* (**168**: 1768).

Tanaka, H. et al. (2009): Effect that "memories colouring paper" gives to slight dementia patient acknowledgment function, psychology function, and side of daily life. *Journal of rehabilitation and health sciences* **7**, 39–42, ESRC. 2009, Retrieved 19th October, 2010.

Fauth E.B., Zarit S.H., Femia E.E., et al. (2009) Behavioral and psychological symptoms of dementia and caregivers' stress appraisals: intra-individual stability and change over short-term observations. *Aging Ment Health*. **10**: 563–73.

USER-CENTRED DESIGN OF PATIENT INFORMATION FOR HOSPITAL ADMISSIONS AND PATIENT EXPERIENCE

Alexandra R. Lang[1], T'ng Chang Kwok[2], Evridiki Fioratou[3], Ivan LeJeune[2] & Sarah Sharples[1]

[1]*Human Factors Research Group, Faculty of Engineering, University of Nottingham, UK*
[2]*Acute Care Centre, Queen's Medical Centre, Nottingham University Hospitals, Trust NUH, UK*
[3]*College of Medicine Dentistry and Nursing, University of Dundee, UK*

This paper describes a user-centred approach to information design in an Acute Medical Unit (AMU). It presents a process for inclusion of clinical staff, nurses and patients in the design of information to be used for improving the efficiency of patient admissions to the wards and to increase patient understanding and satisfaction with the service. Human factors expertise was sought to assess the environment, admissions to the AMU and patient clerking by junior doctors. The paper outlines the challenges of designing for multiple users with varying needs and the intricacies of information design and provision when developing a patient leaflet for use in the NHS.

Introduction

Acute Medical Units (AMUs) in large hospitals are dynamic and busy clinical settings. Admissions to these departments come from both General Practitioner (GP) referrals and from Emergency Departments (ED). Clinical staffs encounter extreme diversity in the patients who present themselves and the medical complaints that they are being admitted for. For patients it can be an extremely stressful experience, being questioned and admitted to hospital for a condition which has yet to be diagnosed. This paper describes a user-centred approach to information design in the NHS. It presents a process for the design of information to be used in this complex clinical environment for the benefit of both staff and patients. The process follows a statement of need from senior clinical staff where human factors expertise was sought to assess the process of admissions to the AMU, patient clerking by junior doctors and recommend solutions to some of the issues being experienced. The paper outlines the process of designing a patient leaflet for use in a busy AMU and the challenges experienced when carrying out user-centred design and meeting the needs of multiple users within in the NHS.

User needs in acute care environments

Acute medicine is the early *"rapid specialist medical and multidisciplinary assessment and treatment"* (Byrne & Silk 2011) of *"patients suffering from a wide range of medical conditions who present to, or from within, hospital and require urgent or emergency care"* (RCP 2007). Acute medical care has been associated with *"lower inpatient mortality, improved patient and staff satisfaction, reduced hospital stays, and increased throughput"* (Wachter & Bell 2012). As such the National Institute of Health and Clinical Excellence (NICE 2007) has promoted the *"improvement of quality, efficiency and consistency of acute medical care"* (RCP 2012).

Doctors and nurses working in these Acute Medical Units (AMUs) are highly skilled and have to deal with a huge variety of patients and medical conditions. It is a challenging environment to work and train in with time pressures and suboptimal environments contributing to a high pressured and varied workload (Williams *et al.* 2013). As such it is recommended that junior doctors experience work in the AMU for blocks of 2–4 months at a time (RCP 2007). A large proportion of this early training will be based around the junior doctor gaining experience of 'clerking' incoming patients (Bircher & Warriner 2009). This involves questioning the patient (and possibly carer) the reason for their admission and 'history taking' about their general health and lifestyle and current medicines. Time taken for the clerking process will be dependent on the patients' health state and stress levels, their healthcare history and the experience of the doctor carrying it out. Several studies highlight the difficulties experienced by junior doctors as they put their learning into practice and have to carry out the clerking procedure. The work done in general surgical admissions wards by junior doctors shows how the documentation may miss important aspects of the history and examination during the busy on-call shifts (Isherwood *et al.* 2013) whilst Gupta *et al.* (2013) describe how junior doctors spend one-third of their time dealing with interruptions which impact completion of the patient history.

For patients, the experience of being sent to the AMU and possibly being told they need to stay in a ward can be a stressful and tiring experience. Ensuring that patient data is accurate and comprehensive is important so that doctors can devise an individually-tailored care plan. The patient pathway of an AMU aims to provide efficient access to immediate care and senior medical opinion (RCP 2007). During the admissions process patients will be seen by a range of clinical staff and will be asked questions repeatedly. This may lead to (interview) fatigue (Frey & Mertons 1995) which can cause lack of engagement and less detailed responses. This publication advocates a patient-centred approach to service and information provision and highlights the benefits of user involvement over consultation with proxy users, considering not only the medical needs but also the psychosocial and emotional experiences of being a patient.

Information provision in AMUs

The NHS Institute for Innovation and Improvement provide guidance for the development of patient information documents (NHS 2008) and state that they should help to ensure that patients arrive on time and are properly prepared, give patients confidence, improve their overall experience, remind patients of what they have already been told, and involve patients and carers in their treatment.

Many proformas have been developed for the improvement of the clerking process for use by clinicians. Faraj *et al.* (2011) describe the benefits of a proforma designed specifically for use by junior doctors admitting elderly patients with trauma whilst Rrtiza-Ali *et al.* (2001) relate the improvements in ease of data retrieval in the clerking process following the use of a proforma. Wethers & Brown (2011) describe how the use of an admission booklet resulted in statistically significant improvement in the completeness of numerous components within clerking, including medical and drug histories, and documentation of risk. The completeness and consistency issues raised in these studies present similar challenges experienced within the local AMU and indicate a potential benefit of 'priming' patients to the clerking questions in advance of their consultation. This priming may improve efficiency and consistency of the clerking procedure and as with the mental health study facilitate the 'history taking' and recording of a more complete drug history. Research into patient information provision has shown that careful design and presentation of proformas can improve patient care and safety and can be important with regards to managing patient expectations (Beisecker & Beisecker 1993) whilst also reducing anxiety for the patient (Sheard & Garrud 2006).

Methodology

The study took place in a large teaching hospital in the East Midlands region of the UK. The remit was to assess the admissions and clerking process and consider staff- and patient-centred solutions to the issues found. A multi-method process was implemented to understand the challenges for AMU admissions and to access a wide range of 'users' – clinical staff, nurses and patients. Figure 1 shows the methods used in the AMU analysis; Contextual Inquiry (CI) and Expert Interviews, and the subsequent Co-design task.

The Contextual Inquiry of the AMU was carried out with a Senior Consultant in Acute Medicine, accompanied by two HF experts and a cognitive psychologist. A range of clinical staff were then interviewed about their perspectives of working in the AMU, their views of the clerking procedures and working practices associated with admitting patients to the clinic. Following review of the data it was considered that, in addition to larger scale organisation modifications (beyond the scope of this paper), preparation of patients through information provision could alleviate some of the issues associated with clerking and admissions. The process for developing this resource took an iterative co-design approach as shown in Figure 1. In addition to the senior, mid level and junior doctors this process also included focus groups with a Patient and Public Involvement (PPI) group and the Nurses Unit Practice Council (UPC) both of whom were invited to review the document following each iteration. In between consultation with different user groups the co-design team (HF researcher, junior doctor and senior consultant) reviewed the document and made changes.

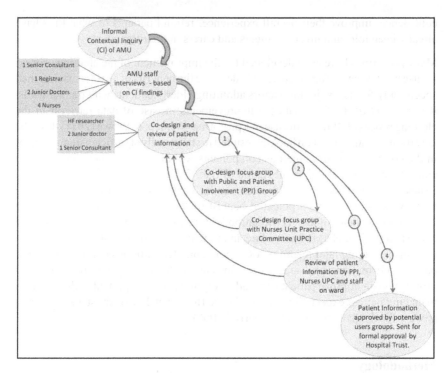

Figure 1.   User-centred Patient Information resource design process.

Results

Problems identified in the CI and Interview stages included the time being taken for clinical staff; particularly junior doctors, to clerk incoming patients. This process was having a negative effect on the waiting times for other patients needing consultation and was exacerbated by two issues. Firstly, the frequency and range of interruptions that clinical staff experience e.g. nurses passing on information for doctors, patients being unwell and requiring further review, requests to speak to relatives and secondly, the lack of preparedness by the patient to answer questions and provide information. This included the articulation of symptoms, comprehensiveness of patient report, missing or false information and the reconciliation of medications provided from the hospital and over the counter. Patients with poor literacy, non English speakers and those with a high burden drug regime may find this particularly difficult.

It was considered by the range of staff consulted that if the admissions process could be made more efficient then some of the wider issues in the unit would be alleviated, e.g. waiting times, duplication of information and tasks and allocation of staff resources and time. Figure 2 describes some of the other issues reported from these scoping tasks. A collaborative decision was made to address this problem through 'priming' patients prior to their clerking consultation. The aim being to help them understand what is expected of them and why and in doing so the preparation would support the clinician in their task whilst simultaneously improving the patient experience. This co-design team (see Figure 1) developed a preliminary information document based on clinical requirements, to be the subject of a PPI focus group. The 8 members of the focus group included a range of patients with personal experience, carers of patients who had been admitted to the acute wards and clinical staff who were no longer practising. The PPI focus group provided insight of the use of patient information by patients and stated requirements which may not have been identified through clinical proxies. The focus group was unanimous in the view that a patient information resource would be a good tool for use in the AMU. Their view was that it would explain to patients and relatives what will happen to them in the AMU and who they will encounter. It was considered that empowerment of patients, enabling them to get all relevant information to hand prior to consultation would improve their satisfaction with the process and that the clerking task would be quicker and more accurate. Figure 2 details some of the other needs raised by this user population. Similarly to the clinical staff, the issue of medicine reconciliation was one that all patients in the PPI group felt to be particularly difficult within the current clerking scenario. It was considered to be stressful, especially for individuals who are being admitted to the AMU for the first time with no prior knowledge of what the doctors will require. As such it was considered that a leaflet prompting the patient to find this information out would be useful in relieving stress and ease the process of acquiring and recording this information.

The next iteration of the information document was presented for assessment and development within a focus group with nursing staff in the AMU. Throughout the activity with this user group it became apparent that their needs of a patient information resource differ significantly from the needs of other clinical staff.

Figure 2 outlines the responses and main points raised within this focus group. Many of the requirements raised by the nurse UPC group are associated with the practical and logistical information that patients require for AMU admission, the provision of which was anticipated to relieve the time pressures on nurses for queries which take them away from their nursing responsibilities. One theme which was particularly significant due to its commonalities with the PPI group was the idea that patients should be empowered through information provision. A particular aspect of which is providing information to the patient in a brief written format to tell them what to expect in the next stage of their care, e.g. whether they require tests (indicating a waiting period before and after the tests have been carried out), if they are being admitted to the ward and why. It was important to the members of the UPC that this information was not only reported in the clinical notes but was made available to the patients as they felt that a verbal explanation could be easily forgotten or misunderstood, an issue also reported by Gupta *et al.* (2013). Following this focus group, the patient document was modified to reflect the needs of the nurse user population, and the resulting leaflet was disseminated to all participating user groups for review and critique. The leaflet was returned with minor recommendations, and with no additional requirements stated.

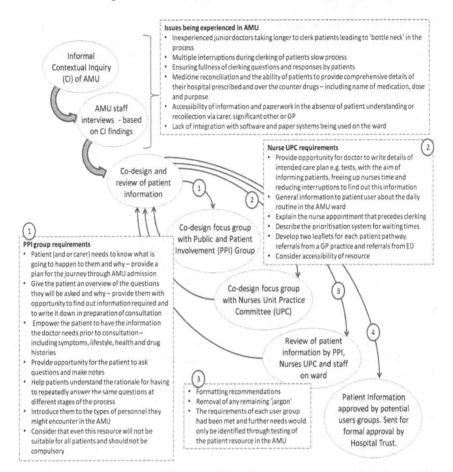

Figure 2.   Process Method with statements of need from user groups.

This process demonstrates a progressive user-centred method for involving all potential user groups in hospital documentation design. Carrying out each stage of development with a different user group and enabling a final holistic review elicited the requirements of all people who might make use of the documents within the AMU. Some statements of need were inclusive of the needs of multiple user groups, however each stage of the process disclosed new needs that had not been considered previously. Nurses and clinicians are sometimes used as proxies for the input of patient perspective. This process demonstrates that clinical proxies are not always able to anticipate the needs of patients and supports the case for user-centred design where all users are involved in the design process fully. Following the request for a human factors intervention into the process of admitting patients to the AMU, it was considered that the patient information developed by a user-centred process could,

- Improve the efficiency of the clerking process and admissions to AMU
- Support junior doctors during clerking the clerking process
- Maximise staff resource and time through 'priming patients'
- Enable patients to understand what information they will be expected to be able to provide to clinical staff and why
- Empower patients for the improvement of their experience

Limitations of this study include the interruptions experienced during the CI process and assessment of the environment, systems and tasks was achieved albeit these intervals. Another limitation of this study was the fact that the final review of the patient information document was carried out by email for the PPI and UPC groups and may not have provided detailed reviews had it been carried out face to face. The final review by clinical staff was carried out on the AMU by the junior doctor during working hours. Participants offered positive comments and limited critique and may not have taken the full time to consider the document due to other priorities. There is potential bias within the PPI and UPC groups, as these representatives of the user populations will typify proactive and confident individuals and therefore may not be able to adequately represent the needs of people who are unlikely to join these kinds of groups. Working with clinical staff and in the AMU provided challenges to the iterative method, working around time constraints and a variety of medical responsibilities. It was easier to involve the PPI and UPC groups as the focus groups were able to be carried out during their regular monthly meetings.

Conclusions

The patient information generated from this process will be useful not only in providing a working solution in a local AMU but also demonstrates in practice a user-centred method of developing patient information. This meets the Department of Health aims of achieving efficient care and providing patient centred care that is respectful and responsive to individual patient preferences and needs whilst also informs, communicates, and educates in order to facilitate patient involvement in clinical decisions and health promotion (IOM 2001). The use of this process by other clinical environments for patient information design could lead to benefits to both staff and patients and the healthcare system in which they are working and being treated.

Acknowledgements

The views expressed are entirely the authors own. The contents of this paper have not been commissioned and have been externally peer reviewed. We would like to express our thanks to the staff and PPI group within the Acute Care Centre of the Queen's Medical Centre, Nottingham UK.

References

Beisecker, A. & Beisecker, T. (1993). Using metaphors to characterize doctor-patient relationships: paternalism versus consumerism. *Health Communication* 5(1): 41–58.

Bircher, T. & Warriner, D. (2009). A career in Acute Medicine." *BMJ Careers*.

Byrne, D. & B. Silke (2011). Acute medical units: Review of evidence. *European Journal of Internal Medicine* 22(4): 344–347.

Faraj, AA., Brewer, OD. *et al.* (2011). The value of an admissions proforma for elderly patients with trauma. *Injury* 42(2): 171–172.

Frey, J. & Mertons Oishi, S. (1995). *The Survey Kit 4. How to conduct interviews by telephone and in person.* Thousand Oaks CA, London, New Delhi: SAGE Publications Ltd.

Gupta, S., Ratnasingham, K. *et al.* (2013). Improvement in clinical recording keeping following the introduction of an admission clerking proforma for acute general surgical patients. *Clinical Audit* 5: 61–66.

IOM (2001) Crossing the Quality Chasm. A New Health system for the 21st Century. Institute of Medicine. March 2001.

Isherwood, J., Patel, M. *et al.* (2013). The positive effects of accurate documentation. *British Journal of Healthcare Management* 19(5).

NHS (2008). Quality and Service Improvement Tools – Patient Information. National Health Service. Retrieved 28/09/13, from http://www.institute.nhs.uk/quality_and_service_improvement_tools/quality_and_service_improvement_tools/patient_information.html.

NICE (2007). *Acutely ill patients in hospital: Recognition of and response to acute illness in adults in hospital* London, National Institute of Health and Clinical Excellence. Guideline 50.

RCP (2007). *Acute medical care. The right person, in the right setting – first time.* London, Acute Medicine Task Force, Royal College of Physicians.

Rtiza-Ali, A., Houghton, CM. *et al.* (2001). Medical admissions can be made easier, quicker and better by the use of a pre-printed Medical Admission Proforma. *Clinical Medicine* 1(4): 327.

Sheard, C. & Garrud, P. (2006). Evaluation of generic patient information: Effects on health outcomes, knowledge and satisfaction. *Patient Education and Counseling* 61(1): 43–47.

Wachter, R. & Bell, D. (2012). Renaissance of hospital generalists. *BMJ* 344(652).

Wethers, G. & Brown, J. (2011). Does an admission booklet improve patient safety? *Journal of Mental Health* 20(5): 438–444.

Williams, D., Hoare, D. *et al.* (2013). Data recording aids in acute admissions. *International Journal of Health Care Quality Assurance* 26(1): 6–13.

RETAINED GUIDEWIRES IN CENTRAL VENOUS CATHETERISATION: AN ANALYSIS OF OMISSION ERRORS

Yi-Chun Teng[1], James Ward[2], Tim Horberry[2], Vishal Patil[3] & John Clarkson[2]

[1] School of Clinical Medicine, University of Cambridge, UK
[2] Engineering Design Centre, University of Cambridge, UK
[3] Cambridge University Hospitals NHS Foundation Trust, UK

Complete intravascular loss of guidewires in patients is an on-going medical concern. This research investigates the guidewire insertion and removal procedure by using a common omission error model by James Reason to identify procedural disposition to omission errors. The research builds on a previous Hierarchical Task Analysis for central venous catheterisation and identifies two crucial points that need further examination with regard to guidewire loss. Suggestions for improved equipment, procedural and system design at these two task steps are given.

Introduction

Central venous catheterisation (CVC) is a prevalent medical procedure that incorporates the Seldinger technique, in which a guidewire is first inserted to direct the subsequent entry of the catheter into the vein (Seldinger, 1953). The standard procedure requires guidewire removal immediately after catheter insertion, but sometimes this fails to occur and the guidewire becomes retained intravascularly. Prolonged guidewire retainment can cause complications such as venous thrombosis and post-thrombotic syndrome without evoking immediate, overt symptoms (Auweiler et al, 2005). The UK Hospital Trust where this research was undertaken recently experienced two such incidents. We previously conducted both a Hierarchical Task Analysis (HTA) and a Human Error Assessment and Reduction Technique (HEART) on CVC to assess the likelihood of future human errors leading to retention (Ward, Teng, Horberry and Clarkson, 2013). This HTA is incorporated in the present analysis which uses an omission error model by James Reason to explore the properties that predispose the omission of steps in CVC, leading to guidewire retention (Reason, 2002).

Method

James Reason's research on omission error assessment identifies factors that contribute to the omission of necessary task steps (Reason, 2002). The features of tasks that predispose their omission include:

1. Steps with a high memory or attention load;
2. Steps that are functionally isolated;

3. Steps that are repeated;
4. Steps that occur after the main obvious goal of the procedure is achieved;
5. Steps in which the object to be acted upon is hidden or inconspicuous;
6. Steps that take place after unforeseen distractions; and finally
7. Steps that occur after weak or unclear prompts.

This list of features was chosen because it offers a means for extensive review of each procedural task. It is easy to apply, provides immediate feedback, and facilitates solution planning. This analysis uses the above range of criteria to examine each task in CVC for properties that encourage their omission.

The HTA previously developed by Ward et al (2013) details the tasks involved in CVC. This research extends the HTA to pinpoint the most error-prone steps in CVC. The present task analysis focuses on and subdivides the steps in CVC that involve catheter insertion and guidewire removal. Steps that encompass two or more attributes for omission errors were deemed to be far more error-prone and were consequently highlighted as crucial moments for guidewire retention.

Results

Hierarchical task analysis of CVC procedure

This analysis extends the HTA by Ward et al into the following sub-steps:

Task	Central venous catheterisation
1.	Pre-catheter-insertion procedures
2.	Catheter-insertion and guidewire-removal procedures
2.1	Grip the catheter
2.2	Grip the guidewire near its proximal end
2.3	Thread the catheter over the guidewire until the catheter encloses the wire
2.4	Change the grip further down the wire towards the patient, and advance the catheter over the wire (this step may be repeated several times)
2.5	Advance the catheter until the distal tip is touching the skin of the patient
2.6	Grip the guidewire at the proximal end and advance the catheter into the vein
2.7	Remove the guidewire
3.	Post-guidewire-removal procedures

Sub-steps under "2. Catheter-insertion and guidewire-removal procedures" were deemed the most relevant for the loss of guidewires and further analysed for attributes that favour task omission. Subtasks 2.6 and 2.7 met several of the criteria on omission error and emerged as the steps most prone to omission in the context of guidewire removal.

Subtask 2.6: Grip the guidewire at the proximal end and advance the catheter into the vein

This step meets several features of tasks that are omission-prone, including numbers 2, 5, 6 and 7 from the list of criteria on omission errors. It is a functionally isolated step: one does not have to grip the proximal end of the guidewire in order to advance the catheter into the vein. Yet without this step, there is a higher chance of the guidewire becoming wholly enclosed within the catheter lumen and drawn into the patient's vasculature as the catheter is inserted. At this point in the procedure the guidewire is largely concealed inside the catheter. CVC sets with relatively short guidewires are more likely to be wholly hidden within the catheter. As a result, the operator is prone to forget to grip the obscured wire at its proximal end. The risk here is further increased if the operator is distracted unexpectedly at this point.

Subtask 2.7: Remove the guidewire

This is a highly omission-prone step as it meets six of the seven omission error criteria listed under methods (1, 2, 4, 5, 6, 7). The task is functionally isolated from catheter insertion and its proper execution relies solely on the operator's memory. It takes place after the main goal of the procedure – intravascular access of the catheter – is attained. Consequently, the operator is preoccupied in securing the catheter and carrying out the post-insertion procedures, and is more likely to forget to remove the wire. Guidewire concealment inside the catheter exasperates guidewire loss at this step. Without the proximal wire protruding from the catheter there is little, if any, cue prompting the operator to take this step. An operator interrupted prior to this task is more likely to assume that the guidewire had already been extracted due to lack of prompting cues.

Discussion

Guidewire retention has been attributed to a wide range of operator-related factors including insufficient staff experience, lack of supervision by someone competent in the procedure, and a general lack of awareness in the potential complications in CVC (Schummer, 2002; Ward et al, 2013). On a superficial level it is probably valid to say that such operator factors in the process increase the chances of guidewire loss. But the underlying design of the procedure, equipment and work organisation are the key factors contributing to human error. Analysis of the CVC process itself demonstrates deficiencies in the system that can be tackled to produce more effective solutions and remove operator error.

The steps requiring the operator to grip the guidewire while inserting the catheter and to remove the guidewire are most prone to omission errors. The analysis has identified several unfortunate aspects of these two steps. The steps are functionally isolated, occur after the main goal is achieved, and the items to be acted upon are hidden and thus provide little or no cues. Unexpected interruptions at any point may seriously jeopardise the completion of these omission-prone tasks. The shortcomings identified in the procedure cannot be addressed satisfactorily by solutions that target overt operator errors (such as additional training and re-educating the operators). Operator mistakes ultimately arise from weaknesses in the system and solutions should ideally target omission-error prone aspects of the CVC procedure instead.

Future research on guidewire retention should shift focus from the errors committed by the operator to changing the procedure itself. This may include implementing reminders prior to tasks 2.6 (advancing the catheter into the vein) and 2.7 (removing the guidewire), modifying specific task steps and engineering the CVC kit to eliminate or minimise the risk of guidewire loss. The wide spectrum of characteristics that favour omission errors as identified in this paper provides fresh angles that can be used to generate new and more promising solutions that better tackle this issue.

Conclusion

Cases of guidewire retention in central lines persistently occur despite repeated attempts to address overt human error factors. This analysis shows that human error in CVC is facilitated by omission-error prone tasks. The two most crucial points for further examination with regard to guidewire loss are catheter insertion and guidewire removal. Implementation of reminders and improvement of equipment and procedural design at these two steps would decrease the likelihood of human error leading to guidewire loss and circumvent total reliance on operator competence.

References

Auweiler M, Kampe S, Zähringer M, Buzello S, von Spiegel T, Buzello W, et al. 2005, *The human error: delayed diagnosis of intravascular loss of guidewires for central venous catheterization*, J Clin Anesth. (November):17(7):562–564.

Reason, J. 2002, *Combating Omission Errors through Task Analysis and Good Reminders*, Quality & Safety in Health Care 11 (1) (March): 40–44.

Schummer, W. 2002, *Loss of the Guide Wire: Mishap or Blunder?*, British Journal of Anaesthesia (January) 88: 144–146.

Seldinger, S.I. 1953, *Catheter Replacement of the Needle in Percutaneous Arteriography: A New Technique*, Acta Radiologica, 39:5, 368–376.

Ward, J, Teng, Y-C, Horberry, T and Clarkson, PJ. 2013, *Healthcare Human Reliability Analysis – by HEART*, In *Contemporary Ergonomics and Human Factors 2013*, edited by Martin Anderson, 287–288 (Taylor and Francis).

NHS MANAGERS' UTILISATION OF INFORMATION RESOURCES TO SUPPORT DECISION-MAKING IN SHIFT WORK

Eleanor Mowbray[1], Patrick Waterson[1] & Ron McCaig[2]

[1]*Loughborough Design School, Loughborough University*
[2]*Cheshire and Wirral Partnership NHS Foundation Trust*

To evaluate the information resources and training available to NHS managers designing shift patterns a literature review and document analysis of Trust policies were used as a foundation for interviews with Ward Managers (n = 7), Senior Managers (n = 5), and e-roster Support Staff (n = 4). Four factors were identified which shape a managers' approach to shift work: Shift work guidance, experience, staff, and performance measures. These inputs can be divided into *Top-Down* (budget and legislation) and *Bottom-Up* (staff, patients, and ward acuity) pressures, both of which must be applied in a balanced and creative way to manage shift work effectively.

Introduction

Shift patterns designed with consideration for employees needs are not only required by law (HMSO 1998) but are critical to managing workload, reducing error, and maintaining staff performance. There is no clear consensus that 12-hour shifts are either beneficial or detrimental compared to a 8 hour three-shift system and questions remain regarding how shift work can be managed effectively, whilst considering staff health and wellbeing and maintaining high standards of care. Although literature regarding shift work management within healthcare is limited, 8 papers were critiqued and were found to advocate innovation in solutions and the consideration of each pattern with respect to the characteristics of the service delivered to patients—for example, ward size and care requirements (Silvestro and Silvestro 2000). An approach that applies empirical research alongside local, contextual knowledge is encouraged (Swan et al. 2012) and collaboration with stakeholders generates acceptance, ownership (Buchanan et al. 2012) and satisfaction (Maenhout and Vanhoucke 2013).

To assess the applicability of these findings within an NHS Trust, 16 managers from Cheshire and Wirral Partnership NHS Foundation Trust (CWP) contributed their views. CWP has undergone two recent changes within shift working: The introduction of an electronic rostering system and 12-hour shifts. The interviews revealed how these have been accepted by managers and their impact on information utilisation, creativity, and job satisfaction. 4 information sources were identified which shape a manager's approach to shift work: *Top-Down* knowledge drawn from

Shift Work Guidance and Performance Measures, and *Bottom-Up* knowledge from Experience and Staff Input.

In terms of information uptake, Ward Managers are inclined to rely on subjective *Bottom-Up* sources and feel pressure from their staff and patients, while Senior Managers focus on *Top-Down* information due to pressures from performance data and legislation. In order to develop the quality of the service, empirical guidance is often overlooked in favour of responding to performance feedback which addresses stakeholders' needs directly. The need to mediate both types of information and take a creative, balanced, and collaborative approach to shift management is evident, but nonetheless challenging.

Methodology

Shift work documents—including CWP's Roster Policy (CWP 2013) and guidance from RCN (RCN 2012) and UNISON (UNISON 2002)—were analysed as a source of expert opinion and to provide the foundation for a series of 14 interviews. The interviews lasted between 10 and 31 minutes and transcripts totalled 32,097 words.

Three interview schedules were developed in collaboration with staff from the Trust, comprising 26, 23 and 28 questions for Ward Managers, Senior Managers, and e-roster support staff respectively.

Schedule 1 (n = 7) was used with Unit, Team, Ward Managers, Clinical Leads, and Senior Nurses who have direct responsibility for rostering staff alongside clinical care duties.

- Schedule 2 (n = 5) was used with Senior Management—Modern Matrons and Clinical Service Managers—who oversee shift management in multiple wards, teams or units.
- Schedule 3 (n = 2) was used with e-roster Support Staff whose role is "training managers and supporting them with understanding how effective or ineffective their rosters are" (e-roster Administrator).

Inclusion criteria limited participants to those with a management role within an NHS Trust which includes responsibility for shift design; however, participants were drawn from inpatient and outpatient services to explore the variation across different types of care—including staff from Adult Mental Health; Older Peoples' Mental Health; Learning Disabilities; Community Teams; Home Treatment; Urgent Care; and Out-of-Hours Services—and to provide coverage of multiple sites within CWP Trust (Chester, Macclesfield, and Birkenhead).

Results

The results are discussed in relation to the use of empirical and local contextual data; working collaboratively with stakeholders; and the need for a creative and unique solution in each set of circumstances.

Empirical and local contextual data

Information uptake varies between managers at different levels of seniority: Ward Managers focus on information inputs from staff and patients while Senior Managers focus on inputs from organisational and political factors (government, legislation and policy). However, at both levels local data from personal experience and internal information (developed within the Trust) was found to have greater weight to reform shift patterns than empirical guidance from the EWTD or HSE.

The importance of mediating these guidelines with local knowledge is demonstrated by the mixed uptake of Trust-wide policies—such as the move to Long Days and e-rostering. This lack of consultation and push for a *"one size fits all"* solution is cited by Waterson (2013, p. 18) as a factor in the failure of the National Programme for Information Technology in Healthcare (NPfIT). However, overreliance on local knowledge was treated with caution by some managers who warned of the potential for unfair and irrational decisions based on the needs of individuals as opposed to the service. The need for balance, *"careful translation"* (Swan et al. 2012, p. 188), and integration of research with local pressures is evident.

Working collaboratively with stakeholders

An approach to shift work which draws on expertise from multiple stakeholders and applies these in a balanced way was identified as the goal among managers. The input from staff as stakeholders was high; their contributions were valued and their wellbeing prioritised.

A creative and unique approach

A creative approach is fundamental to responding to the unpredictable nature of care and greater flexibility would be valued by managers but is restricted by budget, the EWTD, and the e-roster which enforces these. Although frustrating at times, there are many positives to the e-roster (improving fairness; simulating changes; providing evidence; highlighting issues, errors, and EWTD violations; saving time) which staff are embracing with help from the e-Roster Support Team.

Many managers also observed a strong link between successful shift work management and increased job satisfaction, so a creative approach and self-directed research should be encouraged but kept within legal and financial limits.

Conclusions

Overall, CWP's Ward Managers feel they are well supported and the information available to them is sufficient, drawing on a variety of local and universal sources. However, complications arise due to budget and staff allocation which limits the flexibility of the system.

The pressure to push forward the quality of the service means that research-based guidance is often overlooked in favour of performance feedback which corresponds to stakeholders' needs in a direct and measurable way. The interviews revealed that little empirical guidance is used outside the e-roster system (which enforces EWTD legislation) and since its introduction fewer managers have been prompted to look for information on shift work. Instead, Ward Managers are inclined to rely on subjective experience and staff feedback on performance.

The limitations of this type of data were highlighted by the Senior Managers who warned of the risk of bias and short-sighted decisions. Their *Top-Down* approach accounts for many more external pressures and the need to mediate these through collaboration with Ward Managers' *Bottom-Up* understanding of care is evident. This is a challenging balance to strike given the pressures from numerous stake-holders along with the political weight, media interest, and tight budget of the NHS: The thresholds are severe, and greater freedom and authority to work creatively and customise shift patterns is something all managers would value—not only for their own satisfaction, but that of their staff and consequently their patients.

References

Buchanan, D. A., Denyer, D., Jaina, J., Kelliber, C., Moore, C., Parry, E. and Pilbeam, C. 2012, *How Do They Manage? The New Realities of Middle Management Work in Healthcare*, Cranfield University School of Management, Cranfield, UK.

CWP. 2013, "HR 18 Roster Policy." Retrieved 12th February 2013, from: http://www.cwp.nhs.uk/policies/1561-hr18-roster-policy/

HMSO. 1998, *The Working Time Regulations*, SI 1998/1833, HMSO, London, UK.

Maenhout, B. and Vanhoucke, M. 2013, "An integrated nurse staffing and scheduling analysis for longer-term nursing staff allocation problems." *Omega* 41(2): 485–499.

RCN. 2012, "A Shift in the Right Direction." Retrieved 16th October 2012, from: http://www.rcn.org.uk/__data/assets/pdf_file/0004/479434/004285.pdf

Silvestro, R. and Silvestro, C. 2000, "An evaluation of nurse rostering practices in the National Health Service." *Journal of Advanced Nursing* 32(3): 525–535.

Swan, J., Clarke, A., Nicolini, D., Powell, J., Scarbrough, H., Roginski, C., Gkeredakis, E., Mills, P. and Taylor-Phillips, S. 2012, "Evidence in Management Decisions (EMD) – Advancing knowledge utilization in healthcare management." Retrieved 19th November 2012, from: http://www.netscc.ac.uk/hsdr/files/project/SDO_FR_08-1808-244_V01.pdf

UNISON. 2002, "Negotiating on Shiftwork." Retrieved 16th October 2012, from: http://www.unison.org.uk/acrobat/13026.pdf

Waterson, P. 2014, "Health information technology in sociotechnical systems: A progress report on recent developments within the English NHS." *Applied Ergonomics*, 45(2), 150–161.

AS I AM NOW, SO SHALL YOU BECOME

Andree Woodcock

Coventry School of Art and Design, Coventry University, UK

A 3 month participant observation case study of an elderly couple coping with chronic and acute health issues is used to highlight the everyday problems faced by the oldest in society. The increasing number of the elderly, policies and the wish of elderly people to retain independence in their own homes is juxtaposed against the reality of equipment, services and practices which are not fit for purpose. The paper closes with a set of recommendations for greater awareness, interest and funding for research in this area.

Introduction

The Select Committee on Public Service and Demographic Change (2013) opened with the following statement *"The UK population is ageing rapidly, but we have concluded that the Government and our society are woefully underprepared. Longer lives can be a great benefit, but there has been a collective failure to address the implications and without urgent action this great boon could turn into a series of miserable crises."* (p. 7). A rapidly ageing society means older people living longer, often with one or more chronic long-term health conditions. It is predicted that there will be 51% more people aged 65 and over and 101% more people aged 85 in England in 2030 when compared to 2010.

For people aged 65 and over in England and Wales, 2010 to 2030 will witness the following increases: people with diabetes up by over 45%; people with arthritis, coronary heart disease, stroke, each up by over 50%; people with dementia (moderate or severe cognitive impairment) up by over 80% to 1.96 million; people with moderate or severe need for social care up by 90%. Over 50% more people will have three or more long-term conditions in England by 2018 compared to 2008.

Ipsos MORI showed that people are unprepared or unwilling to plan for old age *"assumptions (based on little knowledge), a fear of the unknown, denial, and negative connotations of being a 'pensioner' mean that we put off our financial planning until we are forced to"* or to contemplate and provide for future disability or mental illness, even to the extent of adapting their houses to be suitable for older life. Many assume that the State or their family will look after them. Although most people want to remain independent in their own homes, reports such as that published by the Equality and Human Rights Commission (2011) express concern as to the quality of care provided citing e.g. lack of training, rapid staff turnover, inability of staff to deal with people's needs.

A basic requirement of living independently should be that it in itself does not become a chore, that services and products are accessible, comprehensible and 'joined up'. Woodcock et al (2013) cited research which showed that the elderly who experienced difficulties in everyday life just 'carried on' without looking for other support and that developers neither understand the context of use of equipment, nor support higher level needs or the needs of the whole person.

This research specifically addresses these points by providing insights into the life of an elderly couple of average education, who wish to continue to live independently in their own home, for as long as possible. They represent the oldest old of the population (i.e. over 85 years of old (currently just over 2% of the population)). The study demonstrates why the oldest old become the heaviest users of health and social care services and shows that one of the reasons may be inefficiencies and lack of transparency of health services. The observation period covered a transition in the health of the couple necessitating a move from needing little to moderate nursing and social care.

In this age group just under 40% of men and less than 10% of women live together, either as a married or cohabiting couple (Dunhall, 2008). The predicted expansion of this population requires more research on issues such as the effects of living with chronic disease on (usually the female) partner's health, well-being and sense of identity (e.g. Hagedoorn et al, 2001), models of successful ageing (e.g. Baltes and Baltes, 1990) and the factors which influence it (e.g. acceptance of lower expectations about their appearance and that of their homes, relationships with care workers, Aronson, 2002).

Method

An opportunistic case study (Yin, 1984) was conducted with Joshua and Ivy (both in their mid 80s), over a period of 3 months. The timeframe allowed for the study of everyday life, hospitalization and 'reablement'. Observations were made by two people, with the frequency and duration of observations varying from daily to three times a week. The couple suffered from numerous health problems including diabetes, crumbling spine, macular degeneration, heart disease, arthritis and mobility problems (both use walking aids). Ivy suffers from severe loss of vision and Joshua from Alzheimer's. For most of the observational period, Ivy was the principle care giver, with Joshua retaining leadership of the household. Prior to Joshua's hospitalisation the couple worked co-operatively to maintain their independence and were adapting successfully to old age. Ultimately, Joshua's catheterisation and the steady deterioration of his mental faculties negatively affected their relationship, Ivy's self identity (as caregiver) and their ability to live independently.

The case study represents a situation that is believed to be typical of that of couples, who wish to remain independent in their own homes, but who have a steady decline in health and mobility. Therefore it provides a contemporary account of the everyday difficulties of the elderly which must be recognised by designers of products, health and social care for this group.

Results

The results are grouped under 3 sections, relating to use of non emergency health services, hospitalisation and 'reablement'.

Use of everyday medical services

As people become older, managing health becomes a major task which can control where and how they live. Successful ageing becomes a balance between coping with periods of illness, regaining health and learning to adapt to increasing frailty. In the case of Joshua and Ivy, the couple developed successful coping strategies, compensating for lack of eyesight and mobility and doing household chores co-operatively. Maintaining physiological and psychological well being and reassurance requires liaison with multiple health care providers (e.g. opticians, chiropodists, dentists, local surgery and hospital departments). Joshua and Ivy can have one or two health related appointments every week each of these requires close management, for example:

- Reading and understanding appointments sent through the post. As some of these are made months in advance, or relate to referrals to clinics they are not familiar with, receipt of an appointment may in itself be a cause of relief or stress.
- Successful transference of the appointment to a large desk diary. This necessitates finding the diary, a pen and entering the information correctly under the right date. The information is not always entered on the right day or month. Although Joshua in particular might refer to the diary 3 or 4 times an hour, he does not necessarily know what the current day or month is. Ivy'e eyesight is poor, which means that the appointments may be entered on the wrong day, and the writing may be hard to decipher. Appointment cards are sometimes attached with safety pins to the pages. Even when the appointment is entered on the right date, it may not be noticed.
- Travel to the health provider has to be planned. Ivy can use a scooter to travel to appointments at the local GP, otherwise they rely on family transport and occasionally taxis. Public transport is not used as it is inaccessible, requires transfer to different vehicles, and will still require navigation in the hospital. At the start of the observation period, Joshua and Ivy went to appointments together, at the end, Ivy went to her appointments without Joshua, necessitating advanced planning. With some booked appointments requiring a 4 hour visit, and 10 mile round trip, Joshua would not be able to accompany Ivy (without wandering off) and cannot be left alone.
- Neither were able to communicate well with any of the medical staff, relying on family members to negotiate with gatekeeper staff, make appointments,

understand prescriptions and take them to the right clinics. This process was exacerbated by the lack of joined up health services, which meant that even at the local surgery, patients are not seen by the same doctors. The participant observers, as family members were able to provide sufficient information to enable speedy completion of some visits. Without this intervention, problems would have occurred.

- The local surgery offers a drop in centre for minor ailments, with waiting times of up to 3 hours, in a room of hard furnishings, offering little by way of entertainment. Appointments with assigned doctors can be booked two weeks in advance. Although the surgery has a high number of elderly patients, it does not have a geriatric specialist.

- Very rarely did prescriptions arrive complete or on time, even with automated services. This sometimes necessitated 3 separate visits to the chemist. During the observation period we were subjected to items missed off the prescription list, items from different services having different delivery mechanisms, partly fulfilled prescriptions (which necessitated additional visits to receive a monthly supply of medicines), medicines which had been simply lost. The delivery of prepacked medicines was stopped following periods of noncompliance and a backlog of 3–6 months of unused medicines.

- Noncompliance was a major issue with Joshua. Ivy had recently experienced a life-threatening illness and the importance of compliance had been emphasized. However, her medication changed on a daily basis, requiring her to read instructions twice a day with a magnifying glass and locate very small tablets. Failure of the local surgery to recognise a life threatening condition, has led to distrust in the health service and its failure to take complaints seriously. At one time. Joshua was taking approximately 16 tablets over four time periods. Although Ivy reminded him to take these, the blister packs were difficult to open and the tablets difficult to see, resulting in some being left in the packs. Putting medicines into dispenser packs and monitoring compliance on a twice weekly basis has helped. However, once out of the packs, medicines cannot be administered by health professionals. Medications related to the control of Alzheimer's were steadily resisted Observations and remarks made to the psychiatric nurse revealed that tablets were routinely missed or thrown away, either due to side effects (e.g. dizziness, nausea, drowsiness and hallucinations), misunderstandings about their long term benefits and required build up or the stigma attached to Alzheimer's.

During the initial observation period, the issues outlined above occurred regularly. Joshua and Ivy were still able to living independently, but started to need more support. In particular, it was noted that Joshua's deterioration had a negative effect on Ivy's well being and cheerfulness (reflecting Hagedoom et al's 2001 findings).

Hospitalisation

Joshua was taken ill with a urine infection during a remotely located family holiday. With no local medical services, poor internet connectivity, a non responsive local surgery with irregular opening hours, an emergency surgery was located in a village 10 miles away. Treatment was ineffective and resulted in a call out 12 hours later to an emergency paramedic, who gave enough assurances for a 6 hour drive home to be undertaken, as opposed to waiting the same length of time for hospital treatment.

3 days later, and with no great improvement in his health, Joshua was taken to his local surgery and was sent home to have his bladder emptied by the district nurse. Convinced of inaccurate readings on urine retention the nurse persuaded the doctor that there was a dangerous build up of urine, which she was unable to treat. We drove Joshua to the urology department of the local hospital.

CQS (2013) noted the need for hospitals to recognise the needs of patients with dementia, to record it, and to have training that *"would help staff to identify the person with dementia and reduce length of stay by helping staff to respond effectively to dementia, for example by ensuring their nutritional needs are met."* The Alzheimer's Society recognise the impact of dementia on patient's outcomes; with longer stays, a worsening of dementia and physical health, a greater likelihood of being discharged to a care home instead of their own home and a greater likelihood of antipsychotic medication being used.

Although treated quickly, compassionately and successfully, Joshua was kept in hospital for 10 days for observations, further tests and to allow a care package to be arranged as part of reablement. Fortunately, no further medical complications arose, but the Alzheimers proved difficult to manage, with Joshua having no memory of why he was there, how he had become ill, what the catheter was, what a hospital was, where he lived or when he would be released. As soon as one question was answered, it was repeated again. Problems included:

- Insufficient monitoring of food and fluid intake leading to the need for rehydration (via a drip, which he did not understand) and a weight loss of half a stone.
- Lack of any form of entertainment meaning that Joshua spent most of his day in bed, staring at a wall, unable to walk, read or engage in conversation with other patients. This resulted in stress and anxiety.
- Joshua's lack of memory meant that he did not remember seeing consultants or what was explained to him by the nurses. As nursing staff were rotated, it was difficult to find out what was happening. Information was written on pieces of paper and sellotaped to the wall (e.g. how many more days he had to stay in the hospital, what was wrong with him and what the catheter was).
- Incorrect treatment, resulting in one incident when he was taken to theatre for an examination but given the wrong premeds and 'left to die in a cold basement'.
- The occupational therapist noticed he was soiling himself on a regular basis, and brought this to the attention of ward staff.

Visiting times were relaxed, so we brought in food, newspapers, magazines, games and a CD, all of which were discarded. Conversation revolved around showing holiday pictures, and continual reassurance of the same points. We also provided a sheet of 'Dos and Don'ts' in relation to the catheter in particular. If these were not adhered to, the chances of Joshua being able to transfer back home were reduced.

Reablement and adaptation for special needs

The reablement team were set up by the hospital to encourage independence after hospitalisation. They provided twice daily support and monitored long term needs for 6 weeks, after which other service providers had to be found. Although they provided social care, they do not provide health care (e.g. they can remind the patients to take medicine but not give it to them). Patients with Alzheimer's and catheters fall between social and health care. The team appears to be large, variably trained, with no regular attendants and some coming 90 minutes early.

The occupational therapist admitted that catheterization was normally the tipping point for people with Alzheimer's, and would ultimately require residential care. To postpone this and manage increasingly difficult hygiene issues, the bathroom was converted into a disabled friendly wet room with a bidet toilet; a second bedroom was created so Ivy could have uninterrupted sleep, and the house was decluttered to provide ease of access, cleaning and space for the social care assistants. Problems experienced during this period included:

- Poor installation of bathroom by the accredited supplier, including wrong shower head attached, and a radiator positioned too close to the toilet.
- Decluttering the house enabled Ivy to gain a sense of control and aided cleaning, However, Joshua lost his context, so books and personal artefacts were reintroduced.
- A new television and CD player required careful positioning, attachment to walls and retraining.
- The new bedroom was rejected as an unwanted intrusion on personal life despite it providing better access, easier cleaning, more comfort etc.
- During this time it was noted that Joshua and Ivy were being targeted by nuisance callers, despite being on a nuisance calls register. Further investigations have revealed that the callers are amassing a number of details about the couple, and recognise who is answering the phone. This is a source of worry as Joshua could answer the phone, and may reveal personal information or inadvertently agree to a service.
- Nighttime hunger and wandering remains a problem, which has not been resolved.
- Lack of joined up health and care packages which required Joshua and Ivy being treated as individuals, not a couple and led to different agencies asking for the same information and failing to pass on information or include family members in information gathering, even though this has been explicitly requested and written in instructions.

Although Joshua accepted and understood the catheter as a constant factor in his life, it has not been used properly. As a former mechanic he understands how it works and has accepted that he will be seriously injured if he tried to remove it, However, it is opened constantly, emptied at inappropriate times and locations and not positioned correctly on his leg. District nurses have been called out twice in two months to deal with blockages. Although Ivy has taken some control over its use, she is unable to see the release lever, instructions have not been give to her clearly, and different types of bags are delivered, which are incompatible with the systems they are used to. Although night time bags (a secondary bag, attached to the bottom of the day bag via a long tube) are available, their operation was not well explained, and these are not properly connected. The night bag requires careful management, preferably by someone who does not sleep with a dog and a partner, who lies still and who does not get up at night and drag the bag around with him. This means that even with care workers twice a day to ensure that the bag is empty and set up properly, the catheter is misused resulting in wet bedding two or three times a week and Ivy has to remain vigilant throughout the night and check on its status. Hygiene and cleanliness are obviously major concerns, and have not been resolved properly.

Conclusions

Joshua used to say 'as I am now, so shall you become' when referring to his physical deterioration. The speed of medical advances means that the current generation will suffer from similar health issues to the ones related here. The question is, do we have sufficient understanding of what it means to grow old and are we prepared to make the high levels of investment and sweeping reforms that are needed to improve health products and services. This limited, and personal study illustrates the type of problems experienced by real people in real life, not in laboratories or simulations. Designers have to understand and develop products which will work in this context, with people who have multiple disabilities, for example, those who cannot see well, have limited mobility and memory.

Although much could be learnt from the respectful treatment and care of the elderly in non Western societies, family break up means that many people or couples will have to be resourceful and independent in later life. This may necessitate training of young older people to understand, accept and plan for life changes – for extended periods of old age- to ensure successful aging and coping strategies, and to improve health literacy.

The lack of joined up services and failure to remove systematic inefficiencies should be reconsidered from the perspective of the end user of the services, who are needlessly worried and frustrated in their attempts to get high qualities of service and who are denied a high quality of life.

Lastly, greater consideration needs to be given to the parts played by informal circles of care; family, friends and partners. Not all people want to, or can draw on these resources. With a substantial increase in the number of elderly, new local, sustainable and transferable community based and led initiatives need to developed, such as street or neighbourhood centres where older citizens can support each other, and find advocates.

References

Aronson, J. (2002), Elderly people's accounts of home care rationing: Missing voices in long-term care policy debates. *Ageing and Society, 22*, 399–418.

Baltes, P.B. and Baltes, M.M. (eds), *Successful Ageing: Perspectives from the Behavioural Sciences,* Cambridge, England: CUP 1990.

Care Quality Commission (2013), *Care Update*, Issue 2, Retrieved 21st September, 2013 from http://www.cqc.org.uk/sites/default/files/media/documents/cqc_care_update_issue_2.pdf.

Dunhall, K. (2009), Ageing and mortality in the UK – National Statistician's Annual Article on the Population in ONS (ed), *Population Trends*, No. 134, Palgrave MacMillan, UK, p. 6–23.

Equality and Human Rights Commission (2011), *Close to home*, Retrieved 21st September, 2013 from http://www.equalityhumanrights.com.

Hagedoorn, M., Sanderman, R., Ranchora, A.V., Brilman, E. I., Kempen, G. and Ormel, J. (2001), Chronic disease in elderly couples: Are women more responsive to their spouses' health condition than men? *Journal of Psychosomatic Research*, 51, 693–696.

Select Committee on Public Service and Demographic Change (2013), *Ready for Ageing*, Authority of the House of Lords London: The Stationery Office Ltd.

Woodcock, A., Ward, G., Osmond, J., Unwin, G., Ray, S. and Fielden, S. (2013), Uptake of assistive technology, *Ergonomics Society Conference*.

Yin, R.K. (1984), *Case study research: design and methods*, Newbury Park, CA: Sage.

MUSCULOSKELETAL DISORDERS

ERGOKITA: ANALYSIS OF THE MUSCULOSKELETAL STRAIN OF NURSERY SCHOOL TEACHERS

E.-M. Burford[1], R. Ellegast[1], B. Weber[1], M. Brehmen[2], D. Groneberg[2], A. Sinn-Behrendt[3] & R. Bruder[3]

[1] Institute for Occupational Safety and Health of the German Social Accident Insurance, Sankt Augustin
[2] Institute of Occupational, Social and Environmental Medicine of the Goethe University of Frankfurt, Frankfurt am Main
[3] Institute of Ergonomics of the Technical University of Darmstadt, Darmstadt

The ErgoKiTa study aimed to determine the musculoskeletal strain of nursery school teachers and to identify and evaluate suitable prevention measures. A comprehensive work analysis using objective and subjective methods was performed in nursery schools in Germany. The musculoskeletal strain was determined by means of a comprehensive analysis of postures, forces and movements using the CUELA-system. This paper presents the applied methodology and preliminary descriptive results for the initial data capturing phase for the parameters of knee flexion, trunk flexion, external load moments at L5/S1 and heart rate. Unfavourable postures were identified for both specific tasks and for whole shift values, specifically for trunk and knee flexion.

Introduction

As a result of a change in political policy in Germany, the number of children in the nursery schools aged under-three is increasing. This is expected to have an effect on the physical workload of nursery school teachers, for example an increase in lifting and carrying of children. Only limited research has investigated the stress and strain of occupations in this field (Kusma et al., 2011). Furthermore very few studies have focused on the objective musculoskeletal workload in this profession, with the current situational workload being predominantly unknown. As the facilities are usually designed to be height-appropriate for children, this often results in the nursery school teacher adopting unfavourable body postures to work in this environment (King et al., 2006). Existing means to reduce the physical strain experienced as a result of work-related factors and improve the working posture in this occupation have yet to be evaluated and validated.

The ErgoKiTa study (ErgoKiTa: Ergonomic design of workplaces in nursery schools) was initiated by the Institute for Occupational Safety and Health of German Social Accident Insurance (IFA) to ascertain the current state of knowledge available on the musculoskeletal strain of nursery school teachers; to use objective and subjective methods to ascertain the current situational workload that nursery school teachers are exposed to; and to provide evaluated prevention measures. The research group included experts from the German Social Accident Insurance (DGUV), the Institute of Ergonomics of the Technical University of Darmstadt (IAD), the Institute of Occupational, Social and Environmental Medicine of the Goethe University of Frankfurt (ASU), as well as representatives of the social accident insurance institutions of the involved states of North-Rhine/Westphalia, Rhineland-Palatinate and Hesse.

The study consisted of several phases including a comprehensive review of the available literature on musculoskeletal strain in nursery school teachers, extensive questionnaires to ascertain the current health of the employees as well as question-naires pertaining to the structural parameters associated with work organisation, tasks and work environment that may have an effect on the workload experienced. Furthermore, detailed objective and subjective workload analyses were performed in a smaller sample group in order to ascertain the current musculoskeletal work-load and aid in identifying and evaluating potential preventative measures. Once suitable prevention measures have been identified, these will be evaluated using the same methodology applied to determine the initial workload experienced. The focus of this paper is the objective analyses used to determine the current workload that nursery school teachers are exposed to and preliminary descriptive results for the present situation, prior to the implementation of intervention measures.

Methods

In a previous phase of the ErgoKiTa project, nine nursery schools, representative of the majority of the nursery schools from three different German states, were selected. These nine nursery schools, three from each state, were classified as having a low, intermediate or high intervention need (Sinn-Behrendt et al., 2013). The classification was based on the results of the questionnaire, which assessed the structural and work environment conditions, as well as workplace inspections with the main focus on the available facilities and furnishings. As there is lim-ited research pertaining to the workload within this profession, this classification scheme served as a preliminary step to differentiate and assist with the selection of nursery schools that had a greater intervention need based on the available facilities and furnishings as these factors would most likely effect working posture. These nine nursery schools, specifically selected so that one representative nursery school was selected for each classification group from each of the three German states, were then involved in extensive workload analyses and the intervention study that followed.

In each nursery school the strain and stress of two of the nursery school teachers was assessed with a comprehensive work analysis (n = 18). The participants were all female and had a mean height of 168.50 cm (SD: ±6.09), a mean weight of 72.42 kg (SD: ±14.69), a mean BMI of 25.50 (SD: ±5.00) and mean age of 34.08 years old (SD: ±10.14).

Using the mobile CUELA-system (Ellegast et al., 2010), the musculoskeletal work-load, for each teacher performing her usual duties, was analysed for approximately 4 hours during a normal work day on two separate days. The system determined postures, forces and movements from the data obtained by 10 sensors consisting of ADXL 103/203 3D accelerometers and muRata ENC-03R gyroscopes. Additional trunk movements such as the torsion of the upper body were recorded as a result of sensors located at the upper and lower body being joined with a metal shaft that was connected to a magnetic field sensor (Vert-X 12, Contelec). Loads lifted were assessed using a biomechanical model and pressure sole-insert sensors (Paromed), which allowed for the analysis of ground reaction forces. In addition to this, cardio-vascular strain was assessed by recording heart rate using a Polar heart rate monitor (model RS400) with a sampling frequency of 1 Hz and noise levels were recorded using Brüel & Kjæl noise dosimeter (Type 4448). The participant was filmed for the entire data capturing phase to allow for the assignment of specific tasks, as defined in the study by Kusma et al. (2011), in the data preparation phase. The complete system used to assess workload is depicted in figure 1.

The data obtained from the sensors were analysed as duration percentages of unfavourable postures, loads lifted and the moments and compression forces at the lumbosacral disc L5/S1 (based on biomechanical model calculations) for eight hour work shifts and individual tasks. Additionally the data obtained from the heart rate monitor and noise dosimeter were also analysed in respect to specific tasks as well as for an eight hour work day. Preliminary results presented in this paper

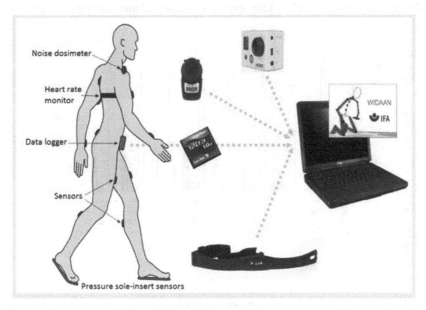

Figure 1.   Set-up of the CUELA-system, heart rate monitor and noise dosimeter used to assess the workload experienced by nursery school teachers in situ.

include results pertaining to the eight-hour work shift for the parameters of knee flexion, trunk flexion, external load moments at L5/S1 and heart rate. Inferential statistical analyses are still outstanding and these results will be published at a later date.

Results

The nursery schools (Kitas) differed from one another with regards to the duration percentages for the different postures for the different tasks as well as for the eight hour work shift.

Nursery schools that were classified as having a high intervention need had a greater percentage of the shift spent in a seated posture where the knee joint was in a position greater than 90°, indicating considerable knee flexion. This is depicted in figure 2. Furthermore this provided an indication of the amount of time the teachers spent seated on children's chairs at children's tables.

Similar results were found for the CUELA posture code, which based on the positioning of the sensors determines the percentage of the duration for different postures. The CUELA posture code indicated that for the shift analysis, the individual nursery schools differed with regards to the percentage of the eight hour working day where tasks were performed in knee-straining postures. Knee-straining postures identified by this code are postures that have been recognised as risk factors for knee pathologies and include squatting, supported and unsupported kneeling,

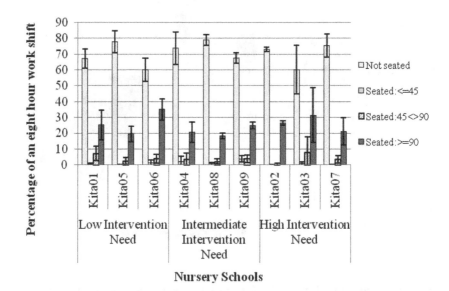

Figure 2.   Percentage of an eight hour work shift spent in a seated posture for different joint interval ranges for knee flexion (left knee joint).

sitting on heels and crawling (BMAS, 2010). Three of the nine nursery schools analysed had over 12% of the shift containing tasks that resulted in knee-straining postures, while the remaining six had straining postures for approximately 5% of the shift.

In all the nursery schools, teachers experienced upper body postures with trunk flexion greater than 20° for at least 16% of an eight hour shift, while in nursery schools classified as having a high intervention need, this number rose to approximately one third of the shift. These results are depicted in figure 3.

The durations of load bearing activities were noticeably higher in nursery schools with greater numbers of children under the age of three, where loads between 5 and 20 kg were carried and recorded to contribute to approximately 5% of the duration of the work shift. The biomechanical model calculations indicated high moments at the lumbosacral disc L5/S1 (>85 Nm) for approximately up to 5.3% of the work shift. These results are featured in table 1.

Further shift analyses revealed that the mean heart rate was below the endurance limit, the limit under which an individual can work without becoming fatigued (Ulmer, 1968), for more than 31% of the shift.

In addition to results for the work shift values, the data obtained during the workload assessment was used to perform a comprehensive task analysis. The task analysis showed variations between the different nursery schools for the same tasks with regards to the durations of knee-straining postures. One example of this was for the

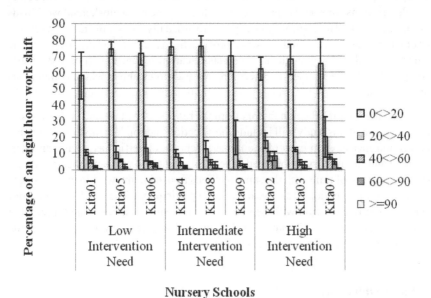

Figure 3.   Percentage of an eight hour work shift for the joint interval ranges for trunk flexion.

Table 1.   Percentage of an 8 hour work shift for the different interval ranges for external load moments at L5/S1 as described by Tichauer (1975).

Classification	Nursery School	0 <= 40 Nm	40 <> 85 Nm	85 <> 135 Nm	>= 135 Nm
Low	Kita01	75.2	24.1	0.7	0.0
Intervention	Kita05	75.0	24.7	0.3	0.0
Need	Kita06	52.4	43.0	4.0	0.6
Intermediate	Kita04	74.3	20.4	5.3	0.0
Intervention	Kita08	68.0	31.6	0.4	0.0
Need	Kita09	37.3	57.4	5.1	0.2
High	Kita02	51.2	44.9	3.8	0.1
Intervention	Kita03	80.6	19.2	0.2	0.0
Need	Kita07	45.0	52.8	2.1	0.1

task "playing", where postures varied from sitting at a child's table to playing on the floor with the child. This task had a greater duration of knee-straining postures in two of the nine nursery schools where the knee-straining component comprised of approximately 11% of the task and for the other seven nursery schools the knee-straining component of the task was not more than 7% of the task duration.

Conclusion

This paper provides preliminary results of the ErgoKiTa Project in the form of objective descriptions for specific parameters indicating the musculoskeletal workloads that nursery school teachers may be exposed to on a daily basis. Further analyses will be conducted and, based on these strain profiles from the shift and task analyses, the nursery schools, as well as the applied classification schemes, will be statistically compared with one another. Furthermore analyses to establish which components of specific tasks are significantly more straining than others will be conducted.

Based on the identified strain factors from the results of the task and shift analyses, the next stage of this study will include the selection and implementation of suitable prevention measures in each of the nursery schools classified as having an intermediate and high intervention need. These interventions will be evaluated after a suitable amount of time with the same methods as used in the workload assessment prior to the intervention phase with the same participants. The results obtain prior and after the intervention will then be compared using descriptive and inferential statistics, in order to draw sound conclusions with regards to suitable intervention measures for the nursery school environment.

Acknowledgements

The project is being carried out under cooperation between the IAD, ASU and IFA and is being financially supported from the research fund of the German Social

Accident Insurance (DGUV). The authors are grateful to all the colleagues at the IFA, IAD and ASU for their help and support with data collection and analysis.

References

BMAS (Bundesministerium für Arbeit und Soziales)., 2010. Merkblatt zur Berufskrankheit Nr. 2112 der Anlage zur Berufskrankheiten-Verordnung. Gonarthrose durch eine Tätigkeit im Knien oder vergleichbare Kniebelastung mit einer kumulativen Einwirkungsdauer während des Arbeitslebens von mindestens 13.000 Stunden und einer Mindesteinwirkungsdauer von insgesamt einer Stunde pro Schicht [Leaflet of occupational disease no. 2112: knee osteoarthritis caused by working while kneeling or similar knee straining with a cumulative duration of exposure of at least 13,000 hours per life and at least one hour per day]. Bek. des BMAS vom 30.12.2009—IVa 4-45222-2122. GMBl 5–6(61):98–103.

Ellegast, R., Hermanns, I., Schiefer, C. 2010. "Feldmesssystem CUELA zur Langzeiterfassung und –analyse von Bewegungen an Arbeitsplätzen." *Zeitschrift für Arbeitswissenschaft* 64 Nr.2: 101–110.

King, P.M., Gratz, R., Kleiner, K., 2006. "Ergonomic recommendations and their impact on child care workers' health." *Work* 26 (1): 13–17.

Kusma, B., Mache, S., Quarcoo, D., Nienhaus, A., Groneberg, D. 2011. "Educators' working conditions in a day care centre on ownership of a non-profit organization." *Journal of Occupational Medicine and Toxicology* Nr.6:36.

Sinn-Behrendt, A., Bopp, V., Sica, L., Bruder, R., Ellegast, R., Weber, B., Brehmen, M., Groneberg, D. 2013. "Klassifizierung von Kindertagesstätten hinsichtlich ihrer (physischen) Belastung anhand struktureller Rahmenbedingungen." *Chancen durch Arbeits-, Produkt- und Systemgestaltung*, 27. Februar–01. März 2013, Krefeld. GfA Frühjahrkongress 2013.

Tichauer, E.R. 1975. "Occupational biomechanics (The anatomical basis of workplace design)." *Rehabilitation Monograph* No. 51. Institute of Rehabilitation Medicine, New York 1975.

Ulmer, H-V. 1968. Ein rechnerisches Kriterium zur Bestimmung der Dauerleistungsgrenze. *Internationale Zeitschrift für Angewandte Physiologie* 27: 299–310.

PHYSIOLOGICAL WORKLOAD AND PERCEIVED EXERTION OF FEMALE LABOURERS IN HARVESTING ACTIVITIES

Renuka Salunke, Susheela Sawkar, Rama Naik, P.R. Sumangala & K.V. Ashalatha

University of Agricultural Sciences, Dharwad-580 005, Karnataka, India

This investigation assessed workload and perceived exertion of female labourers in cutting stalks of sorghum crop in the Dharwad district. The task was performed by 30 apparently healthy female labourers with traditional and improved sickles. Based on the mean working heart rate, cutting stalks of sorghum crop was classified as very heavy for CIAE Bhopal sickle and as heavy for traditional, I-108, and I-104 sickles. Based on the physiological parameters and work output, the improved I-104 sickle was found to be superior to traditional and other improved sickles. However, the women suggested several modifications for sickles for ease operations.

Introduction

Women play a significant and crucial role in agricultural operations, are the pivot around whom the family, society and whole community move. In India, nearly 75 percent of women are from rural sectors, where the impact of science and technology on daily life is limited. Furthermore, women have extremely busy schedule of work on farm and related activities with exclusive involvement in multiple domestic chores. Women perform multiple tasks in a traditional manner due to limited knowledge and skill in application of science and technology to daily living, stretching their time and energy beyond their physical endurance. Thus leading them to low productivity. The introduction of improved agricultural implements may provide high productivity with minimum physiological stress.

Farm women spend long hours, effort and labour in repetitive operations, resulting in fatigue and drudgery. Drudgery is generally conceived as physical and mental strain and fatigue is referred as monotony and hardships experienced while doing job. If appropriate drudgery reducing farm stead implements are made available, rural women may contribute better to productivity by reducing drudgery. Hence, the present investigation assessed the reduction of drudgery by replacing traditional sickles by improved sickles using physiological parameters and perception as indicators.

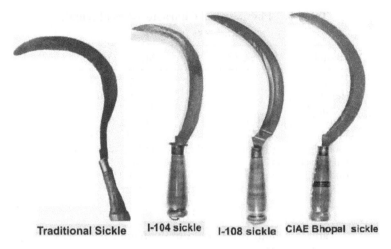

Traditional Sickle   I-104 sickle   I-108 sickle   CIAE Bhopal sickle

Figure 1.   Sickles for cutting stalks.

Methods

The harvesting activity is the major activity in the selected study and female labourers are involved in the harvesting of sorghum crop. An interview schedule was used to collect the demographic profile of the female labourers. Thirty apparently healthy female labourers were considered for the experiment. These were in the age range of 25–35 years, with normal body temperature, blood pressure, heart rate and without any major illness particularly of cardio-respiratory problems during the last five years. Pregnant and nursing mothers were not included in the study. The anthropometric and physiological parameters were recorded with standard methods and computed BMI and energy expenditure. One traditional and three improved sickles were selected for cutting sorghum stalk (Figure 1). The Polar heart rate monitor was used to assess physiological workload of female labourers. Perceived exertion was measured with a five-point scale developed by Varghese et al (1994).

Results and discussion

The physiological characteristics of the female labourers are presented in Table 1. The mean age of the female labourers was 29 years with a mean height of 151 cm. The mean body weight was 48 kg. The computation of Body Mass Index (BMI) (Table 2) revealed the average value was 21. Sixty percent of respondents fell in the normal range, while 10 percent respondents were in the category of chronic energy deficiency and a very few were in the obese group (3%). The aerobic capacity estimated based on heart rate was 1.77 L/min. The mean blood pressure was 113/72. Pulse rate was 75 beats/min and body temperature was 94.6°F, which was in the normal range.

Table 1. Physiological Characteristics of female labourers (n = 30).

Physiological Characteristics	Mean, SD
Age (yrs)	29.03, 3.02
Height(cms)	151.53, 2.70
Weight (kgs)	48.30, 5.05
Body Mass Index (%)	21.02, 2.06
VO_2 max (l/min)	1.77, 0.1
Blood Pressure (Systolic/Diastolic)	113.53/72.81, 126.6/81.4
Pulse (beats/min)	74.96, 9.19
Body Temperature (F0)	94.60, 1.38

Table 2. Body Mass Index of the female labourers.

BMI Classification Garrow et al, (1981)	Frequency (n = 30)
17.0–18.5 (CED Grade I-Mild)	3 (10%)
18.5–20.0 (Low Weight – Normal)	8 (27%)
20.0–25.0 (Normal)	18 (60%)
25.0–30.0 (Obese Grade – I)	1 (3%)
Total	30 (100%)

Table 3. Distribution of subjects according to body type.

Body type	Frequency (n = 30)
Ectomorph (<20)	11.00 (37)
Mesomorph (20–25)	18.00 (60)
Endomorph (>25)	1.00 (3)
Total	30.00 (100)

Table 3 shows that the majority of female labourers (60%) were of the mesomorph body type as per the Quetlets Index followed by ectomorph (36%) and few (3%) were endomorph.

The physical fitness of the sample as per aerobic capacity based on heart rate is in Table 4. Most respondents (57%) were in the good aerobic capacity range (VO_2 max 30.1–37.5 kg min), and 43% were in the very good aerobic capacity range (VO_2 max, 37.6–45.0 ml/kg-min) indicating a healthy state.

The mean working heart rate and energy expenditure was higher while cutting stalks with the CIAE Bhopal sickle (141 beats/min and 13.76 kJ/min, respectively) compared with traditional and improved sickles (Table 5). Based on the mean working heart rate, cutting stalks of sorghum crop was classified as very heavy for

Table 4.   Physical Fitness of female labourers according to aerobic capacity.

Aerobic Capacity (VO_2 max · ml kg/min) (Saha, 1996)	Frequency (n = 30)
<15.0 (Poor)	–
15.0–22.5 (Low average)	–
22.6–30.0 (High average)	–
30.1–37.5 (Good)	17 (57%)
37.6–45.0 (Very Good)	13 (43%)
>45 (Excellent)	–
Total	30 (100%)

Table 5.   The physiological Parameters and rating of perceived exertion of the female labourers in cutting stalks of sorghum crop with selected sickles.

Models of Sickles	Physiological Parameters						
	Mean Working Heart rate (beats/min)	Mean peak Heart Rate (beats/min)	Mean Energy Expenditure (kJ/min)	Mean peak energy expenditure (kJ/min)	Mean TCCW (beats)	Mean PCW (beats/ min)	Mean RPE
Traditional	130.72	134.95	12.06	12.62	1165.72	29.14	3.98
Improved I-104	123.22	129.50	11.51	11.91	920.33	23.01	3.80
I-108	130.43	133.51	12.02	12.50	1138.58	28.46	4.13
CIAE Bhopal	141.40	146.13	13.76	14.51	1427.75	35.69	4.55

the CIAE Bhopal sickle and as heavy for traditional and improved sickles (I-108 and I-104 sickles).

The observations indicated that the sickle curvature was not suitable to hold and slips frequently while cutting, causing bodily imbalance and forward jerk movements to the female labourers. The I-104 sickle recorded least variation in physiological parameters in cutting stalks can be attributed to the sharp and serrated blade and the grip of the handle was convenient for holding the sickle. Based on the above findings, sickle I-104 was considered superior over the traditional and selected improved sickles.

The present results are in conformity with the study conducted by Hasalkar, et al. (2003) and Zend, et al. (2008) that the physiological workload of farm women reduced while weeding and jowar harvesting with improved tools.

The majority of the female labourers (90%) expressed their physical exertion towards cutting stalks with the CIAE Bhopal sickle as very heavy followed by I-108 (80%), I-104 (53%). The traditional sickle was perceived as heavy (73%).

The perception may be due to the smooth running of the cutting stalks and familiarity in using the traditional sickle. The CIAE Bhopal sickle was found to cause more exertion. It could be due to the jerk movements and handle grip itching to the palm, which could be further attributed to weather conditions and several postural changes made while performing the cutting stalks of sorghum crop.

As per the mean peak heart rate response, the female labourers were working above the acceptable limits of workload. This seems to have an impact on the overall physiological stress. Similar results were reported in a study by Sawkar, et al. (2001), who reported the rating of the perceived exertion of the agricultural activities as heavy and very heavy by female labourers.

Conclusion

It can be concluded from the above findings based on physiological parameters and perceived exertion that the improved sickle I-104 was useful in reducing drudgery compared to the traditional and other improved sickles while cutting sorghum stalk. However, there is still scope for improvement by modifying the curvature and length of the sickles.

References

Garrow, G. H. 1987, "Quetelet's index as measure of fatness". *International Journal of Obesity* 9: 147–153.

Hasalkar, S., Budihal, R., Shivalli, R. and Biradar, N. 2003, "Assessment of workload of weeding activity in crop production through heart rate." *Journal of Human Ecology.* 14: 1–3.

Sawkar, P. S., Varghese, M. A., Saha, P. N. and Ashalatha, K. V. 2001, Erogonomic assessment of occupational work load and rest allowances for female agricultural labourers in Dharwad, Karnataka, Published in the proceedings of the conference on Humanizing work and work environment: 11–14.

Varghese, M. A., Saha, P. N. and Atreya, N. 1994, "A rapid appraisal of occupational workload from a modified scale of perceived exertion." *Ergonomics* 37(3): 485–491.

Zend, P. J., Umrikar, S. H., Yeole, S. N. and Kamble, K. J. 2008, "Ergonomic assessment of ear cutter for Jowar harvesting." *Asian Journal of Home Science* 3(1): 34–37.

PREDICTING ISOMETRIC ROTATOR CUFF MUSCLE ACTIVITIES FROM ARM POSTURES AND NET SHOULDER MOMENTS

Xu Xu, Raymond W. McGorry & Jia-hua Lin

Liberty Mutual Research Institute for Safety,
Hopkinton, MA 01748, U.S.A.

Tissue overloading is one of the major contributors to shoulder musculoskeletal injury. The purpose of this study was to develop a set of regression equations predicting the median of normalized electromyography (nEMG) activity of the rotator cuff among the participants from upper arm orientation and net shoulder moments at 95 static postures. The nEMG of the supraspinatus, infraspinatus, and teres minor were measured and the 3-D net shoulder moments were calculated with a biomechanical model. Stepwise regression was used to derive the regression equations. The results showed the r2 of the regression equations were 0.77, 0.75, and 0.63, respectively.

Introduction

Tissue overloading is one of the major contributors to shoulder musculoskeletal disorder (Seitz et al., 2011). To better describe the muscle activities around the shoulder area, previous studies used regression-based methods to predict muscle activities from shoulder joint loading and external forces. Laursen et al. (1998) developed a regression model predicting the electromyography (EMG) activities of thirteen shoulder muscles from hand force, net shoulder moment, and the moment direction at a single static arm posture. In Laursen et al. (2003), this regression method was adapted to a short dynamic movement during cleaning work. Three shoulder muscles were predicted by shoulder net force and moment. The mean correlation between the measured EMG and the predicted EMG was approximately 0.7 among all the participants. In de Groot et al. (2004), a novel regression method was proposed for describing the shoulder muscle activities. It was assumed that the muscle activities can be decomposed to baseline, square component, and cubic component. The contribution of each component depends on the direction of shoulder net force. The results indicated a good fit between the measured EMG and the estimated EMG.

While the use of any regression-based method can effectively predict shoulder muscle activities from shoulder kinetics and kinematics, most such methods have only been established with observations from limited shoulder postures. Since shoulder muscle activity is also strongly related to arm posture (Brookham et al., 2010;

McDonald et al., 2012), extrapolating a mapping from a single posture or task to untested postures may result in errors. Therefore, for a generalized regression-based method predicting the shoulder muscle activities, both shoulder kinetics and shoulder postures should be considered as the predictors.

The purpose of this study is to develop a set of regression equations predicting the median of shoulder muscle activities among participants under various shoulder postures and net shoulder moments. The median describes the average muscle activities among a normal population. The regression equations are then validated using another independent dataset.

Method

Participants and arm postures

The experimental protocol was approved by a local Institutional Review Board. Forty participants (20 females and 20 males, age: 30.9 (10.5), all right-handed), free of acute or chronic upper extremity musculoskeletal disorders, were recruited from local communities. The experiment was first explained to the participants who then gave signed informed consent. An external frame with three rotational degrees of freedom (DoF), which are consistent with the recommendation of the International Society of Biomechanics (ISB) (Wu et al., 2005), was used to guide the thoracohumeral joint (Figure 1). The external frame provided 5 planes of elevation ($\gamma_{TH}1$, 0° to 120° with 30° increments), 4 elevation angles (β_{TH}, −30° to −120° with 30° increments), and 5 humerus axial rotation angles (γ_{TH2}, −90° to 30° with 30° increments) for the thoracohumeral joint. After eliminating unattainable postures, 95 static postures were tested. For all the tested arm postures, the elbow angle was held to 90° flexion.

Apparatus

Activities of the supraspinatus, infraspinatus, and teres minor were monitored by surface EMG. After the skin preparation, bipolar surface electrodes (Ag-AgCl, Noraxon Inc., Scottsdale, AZ, USA) were placed on each muscle (Perotto, 2011). The ground electrode was placed on the middle part of the clavicle. The EMG signals were pre-amplified with a gain of 2000. A 6-DoF load cell (MC3A-1000, AMTI Inc., Watertown, MA, USA) was located underneath the forearm support of the external frame (Figure 1). This force transducer was used to measure the external force and moment on the right arm. The data from the EMG and the force transducer were collected with an A/D converter at 1024 Hz.

A motion tracking system (Optotrak Certus System, NDI, Canada) was used to collect the orientation of the right upper arm, right forearm, the thorax, and the transducer at 100 Hz marker clusters were taped to each body segment. Anatomical landmarks for creating the ISB-recommended anatomical coordinate system (Wu et al., 2005) were digitized by a probe with the participants in an upright standing reference posture, arms at sides. Since the glenohumeral (GH) joint centre could not

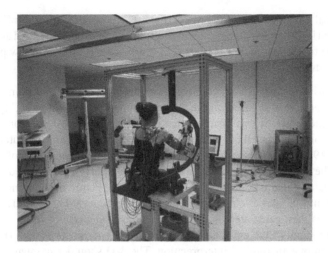

**Figure 1.   Experimental setup. The participants sat within an external frame
and generated various shoulder moments with visual feedback.**

be directly digitized by a probe, the acromion process (ACR) was used instead. The
motion tracking system was synchronized with the EMG and force transducer data.
A customized program (LabView 8.5, National Instruments, Austin, TX) was used
to calculate the GH joint 3-D net moment using the length of the upper arm and the
data collected from the force transducer. The program displayed the 3-D moment
with respect to the humerus coordinate system in each of the three planes (Figure 1)
on a monitor placed in front of the participants to provide visual feedback.

Experiment protocol

Before the experiment, all the static arm postures were randomly ordered by the
plane of elevation first, and then by the elevation angle. Basic participant anthropometry data, such as height, weight, arm length and circumference were measured.
After EMG electrode placement, the participants performed maximum voluntary
contractions (MVC) which activated the monitored rotator cuff muscles. For each
arm posture, the participants generated 3-D shoulder moments as varied as possible for 20 seconds, with the feedback from the real-time net shoulder moment
displayed on the monitor. The participants were asked to generate submaximal
shoulder moments at levels that would not cause them substantial fatigue after the
experiment. A one and one-half minute break was given between the force exertions
for each arm posture.

Data processing

All the EMG signals were band-pass filtered with a 10–400 Hz, 4th order
Butterworth filter (van Dieen and Kingma, 2005) and normalized by the maximum

EMG signal collected during the MVC trials. The normalized EMG (nEMG) was then shifted by 40 ms to compensate for electromechanical delay (Lloyd and Besier, 2003). The transducer data was filtered by a 4th order Butterworth low pass filter at 8 Hz (Gatti et al., 2008).

For each arm posture, the anatomical coordinate systems of the thorax and the humerus were created based on the ISB recommendation (Wu et al., 2005). For the GH joint centre, it was assumed that the GH joint was 10.8% from the ACR on the line between the elbow joint and ACR during the anatomical neutral posture (de Leva, 1996). The mass and the location of centre of mass (CoM) of the upper arm and the forearm were estimated from the previously measured body inertial property data (Zatsiorsky, 2002). The position of the centre of the transducer was derived from the marker cluster placed on the transducer. The shoulder joint net moment was then calculated by inverse dynamic model (Faber et al., 2009). The shoulder moments were calculated in the humerus coordinate system (M_{hx}, M_{hy}, M_{hz}) to improve anatomical interpretation. The 3-D shoulder moment space was then discretized to cubes with 2 Nm each edge. The shoulder moment at each time instant was rounded-up to the closest cube centre before building the regression model.

In the pilot test, it was found that down-sampling the EMG and kinematics data to 10 Hz did not substantially change the regression coefficients of the regression models predicting muscle activities. Therefore, all the data were down-sampled to 10 Hz to improve the computational speed for further regression analysis.

Regression analysis and model validation

A dataset of 25 participants (13 females and 12 males, age: 33.6 (11.8)) was randomly selected from all the participants for building the regression models used for predicting muscle activity. For each posture, if more than 10 participants created a moment in a cube in the discretized 3-D shoulder moment space, the median was then predicted for each muscle.

The potential predictors were: linear terms of frame-defined shoulder joint angles and the 3-D net shoulder moment (γ_{TH1}, β_{TH}, γ_{TH2}, M_{hx}, M_{hy}, and M_{hz}), the quadratic terms of 3-D net shoulder moment (M_{hx}^2, M_{hy}^2, M_{hz}^2, $M_{hx} \cdot M_{hy}$, $M_{hx} \cdot M_{hz}$, and $M_{hy} \cdot M_{hz}$), and the quadratic terms between 3-D shoulder moment and shoulder joint angles ($\gamma_{TH1} \cdot M_{hx}$, $\gamma_{TH1} \cdot M_{hy}$, $\gamma_{TH1} \cdot M_{hz}$, $\beta_{TH} \cdot M_{hx}$, $\beta_{TH} \cdot M_{hy}$, $\beta_{TH} \cdot M_{hz}$, $\gamma_{TH2} \cdot M_{hx}$, $\gamma_{TH2} \cdot M_{hy}$, $\gamma_{TH2} \cdot M_{hz}$). A stepwise regression was then performed for the nEMG of each muscle. The coefficient of determination (r^2) and the root-mean-square error (RMSE) were calculated to evaluate the fitness of the regression model.

The dataset of the remaining 15 participants (7 females and 8 males, age: 26.2 (5.5)) was used to validate the regression model. The medians of the measured nEMG of each muscle were calculated first. The predicted values were estimated from the regression equations derived above. The RMSE between the measured and the predicted values ($RMSE_{val}$) and the percentage error ($Err\%_{val}$) were then used to quantify the quality of the regression model of each muscle.

Results

For the median of nEMG, the r^2 of the supraspinatus, infraspinatus, and teres minor were 0.77, 0.75, and 0.63, respectively. The RMSE were all 0.02. For the validation dataset, the $RMSE_{val}$ were 0.02, 0.03, 0.02, and the $Err\%_{val}$ were 22.0%, 25.8%, and

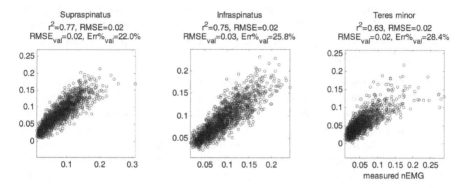

Figure 2. **The measured and the predicted median nEMG for supraspinatus, infraspinatus, and teres minor.**

Table 1. **Coefficient of the predictors for the median of nEMG. An absent coefficient indicates that this predictor was removed by the step regression.**

	Supraspinatus	Infraspinatus	Teres minor
γ TH1	−0.613		0.860
βTH		−0.643	−0.792
γ TH2	0.730	0.609	
Mhx	−0.014	−0.016	
Mhy		0.013	0.030
Mhz	−0.011	−0.012	0.021
Mhx2	0.015	0.018	0.026
Mhy2	0.028	0.078	0.071
Mhz2	0.050	0.063	0.032
Mhx · Mhy	0.042	0.028	
Mhx · Mhz			0.033
Mhy · Mhz	0.018	0.029	0.024
γ TH1 · Mhx	0.005	0.002	0.002
γ TH1 · Mhy	−0.005	−0.006	−0.009
γ TH1 · Mhz	−0.003	0.003	0.010
β TH · Mhx		0.002	
β TH · Mhy			
β TH · Mhz		0.002	−0.003
γ TH2 · Mhx	−0.002	−0.002	
γ TH2 · Mhy	−0.005	−0.006	−0.006
γ TH2 · Mhz	−0.003	−0.002	−0.002
Const	3.802	5.190	2.583

28.4%, respectively (Figure 2). Among all 21 potential predictors, supraspinatus, infraspinatus, and teres minor needed 15, 18, and 15 predictors (Table 1).

Discussion

The goal of the current study was to build a regression equation predicting the median muscle activities of the supraspinatus, infraspinatus, and teres minor among the participants. The results showed the predictors and the response variables were moderately correlated.

The predictors chosen by the stepwise regression revealed some co-contraction of muscle activity patterns. For all the regression equations of the 3 muscles, the coefficients of the squared moments (M_{hx}^2, M_{hy}^2, and M_{hz}^2) were positive. This indicates that as the magnitude of a moment increases, the increased moment will positively correlate to the muscle activity level, no matter whether a muscle contributes to the required net joint moment. Such a finding is partially consistent with previous findings in which shoulder muscle co-contraction was observed (Brookham and Dickerson, 2013; Waite et al., 2010).

There are a few other limitations of this study. First, the force exerted in this study is submaximal to avoid fatigue. Whether the current regression equations will hold with a greater intensity level is not known. Second, only limited arm postures were tested in this study. The current regression equations may not be applicable for predicting muscle activity levels for extreme overhead arm postures. Third, due to the limited number of participants, the gender effect was not examined in this study.

Future studies can apply the current regression equations to the constraints of an optimization-based shoulder biomechanical model. One can use the current regression equations to estimate the approximate average muscle activities given a specific arm posture and net shoulder moment, and add a set of constraints to force the solution of the optimization being close to the estimated muscle activities level. In addition, because high shoulder muscle activity is a risk factor of shoulder musculoskeletal disorders (Bernard, 1997), the current results may be helpful for further developing an injury risk assessment tool for practitioners. However, the criteria of the risk zone of the arm postures and shoulder moments would need to be further investigated.

Conclusion

The current study developed a set of regression equation to predict medians of muscle activities of the supraspinatus, infraspinatus, and teres minor among the participants under various shoulder postures and net shoulder moments. The results showed a good correlation between the predictors and the response variables. Such results can be used to estimate the average rotator cuff muscle activities among a group of people given a specific arm posture and net shoulder moment. Caution should be taken when extrapolating the current results to an untested range of arm postures or net shoulder moments.

References

Bernard, B.P. 1997. Musculoskeletal disorders and workplace Factors: A critical review of epidemiologic evidence for work-related musculoskeletal disorders of the neck, upper extremity and low back. Cincinnati, Ohio: NIOSH.

Brookham, R.L., Wong, J.M. and Dickerson, C.R. 2010. Upper limb posture and submaximal hand tasks influence shoulder muscle activity. *International Journal of Industrial Ergonomics* 40, 337–344.

Brookham, R.L. and Dickerson, C.R. 2013. Empirical quantification of internal and external rotation muscular co-activation ratios in healthy shoulders. *Medical & Biological Engineering & Computing* (In press).

de Groot, J.H., Rozendaal, L.A., Meskers, C.G.M. and Arwert, H.J. 2004. Isometric shoulder muscle activation patterns for 3-D planar forces: A methodology for musculo-skeletal model validation. *Clinical Biomechanics* 19, 790–800.

de Leva, P. 1996. Joint center longitudinal positions computed from a selected subset of Chandler's data. *Journal of Biomechanics* 29, 1231–1233.

Faber, G.S., Kingma, I., Kuijer, P.P.F.M. and van der Molen, H.F., Hoozemans, M.J.M., Frings-Dresen, M.H.W. and van Dieen, J.H. 2009. Working height, block mass and one- vs. two-handed block handling: the contribution to low back and shoulder loading during masonry work. *Ergonomics* 52, 1104–1118.

Gatti, C.J., Doro, L.C., Langenderfer, J.E., Mell, A.G., Maratt, J.D., Carpenter, J.E. and Hughes, R.E. 2008. Evaluation of three methods for determining EMG-muscle force parameter estimates for the shoulder muscles. *Clinical Biomechanics* 23, 166–174.

Laursen, B., Jensen, B.R., Nemeth, G. and Sjogaard, G. 1998. A model predicting individual shoulder muscle forces based on relationship between electromyographic and 3D external forces in static position. *Journal of Biomechanics* 31, 731–739.

Laursen, B., Sogaard, K. and Sjogaard, G. 2003. Biomechanical model predicting electromyographic activity in three shoulder muscles from 3D kinematics and external forces during cleaning work. *Clinical Biomechanics* 18, 287–295.

Lloyd, D.G. and Besier, T.F. 2003. An EMG-driven musculoskeletal model to estimate muscle forces and knee joint moments in vivo. *Journal of Biomechanics* 36, 765–776.

McDonald, A., Picco, B.R., Belbeck, A.L., Chow, A.Y. and Dickerson, C.R. 2012. Spatial dependency of shoulder muscle demands in horizontal pushing and pulling. Applied Ergonomics 43, 971–978.

Perotto, A.O. 2011. *Anatomical Guide for the Electromyographer: The Limbs and Trunk.* Charles C. Thomas Publisher, Limited.

Seitz, A.L., McClure, P.W., Finucane, S., Boardman, N.D. III and Michener, L.A., 2011. Mechanisms of rotator cuff tendinopathy: Intrinsic, extrinsic, or both? *Clinical Biomechanics* 26, 1–12.

van Dieen, J.H. and Kingma, I. 2005. Effects of antagonistic co-contraction on differences between electromyography based and optimization based estimates of spinal forces. *Ergonomics* 48, 411–426.

Waite, D.L., Brookham, R.L. and Dickerson, C.R. 2010. On the suitability of using surface electrode placements to estimate muscle activity of the rotator cuff as recorded by intramuscular electrodes. *Journal of Electromyography and Kinesiology* 20, 903–911.

Wu, G., van der Helm, F.C.T., Veeger, H.E.J., Makhsous, M., Van Roy, P., Anglin, C., Nagels, J., Karduna, A.R., McQuade, K., Wang, X.G., Werner, F.W. and Buchholz, B. 2005. ISB recommendation on definitions of joint coordinate systems of various joints for the reporting of human joint motion – Part II: shoulder, elbow, wrist and hand. *Journal of Biomechanics* 38, 981–992.

Zatsiorsky, V.M., 2002. *Kinetics of Human Motion. Human Kinetics*, Champaign, IL, U.S.A.

SMART ENVIRONMENTS & SUSTAINABILITY

HOW DO MALFUNCTIONS IN 'SMART ENVIRONMENTS' AFFECT USER PERFORMANCE?

Ku Fazira Ku Narul & Chris Baber

*Electronic, Electrical and Computer Engineering,
University of Birmingham*

In this paper, a smart environment is simulated with a projected display is used to support cooking activity in a kitchen of the future. From the point of view of Human Factors, a critical question relates to how users of such a system might cope with its malfunction, either because the system is unable to recognise a person's activity or because it has confused two recipes. This study shows that, even when performing familiar tasks, people can be misled by erroneous information. This suggests that their trust in the advice offered by the system could outweigh their confidence in their own knowledge.

Introduction

In smart environments, digital technologies are integrated into domestic spaces. This could involve, for example, sensors in the room, the cupboards or the tools that are available to use, and projected displays in the walls and work-surfaces. Figure 1 shows an example of an augmented kitchen developed by Bonnani et al. (2005).

The claimed benefits for smart environments in the kitchen, range from supporting the novice cook to providing nutritional advice and guidance to supporting social interactions through cooking and eating. Before developing the hypotheses for this study, we consider some examples of concepts that can be described as ambient kitchens, and how these have been evaluated through user studies. *The Kitchen of the Future* (Siio et al., 2004, 2009) displays recipes pages when users perform

Figure 1. An Augmented Kitchen (Bonnani et al., 2005).

cooking activities: images of the cooking workplace are captured and the cook is able to provide voice memos into multimedia recipes. In *Ambient Kitchen*, projected displays provide guidance to cooks and sensors can detect the activities of people in the kitchen (Olivier et al., 2009). These papers outline the concepts but not report evaluation. Like the *Kitchen of the Future*, the *Living Cookbook* enables people to share their cooking experiences (Terrenghi et al., 2007). A preliminary usability test (using a 'talk aloud' protocol) was performed with four participants (2 men and 2 women) and the results did not reveal any problems with using the prototype, nor did they detail any benefits to be gained from using it. A similar concept, *SuChef* (Paley and Newman, 2009), was tested as a (paper) prototype with 5 participants. While there was no discussion of the way in which the prototype actually supported cooking tasks, people seemed to like the provision of additional material related to the recipes. *Cooking Navi* (developed by the same team responsible for the *Kitchen of the Future*) presents recipes using text, video and audio (Hamada et al., 2005). A preliminary evaluation was performed with 8 participants (2 experienced, 3 intermediate and 3 novice cooks) who were required to cook 2 recipes. Results were positive, with the exception of the question 'did you feel that the provided procedure was appropriate?' (which scored 3.5/10); the authors concluded that this might be because they had imposed timing constraints on the participants rather than allowing them to perform in a more self-paced manner. *CounterActive* (Ju et al., 2001) turns the work surface into a large touch screen for interacting with instructional, step-by-step, projected information. A preliminary test, with two children aged 7 to 10, showed that visual cues were effective in choreographing movement and that the children relied heavily on the videos to demonstrate actions. This suggests that, particularly for people with little or no cooking experience, following a video can be beneficial in performing cooking tasks.

What is apparent from this brief review of papers is that the concept of a kitchen environment which can respond to its users and provide advice and guidance on cooking tasks is becoming increasingly interesting to the pervasive and ubiquitous computing domains. What is equally apparent is that there remains a dearth of studies in how people will interact with such environments (and whether such environments are actually desirable or beneficial). For this paper, our interests lie in the manner in which people cope with malfunctions. In related work, interruptions have been shown to affect cooking tasks (Tran and Mynatt, 2002).

Study

In this study, a functional prototype was designed to 'assist' users performing cooking activity in augmented kitchen environment, projected on top of the table by an LCD projector. Before each task, the ingredients are placed in small, ceramic containers and arranged on the table. This layout provides a convenient structure for the projections which meant that all participants were confronted with exactly the same arrangement with all ingredients and utensils positioned in the same places prior to the start of the trials. Figure 2 shows the surface divided into four sections:

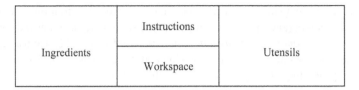

Figure 2.   The layout of the surface for this experiment.

Figure 3.   Ambient interface.

'workspace' (where the user performs the cooking tasks), 'instructions' (containing steps to follow and which ingredients and utensils to use), 'ingredients' (containing 24 ingredients) and 'utensils' (where the equipment and tools are placed).

By tapping the projected icons or by placing their hand on or next to an ingredient or utensil, a coloured disc is projected on top of the object. This disc indicates whether the ingredient is right, or wrong, for that step of the recipe (a light green disc for right ingredient, and a dark red disc when the ingredient is wrong). If the correct ingredient is selected, the name of the ingredient will change to green text at the same time. For the purposes of this trial, the action of the participant is monitored by the experimenter who cues the appropriate information (thus, following a standard 'Wizard of Oz' approach to prototyping). Figure 3 shows the experimental set up in this interface.

Study aims

The purpose of this experiment is to explore how the performance of a cooking task, performed by experts, is affected when the system goes wrong. This study compares user performance under three experimental conditions in an ambient kitchen. These are a control condition with *no system malfunction*, a condition with *simple system malfunctions* and a condition with more *complex system malfunctions*. The *simple system malfunction* involved the projected display highlighting an ingredient as 'incorrect' (red LEDs) when it was the correct ingredient. This could, for example, replicate a situation in which the cueing is to a location rather than to a specific ingredient (and the ingredient is in the 'wrong' place as far as the cueing is concerned). The *complex system malfunction* included a step taken from a different recipe being shown to the participants. This could represent a situation where the automatic prompting confuses one recipe with another.

The cooking task chosen for this study were two Malaysian recipes: **Red Spicy Chicken** and **Fish Grill stuffed with Coconut Chillies Paste**. These recipes represent different levels of requirements in cooking activity (with different number of steps, different types of ingredients and different cooking skills). *Grill Fish* task has fewer ingredients but a higher level of cooking skills, while, *Red Spicy Chicken* has a lower level of cooking skills but a higher number of ingredients and steps to perform.

Two different forms of help were available to participants after the distraction: participants could ask the system to "show" the whole list of steps by tapping the letter "S", or they could ask for an "inventory" by tapping the letter "I". This inventory will lists the ingredients that should been used from the start of original cooking task.

Procedure

The three conditions previously discussed (No system malfunction, simple system malfunction and complex system malfunction) were applied to three (3) subtasks of each main cooking task. This resulted in 9 (3 × 3) different malfunctions for each main cooking task.

The malfunctions were assigned randomly throughout the cooking. In order to ensure that participants did not recall previous cooking system malfunctions there was a gap of at least 5 minutes between each system malfunctions.

Twelve (12) native Malaysian participants were recruited for this study, in which four (N = 4) different individuals participated in each cooking system malfunction. All the participants were aged between 22 and 35 years. Three (3) participants were male, and nine (9) were female. All the participants were familiar with the recipes and had at least three (3) years of cooking experiences. This sample size is sufficient for detecting large effects in user performance with significant power < 0.5 with between subjects using two-way ANOVA repeated measures design.

Participants were informed that during cooking activity, a system malfunction could be introduced, but the details how, when and where it will occur was not given. Once participants had read the instructions, a demonstration was given by experimenter on how to interact with the projected display. Then, participants were given 1 to 2 minutes to familiarise themselves with the display and the required interaction before commencing the experiment. During the experiment, all activity was recorded using a digital camera, and the videos were analysed at the end of experiment. The analysis is based on the analysis of these videos.

Measures

Cooking times were collected in two different forms: *total time* (the amount of time spent performing the experiment, including cooking and malfunction), and *time on system malfunction* (time spent on system malfunction).

Numbers of steps varied between the cooking tasks: redPaste (5), prepareChicken (6), cookChicken (9), chilliesPaste (6), prepareFish (4), cookFish (6).

Errors were evaluated through video analysis and defined as:

- No missing ingredients in each step.
- No missing cooking actions, i.e. missing 'add', 'peel', 'blending'.
- Did not use wrong ingredients.

Results

A series of Two Way ANOVA was conducted on the data to examine the times to complete the task, number of steps completed and number of errors were made for each recipe. In Table 1, * indicates significance at $p < 0.05$, and ** at $p < 0.001$.

In terms of Time, for both cooking task, Post-hoc Tukey tests revealed that participants were significantly faster in the *Control* condition than the *Complex System Malfunction* condition.

Table 1. Results of ANOVA.

Cooking Tasks

	Time
Red Spicy Chicken	$F(2, 27) = 4.370^*$ Complex > control
Grilled Fish	$F(2, 27) = 7.110^*$ Complex > control

Sub-Tasks

	Steps	Error
Prepare chicken	$F(2, 9) = 6.62^*$ Simple > control	$F(2, 9) = 5.41^*$ Simple > control
Cook chicken	$F(2, 9) = 14.78^{**}$ Complex > control Simple > control	$F(2, 9) = 5.15^*$ Complex > control
Prepare fish	$F(2, 9) = 5.57^*$ Complex > control	$F(2, 9) = 9.50^{**}$ Complex > control
Cook fish	n.s.	n.s.

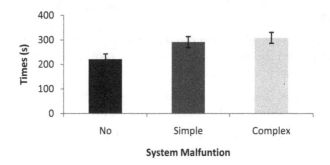

Figure 4. Comparison of average times to complete task under different 'malfunction' conditions.

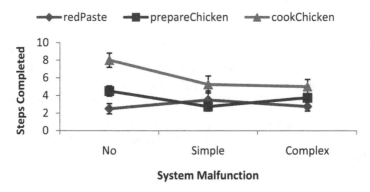

Figure 5. Comparison of number of steps performed under different 'malfunction' conditions for the three 'chicken' tasks.

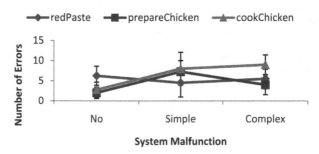

Figure 6. Number of errors made in the 'red spicy chicken' tasks under different 'malfunction' conditions.

In terms of the number of steps completed for the sub-tasks, the results showed that the *System Malfunction* tended to result in fewer steps completed than in the other conditions.

The study also found that the number of errors made in the sub-tasks, were higher in the *System Malfunction* than the *Control* conditions.

Discussion

The study measured the performance of cooking activities by Malaysian participants (who were all assumed to be experts in the recipes and ingredients) when system malfunctions occurred in an ambient kitchen. It was found that system malfunction slows the performance significantly when compared with a control condition in which no malfunction occurs. For example, one participant complained about the cooking task *"I thought I already used this ingredient, why do I need to use it again?"* but she still continued to follow the instruction given. Another participant skipped the 'extra' step after realizing it had confused her *"This is not right, something weird here, I want to skip this step because I have taken this ingredient, why do I need to used twice?"*. This suggests that during performing the cooking activities in the *system malfunction*, participants *might not* be able to differentiate between the steps of the primary cooking task and the 'other' cooking task. This suggests that users have confidence that the system gives *true* information and is guiding them to the accurate completion of the cooking tasks. One participant questioned the about the distraction *"This looks strange, ermm (re-read the ingredients and instruction) well. just follow the system, this is what the system wants me to do"*. Therefore, participants ended up confusing the cooking tasks.

The cooking tasks with more ingredients had more errors than those with a fewer ingredients. The results also show that tasks that have confusion in term of its name or similar ingredients will cause higher errors in the simple system malfunction. This suggests that *system malfunction* involving mixing the right and wrong ingredients can affect cooking performance when similar ingredients are required i.e. sugar vs. salt, turmeric powder vs. curry powder, or galangal vs. ginger.

While projected displays offer the potential to support cooking tasks, this experiment shows that, even quite simple malfunctions, can lead to deterioration of performance. What is interesting in this study is that the participants were all knowledgeable about the recipes being presented and, as such, did not need to follow the instructions presented to them. However, they were misled by the projected display when it was erroneous. This implies that the projection of information was sufficient to capture their attention and to serve as a guide through the steps that they should follow. In terms of ambient displays, therefore, this paper supports the earlier proposal (Baber and Schwirtz, 2010) that ambient systems *could* shift the locus of control away from the user, who then finds it difficult to interpret the quality of their activity or, in this study, to balance their knowledge of a given task with the advice offered to them by the automated system.

References

Baber, C. and Schwirtz, A., 2010. Ergonomics of Ubiquitous Computing: how sensor-based interaction influence user performance in simple tasks, In M. Anderson (ed.) *Contemporary Ergonomics 2010*, London: Taylor and Francis, 192–201.

Bonanni, L., Lee, C.-H. and Selker, T., 2005. Attention-based design of augmented reality interfaces, *CHI '05 extended abstracts on Human factors in computing systems*, New York, N.Y.: ACM.

Hamada, R., Okabe, J., Ide, I., Satoh, S., Sakai, S. and Tanaka, H., 2005. Cooking navi: assistant for daily cooking in kitchen, MULTIMEDIA '05 *Proceedings of the 13th annual ACM international conference on Multimedia,* New York: ACM, 371–374.

Olivier, P., Xu, G., Monk, A. and Hoey, J., 2009. Ambient kitchen: designing situated services using a high fidelity prototyping environment, PETRA'09 *2nd International Conference on Pervasive Technologies Related to Assistive environments*, New York: ACM, article 47.

Paley, J. and Newman, M., 2009. SuChef: an in-kitchen display to assist with "everyday" cooking, *CHI2009*, New York: ACM, 3973–3978.

Siio, I., Mima, I., Frank, I., Ono, T. and Weintraub, H., 2009. Making recipes in the kitchen of the future, *CHI '04*, New York: ACM, 1554–1554.

Siio, I., Hamada, R. and Mima, N., 2009. Kitchen of the Future and its applications, *Human-Computer Interaction. Interaction Platforms and Techniques*, Berlin: Springer, Lecture Notes in Computer Science 4551, 946–955.

Terrenghi, L., Hilliges, O. and Butz, A., 2007. Kitchen stories: sharing recipes with the Living Cookbook, *Personal and Ubiquitous Computing, 11*, 409–414.

Tran, Q.T. and Mynatt, E.D., 2002. *What Was I Cooking? Towards Deja vu Displays of Everyday Memory*, Atltanta, GA: Georgia Institute of Technology, GVU Technical Report GIT-GVU-03-33.

CREATING A LASTING COMMITMENT TO
SUSTAINABLE LIVING

Hermione Taylor, W. Ian Hamilton & Caroline Vail

*Founding Director of The Donation, Wayra UnLtd Academy, Shropshire House,
London, WC1E 6JA*
*Environmental Resources Management, One Castle Park, Tower Hill,
Bristol, BS2 0JA*

The management of energy and resources represents the biggest challenge to contemporary living. So long as people are influenced by
scepticism and denial of manmade climate change they remain reluctant to commit to real and sustained change in their behaviour. The
DoNation project has been designed to engage people in action for
sustainability in a fun, friendly and measureable way. The pledging mechanism behind The DoNation is sponsorship where you give
action instead of cash. Its initial trial was very successful with 216
people pledging actions equating to savings of over 16 t of CO_2. This
paper reports on a trial within ERM in the UK. The evidence shows
that The DoNation can create lasting change in more sustainable
behaviour.

Introduction

The management of energy and resources represents the biggest challenge to contemporary living. But recent research by Poortinga et al. (2013) seems to suggest
that, in the UK at least, public acceptance of man-made climate change is declining. So long as people are influenced by scepticism and denial of manmade climate
change they will remain reluctant to commit to real and sustained change in their
behaviour that could reduce energy and resource expenditure.

Organisations committed to sustainability need support systems that can help people
to engage positively in change. Pledge tools are seen as a useful mechanism to
influence behaviour, gain commitment to more sustainable behaviour and to make
those commitments last. However research (Taylor, 2009) shows that that there
were two main things needed to run a successful pledge program:

1. People: leaders with the ability to influence others on the ground.
2. Tools: to help them record and measure impact, educate, communicate and
   maintain motivation and commitment in the long term.

Some organisations have fantastic leaders and influencers – people with the energy,
enthusiasm, and the right contacts to inspire change on the ground. Usually these
are community-based groups or well-connected networks. But they very rarely have

the resources to build effective tools to disseminate information further, to capture and record the impact that they were making, and most importantly, to maintain motivation and commitment in the long term.

Conversely, many larger organisations – on regional, national and international scales – had set up sophisticated tools to do just this, but very rarely had people on the ground who could connect with individuals on a personal level.

So there was a need for a well-developed tool that could easily be adapted and used by the influencers on the ground. A tool that could be tailored to each group, organisation or campaign, allowing them to own and take claim for the impact they created through it. The DoNation (Taylor, 2011) project has been designed to engage people in action for sustainability in a fun, friendly and measureable way.

Creating The DoNation

The DoNation was designed for use by individuals through 'Donate by Doing'. The pledging mechanism behind 'donate by doing' is sponsorship where you give action instead of cash.

The idea came about whilst the lead author was studying for her MSc. She decided to go on an adventure and cycle from London to Morocco. This was an opportunity to raise sponsorship from others, and support for the 'cause of sustainability'. But she did not like the idea of asking friends for money, and she realised that the environment did not actually need their money as much as it needed their action. So she asked them to support her with action instead of cash – by doing things like cycling to work, eating less meat, or washing clothes at 30°C whilst she was cycling to Morocco.

The response was phenomenal, with 216 people supporting her, saving over 16 t of CO_2 in the process. The idea grew from there, with over 500 people raising sponsorship in this way for challenges of their own over the following 2 years; raising over 6000 pledges of action.

How it works

Around 42% of the UK's carbon emissions are the result of the kind of daily activities of individuals (DEFRA, 2008). This includes how they use hot water, heat their homes, use their cars, use lifts, buy foods, and recycle. In fact the scheme invites people to make pledges on any of around 35 different action areas (as illustrated in Figure 1 below) on the themes of food, travel, home life, lifestyle and work. For instance people can make simple pledges; to use the stairs at work instead of the lift, to boil less water in the kettle, or they can agree to share resources with neighbours.

For example, by washing your clothes at 30°C you could save 45 kg CO_2 a year.

Also, a power drill may only be used for a relatively short time in its life; so does everyone on your street need one? Simple changes like this can create big savings

Figure 1.   Examples of daily activities that produce emissions.

in CO_2 costs. Some of the pledges are larger, for instance committing to lag your loft, which could save around 720 kg of CO_2 every year.

The DoNation pledges are based on simple principles of psychology (as expressed by Thaler and Sunstein, 2009), such as in-group behaviour, peer pressure, personal

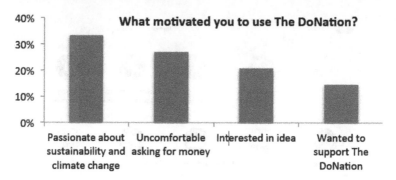

Figure 2.   Social motivations for The DoNation.

challenge, competitiveness and feedback. People make the commitments over a two month period of time, short enough not to be intimidating but long enough to form habits.

Assessing the impact of The Donation

All Doers were sent a questionnaire in January 2013 to find out a bit more about the impact that The DoNation had had on them. There were three surveys: one for Doers who sponsored challenges in 2011 (81 respondents), one for people who had raised sponsorship through The DoNation in 2012 (19 respondents), and another for their sponsors (109 respondents). An overview of the results is presented below.

People raising sponsorship or DoActions

Originally, The DoNation was mainly framed as a sponsorship tool, but during 2012 we saw more people starting to use it as a simple pledge tool for events, competitions, and gift lists. The survey response for this group was small (N = 19) and not very representative of our users as a whole – 66% of our users still use the site to raise sponsorship for a challenge, whereas less than half of the survey respondents had used it in this way. Here are the key points:

- 68% said it made them feel more able to inspire and engage their friends on green issues
- 63% found it easier than asking for money
- 67% found that it motivated them to keep to their challenge and/or to train harder
- Over half of them found their sponsors began talking about sustainability and climate change more as a result.

Although being used in a variety of ways, the key motivation and driver of behaviour change through The DoNation is a social one (see Figure 2 above):

- 77% of respondents had made their pledge in support of someone they knew – a friend, relation, or colleague.

Thinking about this DoAction and your situation at the time of making your
pledge, which of the following statements is most true for you?

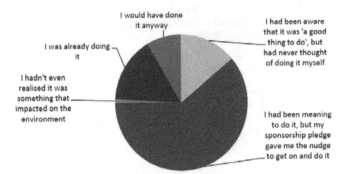

Figure 3.   Linking intentions and actions.

The most common motivation for people making their pledge was to support a
person or organisation, the second most common motivation, experienced by 63%
of respondents, was concern for the cause – climate change and sustainability. This
supports our theory of change, indicating that through the social mechanics of The
DoNation we are able to reach and engage a wider audience than through traditional
information campaigns that rely on people just being committed to the cause.

The real power of The DoNation seems to be in closing the gap between people's
intentions and their actions: 61% of Doers said that they had been meaning to do
the action for a while, but the pledge gave them the nudge to get on and do it (see
Figure 3 above).

Long-term change

One of the most impressive things about The DoNation's impact is how it succeeds
in creating long—term behaviour change (see Figure 4):

- 81% of Doers from 2012 plan to continue doing their DoAction (n = 109)
- 81% of Doers said they are still doing their DoAction over a year later (n = 81).

The DoNation in action with ERM

ERM was impressed with the reported impact of The DoNation. Although initially
developed as an individual pledging tool, The DoNation has been developed into the
Do Good For Business toolkit. As the world's leading sustainability consultancy
ERM places a high importance on living up to the goals we help our clients to
achieve. ERM wanted to run a pilot programme of The DoNation's Do Good for
Business variant to raise the level of personal action on sustainability within our own
organisation. The pilot implementation was planned for the UK business and began
in 2013. The project aimed to get people to do more to improve the sustainability
of their everyday activities.

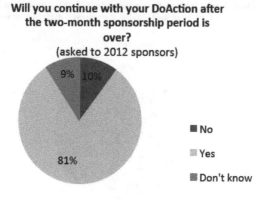

Figure 4.   Commitment to long-term change.

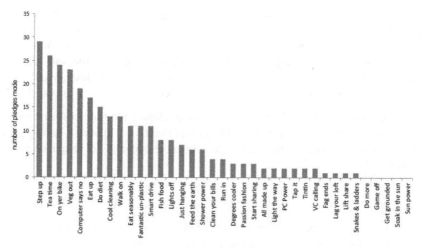

Figure 5.   Popularity of actions chosen.

Results of the ERM pilot study

The implementation project at ERM in the UK has so far resulted in 121 people making 280 pledges that equate to 11,601 kg of CO_2. That is the same amount of carbon as emitted by 23 flights from London to New York, or by driving from London to Timbuktu nine times!

The popularity of the different pledge types is shown in Figure 5 below.

The most popular action has been **Step up**; walking up the stairs at work instead of using the lift. This could be due to a focus on this topic during the launch event.

On yer bike was third most popular pledge, but it had a fairly low completion rate, with only 38% of pledgers completing it (so far). It seems there are some barriers

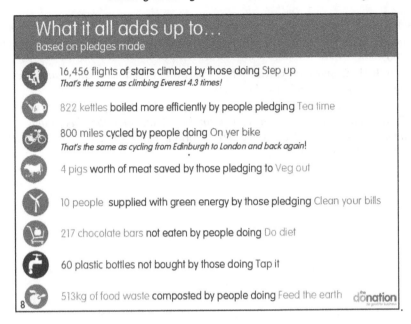

Figure 6.   Actions resulting from pledges.

to action and it could be interesting to explore this further to see what more can be done to help people to cycle to work.

Do More has not yet been pledged. This action is ideal for those who are already doing all the core actions – through this pledge they commit to encouraging others in their team to do something. It might be worthwhile creating incentives for this action in order to increase engagement.

The pledges have resulted in a significant number of practical actions, as shown in Figure 6 below.

63% of your Doers have already confirmed how they got on (many will still be in the 'doing period' so will not have been asked to confirm yet). Of these, 42% said they 'totally nailed it', and 53% managed to do it at least half the time. Overall, 39% of your Doers said that they would definitely carry on with their pledged action beyond the two month period. 17% said they probably would, whilst only 6% said they definitely would not. The project has helped to trigger some really good and potentially long term behaviour changes. These results show that The DoNation demonstrates the success that can be achieved for sustainability from a simple but effective application of psychology.

References

DEFRA, *A Framework for Pro-Environmental Behaviours*. London, UK, Department for Environment, Food and Rural Affairs, 2008.

Taylor, H. *The DoNation* 2011 http://www.thedonation.org.uk/

Taylor, H., *Pro Environmental Behaviour Change and The role of climate change pledge schemes*. MSc Environmental Technology, Imperial College London, 2009.

Thaler, R.H. & Sunstein, C.R., *Nudge: Improving decisions about health, wealth and happiness*, Penguin Books, 2009.

Wouter P, Pidgeon, N.F., Capstick, S. & Aoyagi, M. *Public Attitudes to Nuclear Power and Climate Change in Britain Two Years after the Fukushima Accident*, The Uk Energy Research Centre, Working Paper 19 September 2013: REF UKERC/WP/ES/2013/006.

SYSTEMS: PREPARING FOR THE FUTURE

INTRODUCTION TO HUMAN FACTORS INTEGRATION AND A BRIEF HISTORICAL OVERVIEW

R.S. Bridger & P. Pisula

Human Factors Department, Institute of Naval Medicine, UK

Human Factors Integration (HFI) has its origins in systems theory, which itself was a product of industrialisation and a response to the challenges of building and operating increasingly complex systems. Rapid technical advances caused novelty. It was challenging to design systems with compatible components, including human ones, without prior knowledge, custom or practice in key areas. Selected literature is briefly reviewed in relation to some of the key characteristics of modern HFI, in theory and in practice, and in relation to generic versus system specific human requirements.

Introduction

Human Factors and Ergonomics (HF&E) are concerned with the interactions between people and technology and the factors affecting the interaction. The aim is to make systems work better by improving the interactions. There is a large body of scientific evidence for use in creating optimal conditions for work. In order to exploit this knowledge effectively, processes are required to ensure that system designers and managers have the tools to apply it at the appropriate points in the design cycle. Hereafter, 'Human Factors Integration' (HFI) is used to describe these processes. In the USA, the term 'Human Systems Integration' (HSI) is more commonly used. HSI is rejected here because few man-made systems can operate without people – essential components of all man-made systems, no matter how automated. In this sense, HFI is an essential part of systems design and management.

Systems – A definition

A system consists of a set of elements, the relations between the elements and the boundary around them. Inputs cross the boundary from the environment or from other systems and enter the system: the elements interact on the inputs and with each other to create one or more outputs, which leave the system, enter the environment or become the inputs to other systems. Figure 1 is a generic model – systems may produce products, energy or knowledge or a combination of these and varying amounts of unintended or unwanted products. In practice, the output of the system is the same as its purpose. In the past, the unintended or unwanted outputs of systems (such as waste products in Figure 1) were often ignored, but

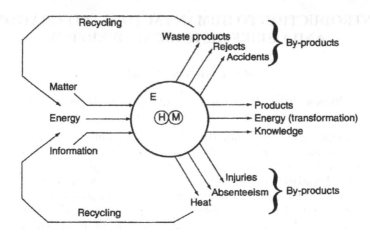

Figure 1.   Example of a System (from Bridger, 2009).

this is an area that is receiving more attention in HFI in relation to macro-system sustainability (Haslam and Waterson, 2013).

HFI is concerned with all of the interactions involving people: the technology they interact with; the inputs and outputs; and the environment they are exposed to when working as part of the system. A 'human-machine system' with one person and one machine can be considered as a starting point for the application of HF&E knowledge. In practice, real systems usually consist of combinations of such systems (several operators and one machine or one operator and several machines) and HFI is often needed to achieve optimal integration of these into a larger system. Optimal integration usually entails some consideration of how the solution to one set of problems in a system will influence the solutions to another set of problems in another system when the two are integrated. Systems theorists in human factors and in other disciplines often state that systems have *emergent properties* (de Greene, 1978). In the present discussion, we might put this differently by saying that systems integration presents designers with *emergent problems*. As elements are bought together, problems of *integration* arise that are not always apparent at the level of the elements or subsystems themselves. HFI is therefore a tool that can be used to overcome problems of integrating people with technology in increasingly complex systems. Leveson (2011) points out that in complex systems, most accidents happen at the boundaries between systems components rather than failure of the components themselves. HFI focuses on one of the major boundaries in any system – the human-technology interface.

Early history of HFI

Although HFI is a relatively recent term, its antecedents can be traced back many years. Early ergonomics textbooks contained a great deal of information about people, gleaned from applied research in anatomy, physiology and psychology (e.g. McCormick, 1957). The book by Singleton (1974) is an exception. This book contains very little information about the core sciences of ergonomics. Singleton's focus was on the 'systems approach' and the activities needed to design systems. He traced the development of the (then) modern technology in several stages. Early machines were built by people with personal experience of the task, which served to integrate many of the HF requirements into the design as tacit knowledge was already integrated into the manufacture. Problems of integration were few because the systems were simple. Next came the era of 'pure engineering' early in the industrial revolution in the 18th century. New machines were made by specialists who were not users or operators themselves. Great benefits accrued, however, because of the increased power made available in comparison with the existing technology. It became apparent, at the end of the 19th century that further progress would not be made simply by building a 'better' machine. The real challenge was to integrate new machines with existing machines and with operators within an overall system of management. One of the first proponents of this 'systems' view was F.W. Taylor (Taylor, 1911) who is famous for saying:

> *"In the past, the man has been the first. In the future, the system must be first."*

Taylor's approach was to simplify jobs to make them easy to learn, thereby increasing the supply of labour; to set a standard (one correct way) of performing a task making predictions of output more accurate; and to establish a bonus scheme to incentivise productivity. Despite the many criticisms of 'Taylorism' over the last century, HFI continues to fail when basic human needs are ignored. Reports of airline pilots and bridge watch keepers falling asleep while monitoring highly automated systems are not uncommon. Such incidents occur because of a failure to properly consider the role of the operator and the tasks allocated in relation to a basic user-requirement – the need for an optimal level of stimulation at work. Failure to provide an optimal level of stimulation may be due to a failure to conduct the allocation of function phase (see below) correctly, resulting in over-utilisation of technology and under-utilising personnel.

Singleton (1974) regarded 'functional thinking' as a key requirement for systems development. Because technological advances provide designers with many ways of implementing a particular system function, the emphasis should be on early identification of all the necessary functions. A function, here, is defined as an activity or set of activities needed for the system to operate. Identification of functions depends, in turn, on a clear understanding of the system's objectives. These objectives follow ultimately from the requirement that has to be met. Specification of requirements is therefore central to the design of systems and to HFI. Definition of functions, in terms of their constituent activities, occurs at the early stages of design and has several advantages: it focuses attention on what the system has to do to fulfill its purpose, leaving the scope for innovation open to later stages. Contractually, it separates responsibility for specifying what has to be done from responsibility for designing the solution (how to do it) to meet the requirements that have been specified. Increasingly long development times of increasingly complex and novel systems were mooted by both Singleton (1974) and Huchingson (1981) as the reason for parallel development of the role of the human operator and the technological solutions, following a functional analysis of system requirements. Thus, processes for the selection and training of personnel are integrated into the design process and not conducted independently or as an afterthought.

HFI in the systems design cycle

HFI activities should take place throughout the lifecycle of the system. In the United Kingdom, the acronym CADMID (Concept, Assessment, Design, Manufacture, In-Use, Disposal) is in common use (e.g. Ministry of Defence, 2012). The main steps are as follows:

Establish project concept and objectives: At the earliest stages, the focus is purely functional – on what is needed not how it is to be achieved. Typically, requirements are identified and defined as succinctly as possible. The choice of techniques to elicit the information (e.g. focus groups, interviews with subject matter experts, user-panels etc.) is itself a key part of the HFI process. The role of operators is determined early in the design process when functions are allocated either to automation or to operators, or flexibly to both. According to Deardon (2000), in those cases where functions can be performed by humans or by automation, the function should only be automated if it is separable from the operator's role and does not interact with it.

Establish Human Factors Integration Strategy: The HFI strategy is essentially the work programme for HF over the project timeline, Identifying the main HF activities, the resources to conduct the activities and the timelines for completion.

Produce Human Factors Integration Plan (HFIP): The HFI Strategy should be included within the HFIP together with details of those responsible for the plan, the main activities of the plan and the methods used to conduct the activities. Inter-relationships with the wider project and any constraints and dependencies should also be made explicit.

Establish Context of Use: The context of use for the solution is established by the solution provider and other stakeholders. This details how the system is going to be used; where and under what conditions; and how in relation to other systems.

Analyse Legacy System Data and Feedback: If an existing system is to be upgraded, requirements specification may be fairly straightforward – a matter of changing the parameter values. If a new system is being developed, information about the operational requirements of the process and the role of human operators may be needed. In both cases, the goal should be to use legacy information to iden-tify and solve any existing ergonomics problems and avoid generating any new ones.

Establish User Population and Characteristics: The physical, psychological and social characteristics of the intended User population are identified, described and documented and communicated to Solution Providers in the form of a Target Audience Description (TAD).

Conduct Early Human Factors Analysis (EHFA): A high-level, comprehensive and balanced analysis of the People-Related aspects of the project, including: Concepts and Doctrine; Equipment; Information; Infrastructure; Logistics; Organisation; Personnel; Training. The output from the EHFA should include a list of key People-Related Considerations including risks, issues, constraints, assumptions, supported by evidence and expert judgement. EHFA occurs as soon as detailed project objectives and constraints are sufficiently mature.

Specification of requirements and acceptance criteria

Requirements specification should not be left to system designers. Subject matter experts and user representatives should be included. In designing a new airport, for example, the requirements include the routes to be serviced, whether international or national flights or both can use the airport, the capacity for aircraft to land and take off and the rate at which passengers can flow through the terminal. Similarly, requirements for successful integration of humans are considered in general terms. There are a number of *generic* human requirements which are applicable to practically all systems (Table 1). These generic requirements can be decomposed into system specific requirements as the system concept is developed. It is in the generation of increasingly detailed and system-specific requirements that subject matter expertise in HF&E is required and this is a critical role for the HF&E specialist on the design team.

MOD (2012) states that all more detailed and system specific requirements will be derived from the generic requirements in Table 1 as the system concept develops.

Table 1. **Generic Human Requirements* and One Level of Decomposition.**

Level 1	Level 2
Employ the right people for the job	Target audience description
Provide a suitable work environment	Equipment usable and safe; Environment meets legal requirements and standards; environment is compatible with tasks
Support work and non work human functions	Tasks compatible with user-expectations and skill; Optimal work loading; Adequate rest, recovery and nutrition
Satisfy User Organisation Constraints	Clear work role; Consult users; Appropriate work patterns; Support from management and colleagues

*MoD 2012, Bridger 2009.

For each of these requirements, acceptance criteria have to be generated to determine whether the requirement has been met in the solution. Acceptance criteria might be in the form of standards or guidelines or they may depend on user-trials or even experiments. This is a second area where formal knowledge of HF&E is required, particularly if trials or experiments are needed. Ergonomists have long held the view that ergonomics works best when it is considered at the earliest stages of design. In terms of the present discussion, specification of requirements is key because if HF requirements are not specified, they are not implemented.

HFI and safety

Leveson (2011) applied systems thinking to safety. All systems consist of: a hazardous process; safety constraints; a hierarchical control system to apply the safety constraints; feedback about the status of the process and the constraints. No control system can be better than the data it receives in the form of feedback about system functioning and the application of constraints. Most accidents happen because: a safety constraint(s) was not provided; a safety constraint(s) was inadequate or misapplied; a safety constraint(s) was not applied; a safety constraint(s) could not be activated; a safety constraint(s) was activated at the wrong time; a safety constraint(s) was removed to soon or applied for too long; a safety constraint(s) was unsafe. All of the above can involve human operators and in all cases, the reasons for the accident can be traced back to the original specification of the requirements, if there was one.

Case study 1: HFI thinking in action

The first author was asked to visit a secure facility wherein the guards were complaining of ill-health due to the heavy burden of personal protective equipment (PPE) they were required to wear on duty. It became clear that procurement had taken place in a piecemeal fashion: there was a wide variety of workplaces and operator roles and hazards were highly distributed and variable. PPE was not regarded as a 'system' itself, or as part of a larger system, and there was an absence of control and feedback at different hierarchical levels. This meant that the conventional approach of traditional ergonomics, based on risk assessment, would be unwieldy and only effective at the time it was applied. A more process-orientated approach focused on HFI and management was taken, resulting in the observations listed below:

1. Numerous PPE assemblies were in use for different operational roles. Several of these exceeded the mass limit set in the appropriate standard (Ministry of Defence, 2008) for a maximum one third of lean body mass. No assembly had been validated against an operational requirement to ensure that degree of protection met the requirement. The assemblies did not offer 'scaleable' protection (adaptability to different security requirements).

2. Officers worked a 12-hour shift with one hour for lunch after six hours on duty. This does not accord with the basic principles of work physiology – when physically arduous work is carried out, short periods of work should be interspersed with short rest periods to prevent the accumulation of fatigue.
3. There was no system of health surveillance in operation: absenteeism due to musculoskeletal ill-health was not monitored; high risk areas had not been identified and the efficacy of any improvements could not be monitored.
4. Human Factors Integration was not apparent in the procurement process. There was a lack of user-consultation and user-involvement.
5. The arrangements currently in place were characterised by low-level complexity that was unlikely to be manageable by a strict top-down policy framework. Rather, the policy should direct those responsible for health and safety to seek local solutions at individual workstations to comply with health and safety legislation.

All of the above can be traced back to either an absence of requirement(s) or violation of a generic HF requirement (Table 1). The solution is not to change the equipment or train the guards, but to improve the control processes that govern how the work is planned, conducted and reported. HFI thinking leads us to consider not only *what* has gone wrong, but *why* and what needs to be done to improve the management of the work.

Case study 2: Review of legacy equipment to propose new requirements

As part of a larger investigation of exposure to vibration and shock in high speed craft, sorties in over twenty craft (currently in use) were made. A variety of basic ergonomic design faults were observed (such as poor seating, a lack of proper seating, poor seat positioning and lack of handrails). Many faults were either hazards themselves or would exacerbate exposure to vibration and shock. It soon became clear that there was a lack of HFI in the procurement process, leading to a lack of HF&E requirements for HSC acquisition. This led to a piecemeal approach to procurement and a bewildering variety of craft with ergonomic hazards of different kinds. Examples of missing requirements, that can all be traced back to generic HF requirements in Table 1 are as follows: All occupants should be provided with a seat, designed for the purpose; All seats should face the direction of travel; Craft intended for use in hostile environments should provide protection for the occupants; Handholds should be located within the zone of convenient reach; The layout of seating should enable ingress and egress without snagging.

Conclusions

HFI is part of systems design and the focus is on what needs to be done rather than how to do it. Requirements specification and acceptance testing are critical for successful integration of HF principles – if a requirement is not specified, it doesn't get done. A user-orientated approach to design has been called for since the inception of ergonomics, yet even today, barriers to the acceptance and integration of HF into systems design still exist (see, for example, Waterson and Kolose, 2010). In complex systems, system-specific HF requirements must be derived from generic HF requirements specified at the beginning. Training in HF&E is required to derive these requirements.

References

Bridger, R.S. 2009. *Introduction to Ergonomics* (CRC Press, Boca Raton, Fl).

Dearden, A., Harrison, M. and Wright, P. 2000. "Allocation of function: scenarios, context and the economics of effort." *International Journal of Human Computer Studies* 52: 289–318.

DeGreene, K. B. 1978. "Force fields and emergent phenomona in sociotechnical macrosystems: theories and models." *Behavioural Science* 23: 1–14.

Haslam, R., and Waterson, P. 2013. "Editorial. Special issue on Ergonomics ans sustainability." *Ergonomics* 56: 343–347.

Huchingson, R. D. 1981. *New Horizons for Human Factors in Design* (McGraw-Hill, New).

Leveson, N. 2011. *Engineering a Safer World: system thinking applied to safety* (MIT Press).

McCormick, E. J. 1957. *Human Engineering* (McGraw-Hill, New York).

Ministry of Defence. 2008. *Defence Standard 00-250. Human Factors for Designers of Systems*.

Ministry of Defence. 2012. *Human Factors Integration for Defence Systems*. Joint Services Publication 912, pp. 78.

Singleton, W. T. 1974. *Man Machine Systems*. (Penguin Education), pp. 174.

Taylor, F. W. 1911. *The Principles of Scientific Management* (Harper and Brothers, New York and London).

Waterson, P. and Lemalu-Kolose, S. 2010. "Exploring the social and organisational aspects of human factors integration: a framework case study." *Safety Science* 48: 482–490.

HFI TRAINING FOR SYSTEMS ENGINEERING PROFESSIONALS

Ella-Mae Hubbard, Luminita Ciocoiu & Michael Henshaw

Engineering Systems of Systems Research Group, School of Electronic, Electrical and Systems Engineering, Loughborough University

This paper will consider what key HFI (Human Factors Integration) skills are needed for Systems Engineers. The paper will use the INCOSE Competency Framework as a guide for the relevant KSAs (Knowledge, Skills & Attributes). The paper suggests some potential barriers to be aware of and outlines strategies for effective learning.

Introduction

Perhaps the most distinguishing feature of Systems Engineering is its interdisciplinary nature. One of the disciplines that must not be forgotten is Human Factors, but how should this be included in the training of systems engineers and how do we overcome the barriers to implementation of such training? It has long been accepted, both in academia and industry, that there is a need for engineers to have an appreciation of human factors, perhaps over and above the need for human factors specialists (Shapiro, 1995).

Understanding how best to improve relationships between ergonomics and a variety of disciplines has been an ongoing consideration, with people considering education from school level, through undergraduate and industry training. However, this may not be a straightforward task. "The relationship between ergonomics and the disciplines it informs has always been tenuous." (Woodcock and Denton, 2001) It is important to consider both what knowledge and skills are needed, but also how and when these should be taught or learnt (Emilsson and Lilje, 2008).

There is overlap of HF and SE and benefits to such an integration (Johnson, 1996, Narkevicius, 2008), but there is no common view of when and how such integration would be best placed (Chua and Feigh, 2011). For students in the earlier years of their learning, introducing human factors within a relevant project is useful. For later training (perhaps postgraduate or continuing professional development (CPD)) inviting students to use a real example from their job can be beneficial. Early and continuing inclusion of human factors considerations will help acceptance and understanding of importance (Woodcock and Denton, 2001).

Identifying relevant skills

INCOSE has developed a competencies framework (INCOSE, 2010), which identifies a range of skills needed by systems engineers and attainment levels for these competencies. The framework was prepared by representatives from Atkins, BAE Systems, Brass Bullet Ltd, DSTL, EADS Astrium, General Dynamics United Kingdom Limited, Harmoni, HMGCC, Loughborough University, Ministry of Defence, Rolls Royce, SELEX Galileo, Thales, Ultra Electronics, University College London and, as such, has good buy in from both industry and academic stakeholders.

The framework groups competencies into three thematic areas: systems thinking, holistic lifecycle view and systems engineering management. Each competence is described and effectiveness indicators are provided to determine levels of attainment within each competency (awareness, supervised practitioner, practitioner and expert). Within the framework, human factors is considered as a specialty.

The contents of the competency framework have been reviewed to determine those areas where there is crossover between SE and HFI. Initial content searches for 'human' and 'user' were performed, followed by a general content analysis for relevant areas.

A total of 8 competencies out of a total of 21 within the framework are identified as associated with HFI (see table 1). Also, whilst the majority of these are within the holistic lifecycle view theme, associations are identified across all three themes.

This confirms that human factors is a relevant, and significant, area for systems engineers. It also helps to give us a basic foundation as to the human factors content which may be necessary for systems engineers. It should be noted that the list of competencies within the framework will not be exhaustive and there will be many other HF skills which would prove beneficial for systems engineers.

A mapping was also performed of the SE competencies to the IEHF knowledge areas (IEHF). It is clear that there is significant overlap/ correlation between these, especially in the IEHF 'people and systems' and 'psychology' areas.

One of the basic learning outcomes for all accredited HE (Higher Education) programmes in engineering is that students "must appreciate the social, environmental, ethical, economic and commercial considerations affecting the exercise of their engineering judgement." (Engineering Council) So human factors education will be of benefit to all UG engineering students. They are also required to have an understanding of design, including "knowledge, understanding and skills to:

- investigate and define a problem and identify constraints including environmental and sustainability limitations, health and safety and risk assessment issues;
- understand customer and user needs and the importance of considerations such as aesthetics;

Table 1.   Relevant Competencies (adapted from INCOSE 2010).

Competency	Description
Systems thinking: systems concepts	"The application of the fundamental concepts of systems thinking to systems engineering. These include understanding what a system is, its context within its environment, its boundaries and interfaces and that it has a lifecycle."
Holistic lifecycle view: systems design – interface management	"Interfaces occur where system elements interact, for example human, mechanical, electrical, thermal, data etc. Interface management comprises the identification, definition and control interactions across system or system element boundaries."
Holistic lifecycle view: validation	"Validation checks that the operational capability of the system meets the needs of customer/end user – 'did we build the right system?'"
Holistic lifecycle view: transition to operation	"Transition to operation is the integration of the system into its super system. This includes provision of support activities … Incorrectly transitioning the system into operation can lead to misuse, failure to perform and customer/user dissatisfaction."
Holistic lifecycle view: determining and managing stakeholder requirements	"To analyse the stakeholder needs and expectations and managing stakeholder requirements."
Holistic lifecycle view: systems design – design for …	"Ensuring that the requirements of all lifecycle stages are addressed at the correct point in the system design."
Systems engineering management: enterprise integration	"Enterprises can be viewed as systems in their own right in which systems engineering is only one element. Systems engineering is only one of many activities that must occur in order to bring about a successful system development that meets the needs of its stakeholders. Systems engineering management must support other functions … and manage the interfaces with them."
Systems engineering management: integration of specialisms	"Coherent integration of specialisms into the project/ programme at the right time."

- awareness of the framework of relevant legal requirements, governing engineering activities, including personnel, health, safety and risk (including environmental risk) issues" (Engineering Council)

Training systems engineers – an example

Within the SE undergraduate (UG) courses at Loughborough University, HFI considerations have always formed part of the compulsory content of both BEng and

Table 2.   Potential barriers.

Barrier	Description
Failure to see importance or link/ necessity for their work	Engineers can often be quite focused in their specific discipline. Systems engineers must have a wider focus than this and see the 'bigger picture'. Getting systems engineers to understand the impact of good HFI (or the implications of ignoring HFI all together). Understanding the reason for learning is key to motivation (Fry et al., 2003). There is some evidence that UG engineers are 'unappreciative of the benefits of ergonomics' (Woodcock and Denton, 2001).
Too busy doing other things/getting on with real work	This can actually be a wider issue within SE, rather than one simply associated with HFI. HFI needs to be considered from early in the lifecycle, however, the benefits of good HFI are sometimes not realised. Such a delay on the return on effort can lead people to focus their efforts or prioritise tasks in other areas.
Somebody else's problem	Linked to both issues identified above, understanding of HFI can be seen as unnecessary because someone else will do it. To use the words of the INCOSE Competencies Framework (INCOSE, 2010), an awareness is needed by all. Understanding the areas that may be impacted or benefit from good HFI is a good starting point, even if the work is then done elsewhere.
Fundamental difference in processes	Engineers are not always comfortable (or even familiar) with methodologies used in social science approaches, and they traditionally do not receive training in these techniques. Engineering students are taught to find an answer, the exact solution, there are rules for them to work to, the world is black or white. Or at least this is typically the case early on in their studies. Students can find it difficult to work on an open ended problem, where there is no defined solution, it is not comfortable for them. When dealing with human factors, we are quite often dealing with shades of grey. Students first need to accept the challenge of this different approach to problem solving before any subject specific learning can begin. "It can be difficult to communicate the findings of human factors specialists into a language that can be interpreted and acted upon by systems engineers." (Johnson, 1996)

MEng programmes (although specific content and delivery will change). There is evidence that options taken (as opposed to compulsory modules) on human factors related modules has decreased since 2005. Such a decrease makes the compulsory elements of teaching on such courses all the more important (if we feel that systems engineers should have an appreciation of human factors). It is not clear whether such a decrease is due to student interest, module availability, general course evolution or some other factors. Understanding such drivers for a particular cohort or group of learners is necessary before any interventions to increase uptake in relevant courses is embarked upon. Within the School of Electronic, Electrical and Systems Engineering at Loughborough University, take up of UG final year, individual projects (dissertations) which have clear objectives or focus related to HFI has been 5% on average since 2008 (it is recognised that some HF considerations should be present in most, if not all, projects and these figures do include students on Electronic and Electrical Engineering and Electronic and Computer Systems Engineering programmes within the school, rather than just Systems Engineering students). There is some evidence of decline in the number of students taking projects with such focus since 2010, although there are many factors which could account or contribute to this.

Potential barriers to acceptance

When teaching HFI skills to engineers, there are some potential barriers, which need to be overcome if the training is to be successful. Many of these link to the consideration of motivating learners, looking at both intrinsic and extrinsic motivation (Fry et al., 2003). Some of these barriers, identified in the example case detailed above, are described below (further work would be needed to determine the wide scale affect of such barriers.

Conclusion – Strategies

In attempting to overcome these challenges, as experienced in various types of education, there are some strategies to be employed.

* Make the training relevant

If you are trying to overcome a barrier to acceptance, the more context relevant you can make the training the better. If the trainees can see direct links (perhaps through the examples you use or the terminology employed) to their work, this will increase uptake and implementation of the ideas and techniques learned. Improving the effectiveness of ergonomics education has been a long-term goal. Making that education appropriate to a particular area of interest or application is powerful for understanding and acceptance – taking it beyond the perception that it is just 'common sense' (Woodcock and Flyte, 1997).

* Make the training applied

HFI teaching as part of wider projects is often successful for acceptance by systems engineers. Depending on their stage of learning they will either be used to the interdisciplinary nature, or they need to learn about that as well. This is linked to increasing the relevance of the training. Ensuring that Systems Engineers have some HF training may help develop a common language, making communication and understanding easier.

References

Chua, Z.K. and Feigh, K.M., 2011, Integrating human factors principles into systems engineering, IEEE/AIAA 30th Digital Avionics Systems Conference, October 2011, ISSN: 2155-7195

Emilsson, U.M. and Lilje, B., 2008, Training social competence in engineering education: necessary, possible or not even desirable? An explorative study from a surveying education programme, *European Journal of Engineering Education* 33:3, 259–269

Engineering Council, The accreditation of higher education programmes, UK Standard for Professional Engineering Competence, http://www.engc.org.uk/ ecukdocuments/internet/document%20library/AHEP%20Brochure.pdf, accessed 29 November 2013

Fry, H., Ketteridge, S. and Marshall, S., 2003, A handbook for teaching and learning in higher education, 2nd edition

IEHF, Knowledge Areas, http://www.ergonomics.org.uk/knowledge-areas, accessed 27 November 2013

INCOSE UK, 2010, *INCOSE UK Framework: Systems Engineering Competencies Framework*, Issue 3, January 2010

Johnson, C.W., 1996, Integrating human factors and systems engineering to reduce the risk of operator 'error', *Safety Science* 22:1–3, 195–214

Narkevicius, J.M., 2008, Human factors and systems engineering – integrating for successful systems development, Proceedings of the Human Factors and Ergonomics Society 52nd Annual Meeting, 1961–1963

Shapiro, R.G., 1995, How can human factors education meet industry needs?, *Ergonomics in Design* 3:32

Woodcock, A. and Denton, H.G., 2001, The teaching of ergonomics in schools: what is happening, *Contemporary Ergonomics*, 311–315

Woodcock, A. and Flyte, M.G., 1997, The development of computer-based tools to support ergonomics in design education and practice, *Digital Creativity* 8:3–4, 113–120

MANAGING THE LIFECYCLE OF YOUR ROBOT

M.A. Sinclair & C.E. Siemieniuch

ESoS Research Group
School of Electrical, Electronic & System Engineering
Loughborough University of Technology

'Robot' for this paper is assumed to be a cognitive device, acting as a co-worker within a team of human workers: a mobile device, with a degree of autonomy, interchangeable prostheses, interacting freely with surrounding humans, in a civilian environment. An exemplar lifecycle is the MoD's CADMID lifecycle, and this paper concentrates on the In-service phase. The approach is from a management perspective; a road-map is provided to acquire a robot, to put it to work, and to support both it and the team during its in-service phase. The emphasis is on what management needs to consider and the structures that need to be in place in order to run this process.

Introduction

The development of robotic technology proceeds apace. However, it is largely driven by engineers thinking in engineering development terms. At some point it becomes necessary to look beyond the purely engineering development effort, impressive as it is, and contemplate the utilisation of robots in society. It is with this aspect that this paper is concerned. First, an exemplar co-working scenario is presented, and then some of the management issues entailed in this scenario are explored.

Much has been made of the potential use of robots in assistive living, both to enable the elderly, and other challenged individuals, enabling them to live at home and relieving the demand on care and medical services. Of potentially more economic significance will be the role of robots in value-creation in society, working in conjunction with human beings in teams. It may be argued that that is happening already on production lines; true, but the robots are encaged, performing repetitive tasks with little autonomous judgement involved. The role for robots in this paper is as a co-worker, carrying out joint tasks with humans, and single tasks in parallel.

Unfortunately, there are no robots in production of significant capability to provide practical experience. For this reason, a scenario approach is used by which a management process could be generated, based on experience in other complex products. Necessarily, this is an abstract process, requiring further elaboration as better understanding is gained.

There are three significant scenarios that can be envisaged as listed below; we discuss one in some detail, and then amend the discussion for the other cases. The scenarios are:

- *Assistive living:* for example, an elderly person (or other reduced-capability person) is provided with a robot to enable that person to live at home and receive moment-by-moment support, complementing that provided by other human-based care services.
- *Co-working in teams:* a team comprising humans and one or more robots work together carrying out joint tasks with humans, and single tasks in parallel with humans. These teams need not be constrained to production facilities; fieldwork teams in the utilities such as gas, electricity and energy, road and rail maintenance and so on could also have robot co-workers. Robots may well be restricted mainly to the 4D tasks; dull, dirty, dangerous and dark, also fetching, lifting and carrying, but they will be doing this in close proximity to human beings, and, perhaps, other robots too. The 'BigDog' robot is a prototype for this role (www.engadget.com/2013/02/28/bigdog-robot-has-an-arm-now-run/).
- *Distributed working:* the Internet of Things case, where a cognitive device is part of a system of systems, perhaps embedded, whose decisions interact with humans in accomplishing a purpose. The robot itself may be a geographically-distributed device.

The scenario chosen for exploration is co-working. For simplicity, we assume there is one robot in a workgroup including several humans, and that this workgroup is functioning within a larger organisation, carrying out maintenance tasks at any of a number of locations, all different. The robot itself is assumed to be mobile, equipped with prostheses (one or more) to enable it to be a tool-user. Furthermore, we assume that the robot has been equipped with sufficient software to enable it to operate, for most tasks, at level 6 on Sheridan's scale (Sheridan 1980, Sheridan 1994, Parasuraman, Sheridan et al. 2000): "allows the human a restricted time to veto before automatic execution [of the decision or task]", or perhaps at level 7 "executes automatically, then necessarily informs the human". It should be noted that this implies that the robot, working among people, necessarily must be capable of ethical behaviour, though perhaps not at the level to be expected of humans (Sullins 2006, Abney 2012, Asaro 2012). This is necessary for its co-workers to trust it when it is working autonomously. For safety reasons (see Table 1), it is assumed that the robot continuously records its working environment and its work, irrespective of privacy concerns.

Furthermore, if for illustration purposes we adopt the MoD CADMID lifecycle as shown in Figure 1, the paper only addresses the second last of these phases.

We may now propose a process for managing the robot over its working life in the organisation, as shown in Figure 2 below. This diagram is predicated on the assumption that the organisation that supports the work group also plans the insertion of the robot(s) into the workgroup. It will be noted that a considerable support effort is required before the robot is put to use within the work group.

Table 1. A listing of co-worker protection requirements.

No	Keyword	Principle
1	Authority 1	Whatever the application, the human party shall have the highest authority.
2	Authority 2	All decisions and action plans originated by robots shall be under the supervision of a human party, who shall take responsibility for these.
3	Authority 3	The last person to issue a command to a robot shall take responsibility for the outcomes; other stakeholders may also be included.
4	Authority 4	Robots shall not 'force' human response.
5	Safety 1	A robot shall not deliberately cause injury, neither physiological nor psychological.
6	Safety 2	Any contact between robots and humans shall be by 'controlled collisions' (e.g. touching, hand-overs, etc.).
7	Safety 3	Any learning undertaken by robots shall be subject to verification and validation (V&V) to ensure continuous compliance with Safety 1 & 2.
8	Safety 4	Periodic inspection of robots shall be undertaken, to ensure compliance with Safety 1, 2 & 3 above.
9	Safety 5	Robots shall not be abandoned by their designers, developers, manufacturers and/or owners until they are withdrawn from service.
10	Safety 6	Where a failure of the cognitive device(s) within a system might lead to a critical safety situation for the user, a back-up system to mitigate the critical safety situation shall be provided.
11	Safety 7	The output of a cognitive device shall not be utilised in any cultural region outside those formally considered during the design or the tailoring during the implementation of the device in a system.
12	Anthropomorphism 1	Unless legally permitted, the design and operation of cognitive devices shall eschew anthropomorphism intended to delude humans interacting with them.
13	Anthropomorphism 2	Designers of robots should avoid anthropomorphism where this might lead to unnatural or socially compromising behaviour, and where it is not clearly required to exercise the system's functionality.
14	Technology addiction 1	Designers of robots should avoid creating systems that could lead to addictive behaviours in their human users
15	Social exclusion 1	Those providing support systems that include robotics should ensure that these do not lead to social exclusion and an increased potential for human neglect.
16	Social exclusion 2	Where users are dependent on robots, the design and employment of these shall not 'trap' users in one location because of insufficient power provision.
17	Representation 1	For those users likely to be physiologically or cognitively dependent on robots, prior to their introduction, discussions with the patients, or their representatives, shall take place.
18	Representation 2	For 'patients' as defined in Representation 1, provision shall be made for the patient to have an ombudsman or other representative for the duration of the care system.

(Continued)

Table 2.   Continued.

No	Keyword	Principle
19	Data protection 1	The capture of data of a personal nature by a robot shall comply with relevant legislation, and shall be agreed with end-users of the system before it is commissioned.
20	Data protection 2	Personal data captured by the robot shall be protected according to current legislation.
21	Data protection 3	Where robots interface to the internet or other information systems, data security shall be paramount
22	Data protection 4	Access to personal data captured by robots shall only be accessible to the person concerned, to those cognitive agents authorised to have access and to other persons given legal permission to have access.

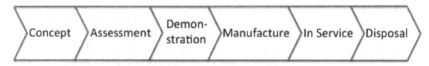

Figure 1.   The CADMID lifecycle, utilised by the UK Ministry of Defence.

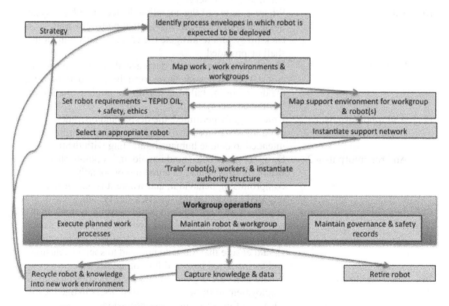

Figure 2.   A high-level process to manage robot co-workers.

It is assumed that the robot is obtained from a supplier; not built in-house. See Figure 3 and the text below for an explanation of TEPID OIL.

Figure 3 below provides a little more detail about Figure 2 above. It outlines the steps from strategy to work group operations in relation to a notional timescale, and

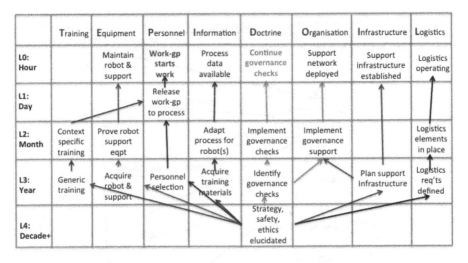

	Training	Equipment	Personnel	Information	Doctrine	Organisation	Infrastructure	Logistics
L0: Hour		Maintain robot & support	Work-gp starts work	Process data available	Continue governance checks	Support network deployed	Support infrastructure established	Logistics operating
L1: Day			Release work-gp to process					
L2: Month	Context specific training	Prove robot support eqpt		Adapt process for robot(s)	Implement governance checks	Implement governance support		Logistics elements in place
L3: Year	Generic training	Acquire robot & support	Personnel selection	Acquire training materials	Identify governance checks		Plan support Infrastructure	Logistics req'ts defined
L4: Decade+					Strategy, safety, ethics elucidated			

Figure 3.   A Mackley diagram to plan the introduction of a robot into a work group.

introduces the 'TEPID OIL' acronym first adopted by the Ministry of Defence in the UK, in the form of a 'Mackley diagram' (Mackley, Barker *et al.* 2007, Mackley 2008). TEPID OIL covers the following aspects of capability:

Training, Equipment, Personnel, Information, Doctrine (in civilian terms, this equates to strategy and policy; for the purposes of this document it also includes ethics), Organisation Infrastructure (for this document, this term includes interoperability), and Logistics

Together, these aspects enable an organisation to deliver a capability.

Figure 3 indicates that the planning may have to start a decade before the work group commences operation, and that many sub-processes will have to happen concurrently, placing considerable emphasis on overall project management. A corollary is that for this to be an efficient process, the management of knowledge will be a key issue. Figure 3 implies a number of lessons, as listed below.

- Planning for the introduction of a robot into a workgroup will require considerable lead time.
- Many processes must happen concurrently; this is driven not just for efficiency reasons, but because of the rapid pace of development of robot technology. Too much time spent on this process likely will result in a near-obsolescent outcome.
- All of these sub-processes are necessary. It is difficult to see how an effective work group could be delivered if any of these is missed.
- While it is strategy that commences the whole exercise, strategy necessarily must be a continuous process, because the outcome is a complex system, and because of the rate of development of technology, as indicated above. The better is the

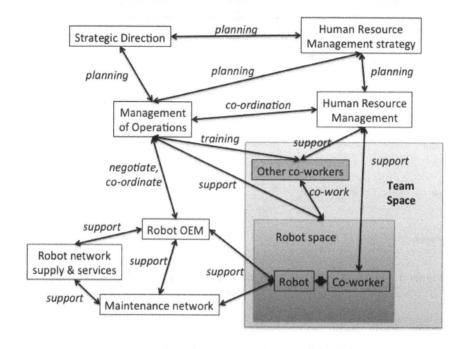

Figure 4.   Network for authority and responsibility to support the work group.

strategic grasp at the beginning, the less likely will be the need for reworking in the sub-processes.

Figure 4 below shows a generic organisational structure that distributes authority and responsibility to run the process in Figures 2 and 3 on a continuous basis.

Figure 4 represents a network for authority and responsibility to support the work group using the processes outlined in Figures 2 & 3. These functions may be distributed across different organisations. At right is a structure to provide human resource support to the individuals in the team as a whole, including those who work directly with the robot. This corresponds to the personnel column in Figure 3. Along the bottom is the sub-process to support the robot, delivering upgrades, maintenance and probably technical support to those managing operations. This corresponds to the equipment column in Figure 3, and is separated out because of the complexities and specialisms of robot engineering. The main diagonal embraces the rest of the columns in Fig. 3, and most of the process in Figure 2. It is also the region in which most of the governance issues will be addressed.

Because of the high-level abstractions in Figures 2 to 4, it is possible to apply these diagrams to the other two scenarios listed earlier, albeit with some changes in emphasis. For example, in the Assisted Living scenario, it is likely that the management of operations will find itself with some technological challenges, implying that some of the robot-orientated authority and responsibility aspects of management would be better placed within the horizontal, equipment supply and maintenance domain. Furthermore, given that a proportion of people would prefer to organise their own care, especially in the early stages, and the likely cost of these robots, it is likely that the best business model would be one in which leasing is the fundamental arrangement to provide care.

Likewise for distributed robotic systems; the problem here is that it is possible that some of the cross-links between the strands in Figure 4 will be contract-based. Because this redistribution might be across jurisdictions, it is likely that some regulatory and standards activity will be required for this scenario.

Finally, we consider some of the safety aspects that must be addressed in the diagrams above, from an ethical standpoint:

It seems evident that there is a long way to go before robots as co-workers become likely in work groups. For other reasons not discussed here and due to liability, personhood, ethics and culture, we believe robots will always be subservient to humans. It is also evident that for members of the IEHF to be involved in this form of HMI, some increase in understanding will be needed.

References

Abney, K. 2012. Robots, ethical theory and metaethics: a guide for the perplexed. In P. Lin, K. Abney, G.A. Bekey. (eds.), *Robot ethics,* (MIT Press, Cambridge, MA), 35–54.

Asaro, P. M. 2012. A body to kick, but no soul to damn: perspectives on robotics. In P. Lin, K. Abney, G.A. Bekey. (eds.), *Robot ethics,* (MIT Press, Cambridge, MA), 169–186.

Mackley, T. 2008. Concepts of agility in *Network Enabled Capability. Realising Network Enabled Capability.* Oulton Hall, Leeds, UK, BAE Systems.

Mackley, T., S. Barker & P. John 2007. *Concepts of agility in Network Enabled Capability. The NECTISE project.* Cranfield, Cranfield University, UK 8.

Parasuraman, R., T. B. Sheridan & C. D. Wickens 2000. "A model for types and levels of human interaction with automation." *IEEE Transactions on Systems, Man & Cybernetics 30(3):* 286–297.

Sheridan, T. B. 1980. "Computer control and human alienation." *Technology Review(October):* 61–70.

Sheridan, T. B. 1994. Human supervisory control. In G. Salvendy and W. Karwowski. (eds.), *Design of work and development of personnel in advanced manufacturing,* (John Wiley & Sons, New York), 79–102.

Sullins, J. P. 2006. "When is a robot a moral agent?" *International Review of information Ethics 6(1):* 24–30.

A FRAMEWORK FOR HUMAN ASSESSMENT IN THE DEFENCE SYSTEM ENGINEERING LIFECYCLE

J. Astwood[1], K. Tatlock[1], K. Strickland[2] & W. Tutton[2]

[1]*MBDA Limited*
[2]*Defence Science Technology Laboratories, UK Ministry of Defence*

The research outlined in this paper investigated the dependencies between Systems Engineering (SE) activities and the application of a Human Factors Assessment Framework (HFAF) developed by Dstl for defence system engineering activities. The research considered how best the HFAF could be applied within the SE Lifecycle with an emphasis on current Human Systems Integration (HSI) best practice within the UK Ministry of Defence (MoD) acquisition cycle. The HFAF lends itself well to each stage of the SE Lifecycle, especially during the early stages where the work undertaken is largely investigative. The HFAF supports the development of requirements, and design specifications, as well as an aid to understand system maturity via trial based assessment.

Introduction

This research was funded by the Defence Science Technology Laboratories (Dstl) and carried out by the Defence Human Capability Science and Technology Centre (DHCSTC) led by BAE Systems, with the purpose of understanding the dependencies between Systems Engineering (SE) activities and the application of the Human Factors (HF) Assessment Framework (HFAF) (Humm et al, 2010). An additional aspect of this research was to identify where the HFAF developed by Dstl could be applied to reduce costs, mitigate risk and ensure the end product/system meets the technical and military requirements. Within this research, an investigation was undertaken to identify the practicality of widening the application of HFAF as it currently stands, and the practicality of expanding HFAF in capability.

Background

In 2009, the Dstl Human Systems Group identified a need for a coherent and consistent approach to assessing HF aspects of Dismounted Close Combat (DCC) Systems, as a result of issues experienced with the quality of HF information available to inform DCC system development. The HFAF provides a high level framework for HF Practitioners to gather good quality HF data needed to support

DCC design and system analysis as well as Human Systems Integration (HSI) activities. To date, the HFAF has been applied across a range of DCC projects, including enabling delivery of Urgent Operational Requirements (UORs) and support to operations research, however, its place within mature systems engineering/acquisition strategy is neither well understood nor formalised. It is the aspiration of senior stakeholders that the HFAF will enable better discrimination between proposed systems/capabilities through a progressive iterative scientific approach in domains beyond DCC.

The HFAF

The HFAF utilises a three level approach which can be considered as the levels of fidelity needed in assessment:

- Level 1: Initial HF assessment – Paper based review of the system concepts, specification or specific parameters of interest, visual inspection and/or functional assessment.
- Level 2: Functional performance assessment – assesses function and performance of a number of users (a representative sample) during a range of simulated military tasks exercised as aspects of the system, using standardised HF data collection methods.
- Level 3: Controlled environment assessment – a lab-based assessment within a controlled environment to exercise all or part of the system. This is undertaken in order to provide evidence to answer specific research questions regarding equipment or soldier performance where more control is required to ensure reliable and robust physiological, psychological or performance-related data are captured.

Aligning the HFAF with SE Lifecycles

While the HFAF has been principally used for down-select activities for UORs, for which it appears to work well as time is of the essence, the research focused on how well the HFAF could work in more systematic system development. With the introduction of the BS ISO/IEC 15288 in 2002 (British Standards, 2002) the discipline of SE established a framework for describing the development life cycle of systems. In order to understand where the HFAF could support the SE Lifecycle a number of SE applications (Model-Based Systems Engineering (MBSE), the UK MoD Smart Acquisition Cycle (e.g. CADMID) and the MoD HSI Process were investigated to determine where the HFAF could align within them.

By reviewing these three applications, and a number of alternative views such as ISO 15288, it was possible to illustrate how HSI can be mapped on to the Systems Engineering Lifecycle (Figure 1), and so provide a basis for identifying where the HFAF can be used. It is also important to note that HF integration models only act as

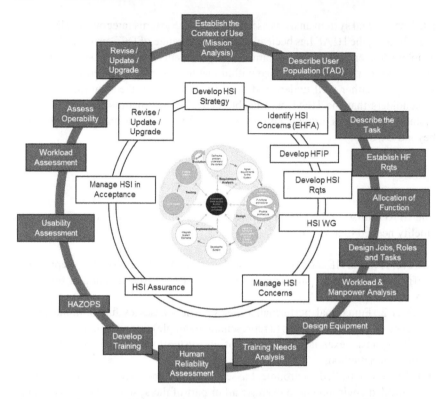

Figure 1.   Typical HSI Activities mapped to SE Lifecycle (See Fig 3) (outer circle outlines technical activities, the inner circle management activities).

guides, since no standard method can remove the need for expert based judgement of what Human Factors work needs to be completed given the specific needs of the system being developed.

Lifecycle touch points

Figure 2 provides an overarching view of the alignment of each of the cycles/processes relevant to the MoD acquisition phases, typically referred to as CADMID (Concept, Assessment, Demonstration, Manufacture, In-Service and Demonstration) that the majority of projects use during the development of defence systems. Note; Figure 2 suggests a firm schedule, the reality is often more fluid.

HFAF utility

Figure 3 provides the Touch Points, or views of, where the HFAF could most usefully be implemented within the SE Lifecycle and how it supports the HSI process.

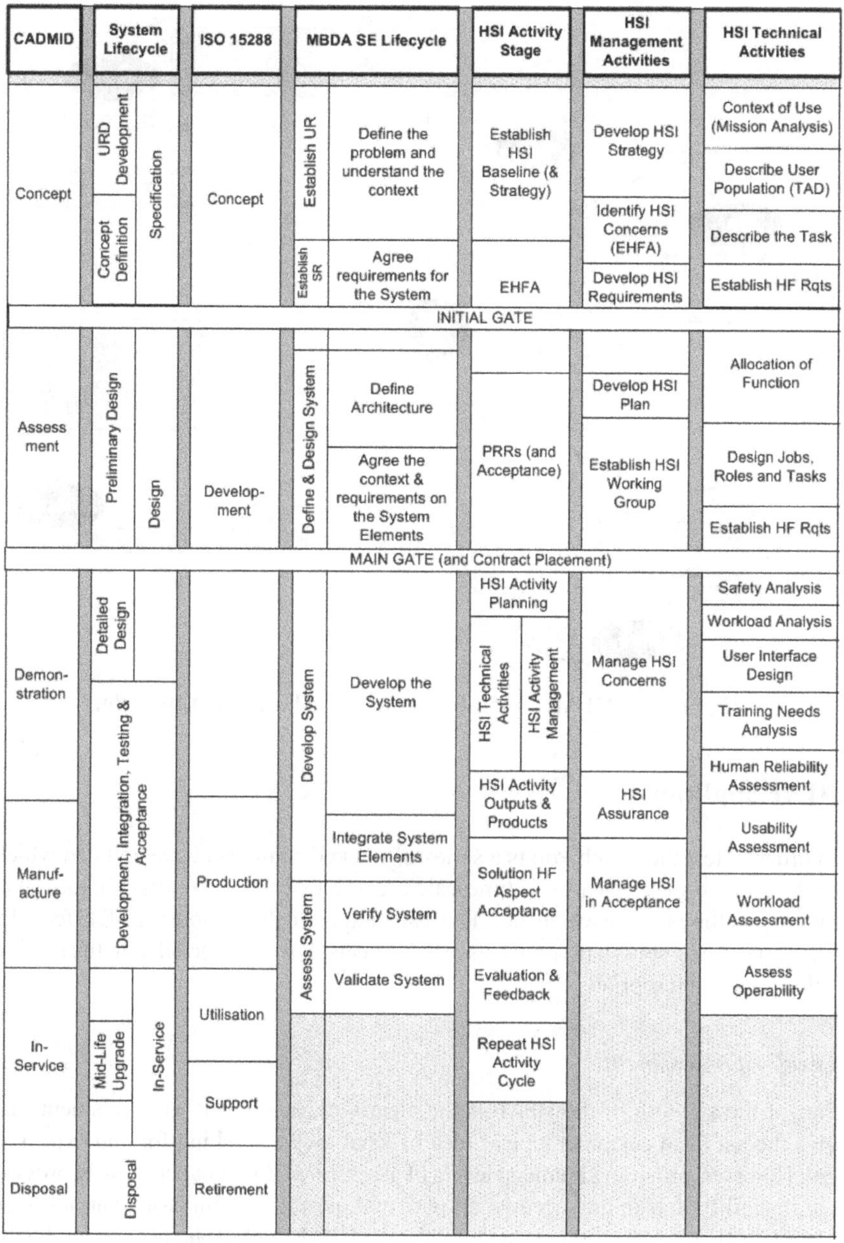

CADMID	System Lifecycle	ISO 15288	MBDA SE Lifecycle	HSI Activity Stage	HSI Management Activities	HSI Technical Activities
Concept	URD — Development / Concept Definition — Specification	Concept	Establish UR — Define the problem and understand the context	Establish HSI Baseline (& Strategy)	Develop HSI Strategy	Context of Use (Mission Analysis)
						Describe User Population (TAD)
					Identify HSI Concerns (EHFA)	Describe the Task
			Establish SR — Agree requirements for the System	EHFA	Develop HSI Requirements	Establish HF Rqts
INITIAL GATE						
Assessment	Preliminary Design — Design	Development	Define & Design System — Define Architecture	PRRs (and Acceptance)	Develop HSI Plan	Allocation of Function
			Agree the context & requirements on the System Elements		Establish HSI Working Group	Design Jobs, Roles and Tasks
						Establish HF Rqts
MAIN GATE (and Contract Placement)						
Demonstration	Detailed Design — Development, Integration, Testing & Acceptance	Production	Develop System — Develop the System	HSI Activity Planning	Manage HSI Concerns	Safety Analysis
				HSI Technical Activities / HSI Activity Management		Workload Analysis
						User Interface Design
						Training Needs Analysis
						Human Reliability Assessment
Manufacture			Integrate System Elements	HSI Activity Outputs & Products	HSI Assurance	Usability Assessment
			Verify System	Solution HF Aspect Acceptance	Manage HSI in Acceptance	Workload Assessment
			Assess System — Validate System	Evaluation & Feedback		Assess Operability
In-Service	Mid-Life Upgrade — In-Service	Utilisation		Repeat HSI Activity Cycle		
		Support				
Disposal	Disposal	Retirement				

Figure 2. Alignment of CADMID, System Lifecycle, HSI Activities and Key Documents.

From the touch points it is evident that the HFAF lends itself well to each of the stages of the SE Lifecycle. In particular during the early stages, the HFAF supports the development of requirements, and design specifications, as well as aiding the project to better understand system maturity via trial based assessment.

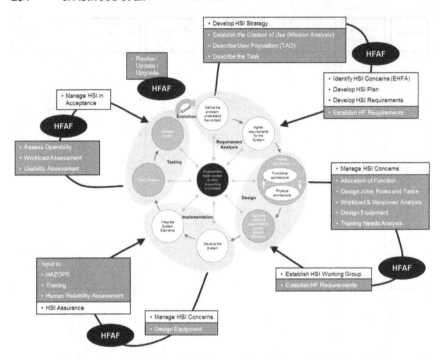

Figure 3.   HFAF Touch Points with HSI and SE Lifecycle.

HFAF application

In order to test the Touch Points a series of worked examples were collated which cover a range of applications, timescales, complexity and maturity factors, that investigate the extent to which the HFAF can be applied throughout the SE Lifecycle. The worked examples had some classified aspects, but are referred to in the section below where appropriate.

Levels of assessment

Part of the strength of the HFAF is the clearly defined levels of assessment that give the user and customer a clear idea of what they are asking for and expect to get. However, providing bounded levels of assessment does not necessarily provide the flexibility to take into account all potential uses. The main issue that has been identified that requires addressing to make the HFAF applicable throughout the SE Lifecycle in complex programmes is that the current Levels of assessment do not lend themselves directly to all potential applications. Examples of this are outlined below (names of the system cases used to test the touch points have been anonymised due to classification sensitivities):

- Worked Example: System Case 3 – Lab based experimentation was carried out at an early stage of the design phase to address specific questions (but of a very limited capability) and also to gather subjective comments from end-users about the general feel of the system. At first this case may be viewed as a HFAF Level 1 assessment but in a lab-based environment under semi-controlled conditions; it was unlikely to have been a HFAF Level 3 assessment due to the relaxed methodology for gathering subjective data about the general issues in the system.

* Worked Example: System Case 2 – Level 3 type experimentation was carried out in the mature stages of the project. This period of controlled experimentation included collating evidence for validation and verification, gathering subjective information for future development via Training Sessions and After Action Reviews, and also allowing for a more free-form exploratory element that allowed for development of the CONOPS for the system. Some of the more free-form, subjective elements could have been captured by a separate Level 1 type assessment.

The complexity in understanding what level(s) of assessment is required is one which HF practitioners are capable of addressing and therefore should be able to utilise the HFAF in the spirit in which it is intended. One of the strengths of the HFAF is as a tool for communication between stakeholders, and using the framework in this flexible manner has the potential for misunderstanding particularly if applied in a contract. The description of the current HFAF levels of assessment may be generalised to take into consideration this issue. It has become clear, however, that the HFAF guidance provided to users will need to take into account a number of dimensions, such as; task type, project phase, equipment maturity, output required and project complexity, in order to provide sufficient information regarding the scope of work and to align with the wider SE Lifecycle.

Output required

One of the key factors when considering the type of assessment that is required is the specific question(s) that need addressing. Questions can be broad-scoping (to address general ideas and concepts) or focussed (to investigate specific capabilities), and also have the potential to be exploratory (for example, in the development of a Concept of Use Document (CONUSE)) or to provide outputs for a specific need (for example, in support of Verification & Validation (V&V)). The key consideration is to determine what level of evidence/output is acceptable, i.e. what is the burden of proof and what level of scrutiny will the output have to withstand?

Assurance & acceptance

It was found that HFAF could provide a standardised format for key assessment points within a project life-cycle. This standardised format allows for good comparison across ideas and concepts within a project and also potentially for comparison across projects (where security allows).

The quality of the SE Lifecycle (and therefore the HSI process within this) conducted within a project relies on two key themes:

- Assurance, i.e. the customer (MoD) being confident of the quality of the process undertaken; and
- Acceptance, i.e. the customer (MoD) being content through Integrated Test Evaluation and Acceptance (ITEA), and V&V activities that the system being developed is fit for purpose.

Both of these themes need to be taken into consideration when understanding where the HFAF fits into the HSI process. It is the Assurance process that will ensure assessment takes place at the appropriate time and demonstrates the risk reduction process; however, ultimately the assessment needs to assist the Acceptance process demonstrating the output meets the requirement.

Assurance and Acceptance are addressed in many ways throughout the project lifecycle ensuring large scale, complex projects remain on track. Specific reviews are often carried out at key milestones e.g. Initial Gate, Main Gate, Critical Design Review; taking into consideration the results of assessments and how decisions were reached.

In order to support the Assurance process, the HFAF would benefit from inclusion of a structure to support the needs of each of the phases of the SE Lifecycle and also in tying up the phases of the SE Lifecycle together (i.e. provision within the framework to support connection to requirements and V&V activities).

From the analysis of the HFAF performed here, as far as the Human Component of Capability (HCoC) is concerned, it would seem that the HFAF has significant potential to support acceptance processes, like the ITEA process. Undertaking assessments via a uniform framework such as the HFAF supports the needs of multiple stakeholders in the programme development. This focuses the consideration of the outputs (e.g. for the purposes of integration) on the results, without the need for undue concern regarding the methodology that was undertaken.

Conclusions and recommendations

A number of adaptations to the HFAF were the primary output from this research, and are being undertaken in the current MoD research programme. In addition, the adapted framework is being considered as a guide document in the future suite of HSI publications, alongside future updates to Defence Standard 00-250, Human Factors for Designers of Systems (UK MoD, 2008), ensuring that there are applied materials for HF practitioners and acquisition professionals in MoD and industry. The key adaptations recommended included:

- Review guidance material within the HFAF to ensure the scope of applications is considered to avoid the guidance becoming too complex, consider developing separate applications for different domains (e.g. Equipment/Cognitive/Social & Organisational). In particular, make careful consideration of how best to present the HFAF to the non-HF practitioner (i.e. military SME and acquisition professionals).

- Ensure level of assessment is based on project maturity (not phase of project lifecycle) and linked to requirements and V&V activities.

Acknowledgments

The research reported in this paper was supported by Dstl contract DSTLX-1000069524.

References

British Standard ISO 9241-210:2010, *Ergonomics of human-system Interaction, Part 2 Human-centred design for interactive systems.* (British Standards)

British Standard ISO/IEC 15288:2002, *Systems Engineering – System life cycle processes.* (British Standards)

Human Factors Integration Defence Technology Centre: *The MOD HFI Process Handbook*, Edition 1, August 2007. (Human Factors Integration Defence Technology Centre)

Humm E, Tutton W and Woolford K, *Human Factors Assessment Framework Working Paper*, Issue 1, March 2009 – February 2010. (UK Ministry of Defence). UNCLASSIFIED

INCOSE-TP-2003-002-03.2, *Systems Engineering Handbook, A guide for System Life Cycle Processes and Activities*, January 2010. (INCOSE)

UK Ministry of Defence, Joint Services Publication 912: *Human Factors Integration for Defence Systems*, Version 2.0, 22nd April 2013. (UK Ministry of Defence). UNCLASSIFIED

UK Ministry of Defence, Defence Standard 00-250, *Human Factors for Designers of Systems.* Issue 1, 23 May 2008. (UK Ministry of Defence). UNCLASSIFIED

TRANSPORT: JOURNEY EXPERIENCE

WILL AUTONOMOUS VEHICLES MAKE US SICK?

Cyriel Diels

Coventry School of Art & Design, Department of Industrial Design, Coventry University, Coventry CV15FB, UK

Autonomous vehicles have the potential to radically change the way we use and interact with our cars. Current thinking assumes that drivers will engage in non-driving tasks and, accordingly, future vehicle design may look dramatically different. However, the use cases envisaged are also known to exacerbate the incidence and severity of carsickness. This paper will discuss these scenarios with reference to the aetiology of carsickness and suggest design constraints to facilitate acceptable future autonomous vehicle design.

Introduction

Maturation, integration and affordability of enabling technologies have turned automated driving into a reality. We have seen Google's driverless car clocking up thousands of accident-free miles and now several US states, the UK and Japan have passed laws permitting (human-supervised) autonomous cars on their roads for R&D purposes. Several car manufacturers including GM, Mercedes, and Nissan, also recently announced their intention to offer semi-autonomous vehicles by 2020. These vehicles provide dual-mode operation whereby, on demand, longitudinal and lateral vehicle control can be handed over to the vehicle. The system essentially combines full range adaptive cruise control with automated lane keeping applying steering actions using electrical power steering. On-road trials are currently also underway to evaluate so-called platoon driving, i.e. the grouping of vehicles maintaining a short time headway achieved by using a combination of wireless communications, lateral and longitudinal control units, and sensor technology. Current concepts under consideration assume a system whereby the platoon is led by a trained, professional driver whilst the following vehicles are driven fully automatically by the system (for an overview of vehicle automation see SMART (2011).

By taking the driver out of the loop, automated personal mobility has the potential to be more efficient, safer, and greener (e.g. Robinson et al. 2010). At the same time, it allows drivers to engage in non-driving tasks. With vehicle control in the hands of the automated system, the driver, now passenger, can sit back and relax, have a coffee, check emails, read the morning paper, or swivel his or her chair and have a face-to-face conversation with other passengers. Vehicle interiors will be designed to become more like social, work, and entertainment spaces.

Besides the critical aspect of liability, there are several human factors issues that require a better understanding to ensure the successful introduction of vehicle automation. Current research activities focus around the topics of transfer of control, situational awareness, HMI design, mixed traffic conditions, system trust, reliability, and user acceptance. There is one aspect, however, that thus far has appeared to have gone unnoticed: *carsickness*.

Susceptibility to carsickness varies widely but it has been found that around 60% of the population has experienced some nausea from car travel, whereas about a third has vomited in cars before the age of 12 (Griffin, 1990). Although the ultimate manifestation of motions sickness is vomiting, it is typically preceded by signs and symptoms such as nausea, headache, fatigue, and drowsiness, which may linger on for hours (Griffin, 1990).

Coincidentally, the use cases that are being envisaged for automated driving are also those we know to lead to increased levels of carsickness. First, automation alters the driver's function from an active to a passive, monitoring one. Secondly, occupants are assumed to engage in non-driving tasks taking the eyes off the road ahead. Finally, flexible seating arrangements may involve rearward facing seats. In the context of carsickness, the common denominator across above scenarios or use cases is the occupants' inability to predict sufficiently accurately the future path of the vehicle which is known to be a main determinant of sickness (e.g. Golding & Gresty, 2013). Following a brief introduction to the aetiology of carsickness, the different use cases and their exacerbating effect on carsickness will be discussed below.

Aetiology of carsickness

Motion is primarily sensed by the organs of balance located in the inner ear and our eyes. Motion sickness can occur when these motion signals are in conflict with one another or when we are exposed to motion that we are not accustomed to (Reason, 1975; Oman, 1982). It can be caused by a wide variety of motions of the body and the visual scene and is a common problem in travellers by car, train, air, and particularly sea. Seasickness may happen whilst being below deck where a clear view of the visual scene outside the ship is lacking. Under these conditions, motion sickness occurs because the movements of the ship, as perceived by the organs of balance, are in conflict with the motion perceived by the eyes, which indicate a static visual surround.

Sickness can however also occur when we are exposed to motion that, from an evolutionary perspective, we are not used to. Our bodies are not accustomed to low frequency oscillating motion. Sea and airsickness, for example, are mainly caused by slowly oscillating vertical motion. Carsickness, on the other hand, is associated with horizontal accelerations (sway) caused by acceleration, braking, and cornering (Guignard & McCauley, 1990; Turner & Griffin, 1999).

With regard to carsickness, it is linear accelerations (sway) in the low frequency bands (0.1-0.5 Hz) that are most relevant and their effects increase as a function of duration of exposure and the intensity of acceleration (Turner & Griffin, 1999). Apart from the route itself, carsickness heavily depends on the way the car is driven. An aggressive driving style involving plenty of accelerating and braking is therefore more likely to result in carsickness. A study on suburban car journeys reported that the fore-and-aft and lateral acceleration motion patterns were similar over the lower frequency range and were provocative in inducing motion sickness. These low frequency fore-and aft and lateral oscillations are more dependent on the driving behaviour of the driver than the characteristics of the vehicle (Griffin & Newman, 2004).

Characteristics of the vehicle mainly affect higher frequency motion. Similarly, road surface quality also affects the high frequency motion vibrations. From this it follows that road surface quality and suspension affect riding comfort, but do not induce carsickness. Exceptions to this rule are cars with particular soft suspensions. In general, when the suspension frequency is below 1 Hz, the likelihood of carsickness significantly increases (Turner & Griffin, 1999). Cars with stiffer suspensions are therefore less likely to lead to cars sickness. Larger amplitudes of lateral (sway) are particularly provocative. As the amplitude of sway tends to increase towards the rear of vehicles (cars and buses), rear seat passengers are particularly prone to car sickness, especially under conditions where external visual views are limited (Turner & Griffin, 1999).

Carsickness and autonomous vehicles

The novel use cases that are being envisaged for autonomous driving are also those we know to significantly increase the incidence and severity of carsickness. First, automation alters the driver's function from an active to a passive, monitoring one. Secondly, occupants are assumed to engage in non-driving tasks taking the eyes off the road ahead. Finally, flexible seating arrangements may involve rearward facing seats. In the context of carsickness, the common denominator across above scenarios is the occupants' inability to sufficiently accurate predict the future path of the vehicle which is known to be a main determinant of sickness (e.g. Golding & Gresty, 2013).

Changing roles: From driver to passenger

With longitudinal and lateral vehicle control automated, the driver is no longer required to actively engage in the driving task. In dual-mode systems where the driver has the choice to drive the vehicle manually or hand over control to the automated system, the driver may still be required to monitor vehicle status to allow for manual override in case of emergencies. In effect, however, the driver becomes a passive passenger.

It is commonly reported that drivers of cars, pilots of aircraft, or Virtual Reality users in control of their own movements are usually not susceptible to motion sickness despite the fact that they experience the same motion as their passengers (Geeze & Pierson, 1986; Reason & Brand, 1975; Stanney & Hash, 1998). This moderating effect of control on the generation of motion sickness symptoms has typically been attributed to the presence of muscular activity. When we initiate a movement, a copy of the movement command sent out by our central nervous system (CNS), referred to as an "efference copy", is used to perform a simulation of the expected results (output or "reafference") of the command. The expected reafference is then compared with the actual sensed reafference within an internal model in our CNS. If there is a discrepancy, for example, a movement command normally used to move our finger to our nose does not produce the intended arm movement due to additional exercise weights added to the wrists, the internal model is updated. In this case, the efferent signal is increased to account for the increased resistance. Taking the weights off subsequently results in arm *overshoot* and thus requires a further recalibration of the internal model. The presence of an efference copy to activate an internal model is thought to facilitate this habituation process. With reference to motion sickness, those in control can benefit from this mechanism to a larger extent and are generally found to desensitise or habituate much faster (Oman, 1982; Reason, 1978; Reason & Benson, 1978; Reason & Brand, 1975; Rolnick & Lubow, 1991; Stott, 1990). Oman (1991) argued that motion stimuli are relatively benign when individuals are able to motorically anticipate incoming sensory cues. However, a fundamental question is whether this anticipatory mechanism is only activated when the perturbation is self-produced or whether this mechanism is also set in motion in case the perturbation is made predictable by sensory information.

An anticipatory mechanism has been explicitly incorporated in the Subjective Vertical-conflict model or SV-conflict model developed by Bles and colleagues (Bles et al., 1998). As in the classical sensory conflict theory (Oman, 1982; Reason, 1978), self-initiated movement results in an efference copy of the command signal sent to the internal model. This subsequently predicts how the body will react, what the sensor responses will be, and which motion and body attitude is to be expected. In the SV-conflict model, however, an anticipatory mechanism is incorporated so that even during imposed passive motion, the internal model is also activated as long as this motion can be anticipated based on sensory information. Therefore, the SV-conflict model predicts that not only drivers but also passengers sitting next to the driver to be less prone to motion sickness, provided passengers have a clear view and looking at the road ahead (Bles et al., 1998; Bles et al., 2000). Note, however, that this does not preclude particularly sensitive passengers from getting sick.

Engagement in non-driving activities

Automated vehicles allow the driver to engage in non-driving activities. It is highly probable that popular activities may include reading, checking one's emails, or engaging otherwise with nomadic or integrated infotainment systems such as in-vehicle displays, laptops, video games, or tablets. On the basis of the sensory conflict theory of motion sickness, one would expect to see an increase in carsickness under these conditions. Similar to reading a map or book whilst driving, the (static or dynamic) image displayed on displays will not correspond to the motion of the vehicle which ultimately may lead to carsickness. This will be particularly true for downward viewing angles or displays that prevent a clear view from the road ahead or horizon. Note therefore that see-through displays may provide one possible solution to minimise the impact of incongruent motion cues.

Research indicates that in-vehicle entertainment systems indeed increase the likelihood of carsickness. Cowings et al. (1999) reported a negative impact on crew performance and health when subjects attended to visual computer screens while the vehicle was moving. More recently, in a study by Kato and Kitazaki (2008), 20 people were driven around for 30 minutes whilst sitting in the backseat either watching the road ahead, or a rear-seat display showing written text. During each of the two drives, the participants were asked to verbally rate their motion sickness on a motion sickness scale that ranged from 0 (*"No symptoms, I feel fine"*) to 6 (*"moderate nausea, I want to stop"*). As expected based on the conflict between the motion sensed by the visual and vestibular system, results confirmed that watching the in-car screen led to significantly higher levels of carsickness.

Flexible seating arrangements

An idea that can be traced back to at least the 50's, autonomous vehicles are considered to provide an opportunity to facilitate social interaction. Numerous concepts for autonomous vehicles suggest flexible interior layouts, which frequently involve swivelling chairs allowing the driver and front passenger to turn to the rear passengers. In the light of the previous sections, it becomes apparent that facing rearwards may not only lead to conflicting sensory information provided by the visual and vestibular system, it also reduces the ability to anticipate the future motion path. Consequently, alternative layouts with rearward facing seats will almost certainly lead to increased levels of carsickness.

Surprisingly, there appears to be no published data to support this contention. This is even more surprising given the fact that rearward facing seats are standard in trains. UK train operators offer customers the option to choose the preferred direction of travel when purchasing pre-booked tickets. This would imply a significant proportion of the customer base to have a preference to travel forward facing. This is in agreement with the anecdotal evidence which suggests that passengers prefer not to face rearwards in order to avoid motion sickness. Facing forwards allows the passenger to anticipate the train's motion to a larger extent than facing backwards even though the available visual information in trains will be limited. Unlike drivers, train passenger will not be able to see the Focus Of Expansion (FOE) which refers to the most informative part of the observers' visual field with regard to the direction of travel.

Design implications for autonomous vehicles

The above discussion points towards two fundamental principles that need to be taken into account to prevent carsickness: (1) avoid sensory conflict where possible, and (2) maximise the ability to anticipate the future motion path. When applied to the design of autonomous vehicles and its anticipated use cases, the following design guidelines are suggested.

Forward and sideways visibility should be maximised. Ideally, occupants have a clear view of the road ahead. However, under conditions that this view is compromised, any visual information (i.e. optic flow) that correctly indicates the direction of travel will reduce the amount of sensory conflict and enhance the ability to anticipate the motion path. The design should therefore aim for maximum window surface areas or Day Light Openings (DLO), minimal obstruction by A-, B-, and C-pillars, and low belt lines or seats of sufficient height to ensure passengers ability to look out of the vehicle. New lighting technologies such as OLED (Organic Light Emitting Diodes) may provide the possibility to provide simulated optic flow patterns inside the vehicle. See-through displays, such as head up displays will reduce the impact of incongruent motion cues. Future research may also explore the effectiveness of using visual, auditory, and/or tactile cues to provide an artificial horizon and signal the future motion path. With regards to seasickness and airsickness, artificial spatial or motion cues have already been shown to alleviate sickness (e.g. Rolnick & Bles, 1989; Tal et al. 2012). The extent to which these techniques can be extrapolated to the automotive field has yet to be determined.

Finally, the occurrence of carsickness in autonomous vehicles will be dependent on the driving scenario. Our organs of balance are in essence biological accelerometers and this means that they are sensitive to accelerations only (Howard, 1982). As a corollary, sensory conflict and hence the likelihood carsickness from occurring, is significantly reduced when travelling at constant speed. The organs of balance signal the body to be stationary and therefore any stationary scene as sensed by our eyes will be perceived as congruent. Under conditions of constant motion, i.e. no lateral or longitudinal accelerations, carsickness is less likely to occur. With respect to the implementation of autonomous systems this would suggest that future levels of carsickness may be manageable provided the automation is not applied under traffic conditions that involve high levels of accelerations as typically observed in urban or rush hour motorway traffic.

References

Bles, W., Bos, J. E., de Graaf, B., Groen, E. and Wertheim, A. H. 1998, Motion sickness: only one provocative conflict? *Brain Res. Bull.* 47(5), 481–487.

Bles, W., Bos, J. E. and Kruit, H. 2000, Motion sickness. *Current Opinion in Neurology* 13, 19–25.

Cowings, P. S., Toscano, W. B., DeRoshia, C. and Tauson, R. A. 1999, *The effects of the command and control vehicle (C2V) operational environment on soldier health and performance (ARL-MR-468,* Aberdeen Proving Ground, MD: U.S. Army Research Laboratory.

Geeze, D. S. and Pierson, W. P. 1986, Airsickness in B-52 crewmembers. *Mil. Med.* 151(12), 628–629.

Golding, J. F. and Gresty, A. 2013, Motion Sickness and Disorientation in Vehicles. In A. Bronstein (Ed.), Oxford Textbook of Vertigo and Imbalance. Oxford: Oxford University Press.

Griffin, M. J. 1990, *Handbook of Human Vibration*: London Academic Press.

Griffin, M. J. and Newman, M. M. 2004, Visual Field Effects on Motion Sickness in Cars. *Aviation, Space, and Environmental Medicine* 75(9), 739–748.

Howard, I. P. 1982, *Human visual orientation*. Chichester, Wiley.

Kato, K. and Kitazaki, S. 2008, Improvement of Ease of Viewing Images on an In-vehicle Display and Reduction of Carsickness. *Human Factors in Driving, Seating Comfort and Automotive Telematics, 2008 (SP-2210, SAE Technical Paper Series 2008-01-0565*.

Oman, C. M. 1982, A heuristic mathematical model for the dynamics of sensory conflict and motion sickness. *Acta Otolaryngol. Suppl.* 392, 1–44.

Reason, J. T. and Brand, J. J. 1975, *Motion Sickness*. London, New York, San Francisco: Academic Press.

Reason, J. T. 1978, Motion sickness adaptation: a neural mismatch model. *J. R. Soc. Med* 71(11), 819–829.

Reason, J. T. and Benson, A. J. 1978, Voluntary movement control and adaptation to cross-coupled stimulation. *Aviat. Space Environ. Med.* 49(11), 1275–1280.

Robinson, T., Chan, E. and Coelingh, E. 2010. Operating Platoons On Public Motorways: An Introduction To The SARTRE Platooning Programme., Seventeenth World Congress on Intelligent Transport Systems, 25–29 October 2010, Busan, Korea.

Rolnick, A. and Bles W. 1989, Performance and well-being under tilting conditions: the effects of visual reference and artificial horizon. *Aviat. Space Environ. Med.* 60(8), 779–85.

Rolnick, A. and Lubow, R. E. 1991, Why is the driver rarely sick? The role of controllability in motion sickness. *Ergonomics* 34(7), 867–879.

SMART 2011, Definition of necessary vehicle and infrastructure systems for Automated Driving. Final report SMART 2010/0064, European Commission.

Stanney, K.M. and Hash, P. 1998, Locus of User-Initiated Control in Virtual Environments: Influences on Cybersickness. *Presence* 7(5), 447–459.

Stott, J. R. R. 1990, Adaptation to nauseogenic motion stimuli and its application in the treatment of airsickness. In G. H. Crampton (Ed.), *Motion and Space Sickness* (1 ed., pp. 373–390, Boca Raton, Florida: CRC Press.

Tal, D., Gonen, A., Wiener G., Bar, R., Gil, A., Nachum, Z. and Shupak, A. 2012, Artificial horizon effects on motion sickness and performance. *Otol Neurotol.* 33(5), 878–85.

Turner, M. and Griffin, M. J. 1999, Motion sickness in public road transport: the relative importance of motion, vision and individual differences. *Br. J. Psychol.* 90(4), 519–530.

CAR AND LIFT SHARING BARRIERS

A. Woodcock, J. Osmond, J. Begley & K. Frankova

Integrated Transport and Logistics Grand Challenge, Coventry University

The current emphasis on congestion, pollution, quality of transport and the spiralling costs of energy has led to a growth of interest in social forms of transport, such as car and lift share schemes. Together with greener fuel and use of technology, it is possible to widen transport choices available. However, there has been user reluctance to join car/lift share schemes, even though these would offer benefits: lower costs, less congestion and carbon reduction. A pilot study was undertaken to understand the UK current situation and to inform debate about future developments in social transport.

Introduction

The importance of car ownership was highlighted in the 2010 UK National Travel Survey, in which 64% of respondents stated that their main mode of transport was either car or van, as a passenger or driver. Access to personal transportation is seen as 'an essential gateway' to critical services and social interaction, with those owning vehicles travelling nearly twice as much as non-car owners.

Private car availability brings numerous benefits, such as mobility, freedom and convenience, but its continued growth is at odds with sustainable development. It also brings a variety of challenges; congestion, parking, ownership costs, use and maintenance, detrimental land usage. A culture of single driver occupancy has developed and in Britain occupancy rates have largely remained static since the early 1990s, at approximately 1.6 occupants per car per stage of each trip. In 2010, 61% of each stage of a car trip was undertaken by a single driver.

The aim of this research was to provide a contemporary picture of the potential for car/lift sharing schemes as alternatives to personal motor ownership and to recommend how such schemes may be made more attractive to consumers.

Definitions: for the purposes of this paper, car sharing refers to schemes in which a vehicle is owned by a person who is not the driver, for example a workplace scheme, car hire/rental scheme or community-based schemes (car pooling/bicycle sharing as an integral part of housing scheme). Lift sharing can include both informal/formal schemes in which a vehicle owner shares (free, to split costs or for profit) the vehicle on an ad hoc basis, by prearrangement or via portals/social media.

Car sharing

Car sharing offers alternatives to vehicle ownership by providing access to a car when other modes of transport are not possible or are inconvenient (Katzev, 2003). It exists in the space between taxis and rental cars. Members can access cars for brief trips, some as short as a half hour (Steininger et al, 1996; Katzev, 2003) with a less 'formal' booking procedure than for rental cars.

The benefits of car sharing can be placed into four main categories: for transportation: it can reduce vehicle ownership and stimulate uptake of more sustainable and active forms of transport; environmentally, reduced car ownership may decrease the miles/kilometres travelled and thus reduce traffic congestion and lower greenhouse gas emissions; for land use, fewer cars means less demand for parking spaces; for social effects, households benefit by avoiding full cost of ownership and can also realise the true cost of travel.

Motivating factors which increase car sharing membership (Steininger et al, 1996; Meijkamp, 2000; Katzev, 2003) include financial savings, environmental concerns, public transport inconvenience, lack of access to/age of car, residential parking problems, significant lifestyle or mobility changes, differentiated car choice, eligibility for car share from a variety of locations.

The competitive edge of the private car is the main obstacle faced by car sharing organisations. Car ownership may be linked to lifestyle or identity and this may be seen as more valuable to the owner than cost savings. However, this could be addressed by policies that '*work towards increasing the number of people for whom the shared use of cars is a significant promoter of their identity*' (Prettenthaler & Steininger, 1999: 450). Other, more practical concerns include unavailability of car when needed, last minute booking, advanced planning and reservation, distance to closest car sharing vehicle, clean car policies, paperwork, membership requirements etc (Meijkamp, 1998; Katzev, 2003).

Loose et al (2006) investigated demand for car sharing with 1000 German respondents and revealed a considerable lack of awareness of car sharing, with 53% unable to explain the term. Awareness was above average in cities with over 500,000 inhabitants and in those with higher incomes and education. However, the majority were unaware of local car sharing offers.

There is also an emphasis on the role of public transport in assisting car sharing (Wagner and Shaheen, 1998, Bonsall 2002) as it '*becomes the backbone of daily mobility for many car sharing users*' (Loose et al, 2006: 368). The non-availability or inconvenience of rural public transport may leave car transport the only feasible option. Car sharing and public transport should complement each other, and schemes should establish partnerships to benefit both organisations.

Loose et al (2006) found that those more likely to engage in car sharing tended to be women, cyclists, regular public transport users, public transport season ticket holders and those with car sharing experience. They claimed that '*among the 26–35 year olds, the proportion of potential car sharing customers is especially rare*'.

This may be due to starting careers/families creating a higher interest in car owner-ship. Overall, car sharers are more likely to be younger-middle aged women, highly educated, relatively environmentally conscious with slightly higher incomes. Fur-ther, a 'balanced' mix of users is important as neighbourhood users typically reserve vehicles on evenings/weekends, and business users from Monday to Friday, dur-ing daytime hours. Hence, the two user groups complement each other to ensure maximum usage of vehicle fleets.

Lift sharing

The DfT (2005: 18) admit that there is *'a great variety of perceptual barriers to car sharing [sic], all of which can be tackled'* and to address this there should be the opportunity for trial periods before commitment. Some of the concerns include: fear of sharing with strangers, personal safety and security issues; safety of cars parked on the street; poor driving/speeding; sharing payment and travel costs; gender of car sharing partner; trust and security; interaction within the car.

In common with car sharing, one of the barriers is limited knowledge of these schemes. Kearney and De Young (1995: 652) claim that *'lack of information ... the desire to avoid uncertainties or unfamiliar situations play significant roles in the decision not to adopt alternatives to solo commuting'*. Therefore, information on what to expect and clarity in promotional materials are essential.

Teal's (1987: 204) review claims that the propensity to carpool is *'affected more by attitudinal factors than socio-demographic attributes.'* Household vehicle avail-ability plays a role. Individuals are unlikely to stop solo commuting due to the inconvenience/inflexibility of lift sharing and restrictions on independence (Teal, 1987; Bonsall, 2002). Five out of nine of all commuters tend to take relatively inexpensive car trips and thus are not motivated to carpool. The solo driver domi-nance also suggests that carpooling and public transport are seen as inferior choices. Bonsall (2002) argues that informal lift sharing occurs when car ownership is low and public transport unsatisfactory, implying a multi-faceted relationship between carpooling and public transport.

Populations likely to engage in organised lift sharing have no developed network of lift sharing and have the following characteristics (Bonsall 2002); close-knit communities; low car ownership areas/poor public transport/absence of local facil-ities; areas where trip patterns are amenable to lift giving; areas with high parking charges and effective incentives for car sharers.

Potential markets

The following markets have been identified for both car and lift sharing.

Rural markets: Some rural neighborhoods run car sharing schemes or demand responsive transport, and family, friends/neighbours providing lifts are significant, particularly to the young and elderly, resulting in fewer public transport journeys. Gray et al (2006) explored the relationship between social capital, rural mobility and social exclusion, concluding that lift giving is common in tight-knit communities and should be supported with transport subsidies.

Young people and lift sharing: Young people in rural areas are dependent on others to provide lifts to work or public transport stops. Problems with late night transport provision have implications for social life. Dependency on parents' means autonomy may be compromised, so a lift-sharing scheme may be a better option. (Storey and Brannen, 2000: 2).

On-line lift sharing portals: Sharples et al (2012) in a study investigating user attitudes/requirements for technologies to support car sharing, found that flexibility and ease of use of online portals were important: '*If technologically-enabled solutions to socially-connected travel are to be realised, then human factors has a central role to play in involving users, and the factors that constrain, shape and motivate their travel choices, in the design and implementation process*' (9).

Method

An on-line survey and interview study were conducted over 6 months, 2012/2013 and 209 surveys were fully completed and analysed. The majority of the responses (86%) were from those who did not car/lift share (r = 174) followed by 15% (r = 31) who did. This is despite a large proportion (36%) of responses from Coventry University, which has its own car/lift share scheme. This seems to indicate that the majority do not car/lift share even if they have access to a scheme. Differences between the respondents are presented in Table 1.

Table 1.   Results.

	DO NOT participate in car/lift share	DO participate in car /lift share
Sample size	R = 174, gender split relatively equal. The largest age range was 36–45, followed by 26–35 and 46–55 at 19% respectively. Eighteen per cent were 17–25; 14% were 55–64. They were typically white British (82%); 92% did not declare a disability and (70%) lived in an urban area.	R = 31; none used car sharing schemes. 58% selected lift-sharing; 42% ad-hoc arrangements. Respondents were white British (74%), did not declare disability (94%); lived in an urban area. (61%). Mainly female (68%): age range 45–55 (32%); 26–25 (23%).
Scheme	44% did not work in an organisation which offered a scheme: 36% did; 21% were unsure.	Of those who participated in a regular scheme, 50% worked for an organisation that provided one but many were vague about details.

(Continued)

Table 1.   (Continued)

	DO NOT participate in car/lift share	DO participate in car /lift share
Non drivers	Cost (learning, purchase and insurance) were deterrents. Access to family/friends who provide lifts and availability of other forms of transport were important. Benefits of non-driving were reported as more exercise, less stress, environmental benefits, low costs and increased productivity (working on train or bus). Drawbacks included bad weather, delays, prices, safety issues, deserted stations, lack of public transport to some destinations and travel time needed when using different transport during same journey.	For non-drivers the costs of buying a car/insurance important to not driving. Two out of three felt that the cost of learning to drive, access to other forms of transport and 'family/friends drive me' were important. Of equal importance were 'busy/congested roads' and 'environmental reasons'. Only one out of three was 'not interested in driving or 'too busy to learn', and of low importance were 'too young', 'too old', 'physical difficulties/disabilities' and 'put off by theory/practical test'. These respondents used a mixture of transport: bus, bicycle, car or train.
Drivers	93% could drive, with majority (74%) driving for 10 years or more. 83% owned a private vehicle and stated that the biggest advantage was freedom and convenience. 63% cited the cost of car ownership as a major issue.	90% were drivers, with 74% driving for 10 years plus; 86% owned a private vehicle. Most cited advantage was freedom/convenience (41%), followed by 'better/cheaper than public transport'. The most cited disadvantage was cost (76%).
Participation	Drivers/non-drivers – main reason for non participation: lack of spontaneity/mismatching work patterns (35%), lack of awareness (24%), privacy, driving habits, location, cost.	Primary motivation – to get to work (67%); secondary motivation – to get to an educational institution, to work, or go shopping (all at 18%).
Costs/benefits	Costs – inflexibility (45%), reliance on others (26%) and lack of privacy (17%) Benefits – cost savings/environment (16%)	*Lift sharers*: Costs – ties to someone's schedule; increased journey length; others' location; no dedicated/discounted parking Benefits – 67% cost savings (petrol/parking); greater flexibility and freedom; social aspects; space to solve work related issues; environment. *Ad hoc lift sharers*: Costs – less flexibility in departure and arrival, additional journey length, lack of a car 'going the way I want to go'. Design of sites a barrier – 'clunky to use, passwords and log-ins do not work smoothly, lots of people can't be bothered to join.' Benefits – company, cost savings.

Results

In open-ended questions, 71% of those not currently lift/car sharing fell into the '*it is a good idea, but* ...' theme, indicating that if a scheme was practical, it might be considered. However, lack of practicality – less flexibility/independence, rural living, problematic work patterns, child-care needs – precluded use of schemes, alongside a lack of awareness of how such schemes work. Comments included: 1) *"Although I am vaguely aware of [a] scheme – it has not been well advertised and I don't know how to investigate using it."* 2) *"Any kind of car share system has to have minimal task loading to make arrangements – no 'systems' you log onto to ask for lifts, no apps, no ambiguity on payment rates."* 3) *"Segregate the car parks by destination like a bus terminal and people can ask for lifts or informally approach people going to the same place they live."*

Therefore, the recommendation is that a concentrated publicity campaign for any newly introduced scheme is undertaken well before a launch.

Twenty-three follow up interviews took place, which included eleven interviewees who did car/lift share, ten interviewees who did not, one technology platform provider and one car/lift sharing organization. Barriers included: wariness of sharing with strangers e.g. opposite gender, conflicts about internal environment (e.g. radio) and poor driving skills; mismatch of work schedules, including partner reliability; reluctance to give up car ownership and autonomy; lack of awareness of schemes and how they work; overcomplicated joining procedures.

The overall recommendations from the interviews was that all schemes must offer a simple technology platform that evidences clarity of purpose – including benefits, safety measures and incentives (such as a cost calculator as a possible trigger point for change) which should be evidenced in all promotional materials.

Conclusion

The importance of car ownership highlighted in the UK National Travel Survey (2010) has been borne out by this study into people's perceptions/use of social transport with the majority of responses indicating that social transport is seen as a poor alternative to the convenience and independence of car ownership.

However, there were indications that some would consider alternative transport if information and benefits of schemes were clearly promoted. For example, highlighting the cost-saving possibilities of using social transport could prove to be a trigger point in changing behaviour. Of greater interest is community based social transport, where home-buyers choose a car sharing integrated lifestyle. Such schemes have not been very successful to date, but this may change due to urban regeneration policy drivers, fuel increases, fuel availability reduction and widening austerity. These require increased investment in public transport and new forms of mobility: linking to initiatives such as HS2 will also be key.

There were also clear indications that the technology used to promote social transport schemes had a pivotal role to play in attracting/keeping potential users. Some participants were put off by clunky technology and others were wary of car sharing with strangers. A simple to use interface and rating system could address trust, and comfort and compatibility between social transport users.

In conclusion, EU transport policy suggests a greater push for integrated schemes/more sustainable transport of which car/lift sharing is an example. The market will continue to support start-ups and larger companies, but as the existing platform software tends to be from the same provider/all schemes need a large pool of viable customers, buy outs/amalgamations will become common. Key players are emerging and there is potential for taxis and car hire/leasing companies to enter this arena as a natural extension/evolution of their business.

For OEMs, potential lies in providing a diverse range of vehicles to suit different journey types and market sectors. Additional attention should be given to in-vehicle telematics, integration of safety features (e.g. car tracking and driver/passenger recognition) and recharging/refuelling capabilities. Car pooling will also demand greater attention be placed on HMI, rapid recognition of driver characteristics, preferences and driving styles. Changing models of car ownership mean that drivers may be unfamiliar with layout, operation and car performance. OEMs need to consider long term effects of a changing pattern of car ownership as well as wide range of factors related to sustainability.

Acknowledgement

This research was supported by the Transport KT, and OandS Technologies, with the help of Coventry University, SME Engagement Fund.

References

Bonsall, P. 2002, *Car share and car clubs: Potential impacts – Final report*, prepared for the DTLR and the Motorists Forum, University of Leeds: Institute for Transport Studies

Cairns, S. 2011, *Accessing Cars: Different ownership and user choices*, Royal Automobile Club

Department for Transport, 2005, *"Making car sharing and car clubs work: A good practice guide"*

Gray, D., Snow, J. and Farrington, J. 2006, "Community transport, social capital and social exclusion in rural areas", *Area*, 38 (1), 89–98

Katzev, R. 2003, "Car sharing: A new approach to urban transportation problems", *Analyses of Social Issues and Public Policy*, 3 (1), 65–86

Kearney, A. R. and De Young, R. 1995, 'A knowledge-based intervention for promoting carpooling', *Environment and Behaviour*, 27 (5), 650–678

Loose, W., Mohr, M. and Nobis, C. 2006, "Assessment of the future development of car sharing in Germany and related opportunities", *Transport Reviews*, 26 (3), 365–382

Meijkamp, R. 1998, "Changing consumer behaviour through eco-efficient services: An empirical study of car sharing in the Netherlands", *Business Strategy and the Environment*, 7, 234–244

Prettenthaler, F. E. and Steininger, K. W. 1999, "Analysis – From ownership to service use lifestyle: The potential of car sharing", *Ecological Economics*, 28, 443–453

Sharples, S., Golightly, D., Leygue, C., O'Malley, C., Goulding, J., and Bedwell, B. 2012, "Technologies to support socially connected journeys: Designing to encourage user acceptance and utilization" in D. de Waard, N. Merat, A.H. Jamson, Y. Barnard, and O.M.J. Carsten (Eds.) (2012). Human Factors of Systems and Technology (pp. 1–11). Maastricht, the Netherlands: Shaker Publishing

Steininger, K., Vogl, C. and Zettl, R. 1996, "Car-sharing organizations: The size of the market segment and revealed change in mobility behavior", *Transport Policy*, 3 (4), 177–185

Storey, P. and Brannen, J. 2000, *Young people and transport in rural areas – summary report*, Joseph Rowntree Foundation

Teal, R. F. 1987, "Carpooling: Who, how and why?" *Transportation Research* A, 21, 203–214

Wagner, C. and Shaheen, S. 1998, "Car sharing and mobility management: Facing new challenges with technology and innovative business planning", *World Transport Policy and Practice*, 4 (2), 39–43

MEASURING QUALITY ACROSS THE WHOLE JOURNEY

Andree Woodcock[1], Nigel Berkeley[1], Oded Cats[2],
Yusak Susilo[2], Gabriela Rodica Hrin[3], Owen O'Reilly[4],
Ieva Markuceviciute[5], & Tiago Pimentel[6]

[1] *Coventry University, UK*
[2] *KTH Royal Institute of Technology, Sweden*
[3] *Integral Consulting, Romania*
[4] *Interactions Ltd, Ireland*
[5] *Smart Continent, Lithuania*
[6] *VTM, Portugal*

Many countries are looking to public transport to alleviate problems of congestion and pollution and increase sustainability. In order to develop a large modal shift in traveller behaviour, transport providers and planners need to deliver a high quality passenger experience. This paper firstly introduces the EU funded METPEX project, the aim of which is to develop a Pan European tool to measure the quality of the whole journey experience, and secondly discusses the results of stakeholder interviews in the UK, which show where such a tool might fill gaps in existing knowledge of passenger behaviour and mobility requirements.

Introduction

The Lisbon Strategy set a goal for the European Union to become the most competitive and dynamic knowledge-based economy in the world, capable of sustainable economic growth with more and better jobs and greater social cohesion. Public transport has a central role to play in ensuring equitable access to social, economic, educational and health services. Its effective use is also seen as key to reducing urban congestion and greenhouse emissions. However, public transport is regarded by many as an inferior form of transport. If public transport is to realise its full potential, a modal shift in traveller behaviour is required, from private to public transport. This will contribute to a significant reduction in the annual costs of road accidents, congestion, energy consumption and pollution, thus releasing funds for economic development, whilst meeting new political challenges such as climate change, energy policy, air quality legislation and the difficulties of tackling congestion.

The question of how to enhance mobility while at the same time reducing congestion, accidents and pollution is a common challenge in all major cities in Europe. Efficient and effective urban transport can significantly contribute to achieving objectives in a wide range of policy domains. The success of policies and policy objectives that have been agreed at EU level, partly depends on actions taken by national, regional and local authorities.

Although much research has been conducted on the integration of transport modes and travel information on the one hand, and travel behaviour and demand analysis on the other, these two streams of knowledge have not had adequate interaction hitherto. Furthermore, key stakeholders such as Transport Authorities, urban and regional local government and transport operators need to have access to standardised methods and results in ways that can inform policy and analysis, enabling the implementation of integrated transport systems and increased accessibility.

A key factor for all transport operators, regardless of mode, is the quality of the passenger experience as this in part effects the uptake of the service. However, it is naive to assume that the quality of the vehicles and service is sufficient in itself to encourage this modal shift without a link to policy and urban planning.

METPEX (ww.metpex.eu), is a three year, FP7 funded project developing a Pan European tool to measure the quality of the passenger experience. The European consortium draws its membership from UK, Ireland, Portugal, Spain, Italy, Greece, Lithuania, Rumania, Sweden and Switzerland, encompassing a wide range of transport scenarios. The focus of the project is on the measurement of the quality of the passenger journey (door-to-door), including private or individual forms of transport such as walking, bicycling and car sharing.

The primary objective is to develop, validate and evaluate a standardised tool to measure passenger experience across whole journeys, to develop best practice and benchmark services in a standardised manner. The information derived from the use of the METPEX toolset will ultimately be used to inform policy makers in providing inclusive, passenger-oriented integrated transport systems that are accessible by all citizens. Specific objectives include:

1. Developing an integrated approach to the measurement of the whole journey passenger experience that takes into account human (physiological, perceptual, cognitive, sensory and affective) socio-economic, cultural, geographic and environmental factors.
2. Assessing the costs of 'inaccessible transport' for different sectors of society (such as those from low income groups, rural communities, the elderly, disabled and those with lower levels of literacy).
3. Assessing the extent to which the measurement of the passenger experience can be used to drive innovation and attention to transport quality from the customer's perspective in the transport industry.
4. Evaluating the passengers experience from different regions of Europe and to support the integration of regional transport networks into an European transport network
5. Facilitating the harmonization of travel behaviour research and analysis across European Union Member States.

Providing an accessible transport service for all is important to ensure people are not excluded from reaching places of employment and health, education and leisure services, and in maintaining a high quality of life for Europe's diverse communities. However, different travellers have different needs and priorities. Some individuals have more complex travel-activity patterns than others as they participate in daily activities at different times and in different locations (e.g. Golledge and Gärling, 2004).

Whilst guidelines and standards aimed to accommodate the different needs of different travellers have been established, there is still a lack of knowledge on what is really valued by different groups of travellers who use different travel modes, and the requirements of those who do not use public transport. Moreover, previous studies often ignore the impact of the access and egress legs on the overall travellers' journey satisfaction. Taking a holistic approach to the study of the passenger experience and journey satisfaction, not only from users' perspective but also stakeholders', will provide an important bridge between action and intention to use more sustainable travel mode (Friman et al., 2011).

Prior to developing the METPEX tools (e.g. mobile app, on-line questionnaire, focus group and interview schedules) to measure the overall quality of the passenger experience, it was important to build on knowledge already gained in previous projects (such as those conducted under the CIVITAS programme) and to understand what information is currently recorded by transport planners, authorities and operators and where the gaps are. From these studies a comprehensive set of variables could be developed from which tools could be derived.

In order to understand the current state of the art with regard to the measurement of the whole journey experience, a series of semi structured interviews were held with stakeholders across Europe to ascertain what journey information was currently being conducted. The results presented in this paper, provide an overview of this research, with especial reference to the situation in the UK.

Data collection

The stakeholder interview survey was launched in April 2013 for 6 months across ten cities – Bucharest, Dublin, Grevena, Rome, Stockholm, Turin, Valencia, Coventry, Vilnius and Zurich – and one European body – the European Disability Forum (see http://www.edf-feph.org/). Stakeholders include local transport authorities, transport operators, government bodies, municipalities, passenger interest groups, those responsible for different aspects of the network and major regional employers.

A standardised interview schedule was drawn up to enable control of data gathering across the different countries. Questions related to the extent to which the stakeholder was informed about the perceived quality of service (such as vehicle and station comfort, network comprehension); the importance of such information; their knowledge about the passenger journey (e.g. origin, final destination, travel and waiting times); the importance they placed on that information; accessibility of the service, infrastructure and network (including travel information and ticketing provision, stations, vehicles and interchanges) for those with reduced mobility; the likely impact a project such as METPEX might have in increasing customer trust and patronage, operating costs, policy and employee satisfaction; and interest in benchmarking.

The interview also gathered stakeholders' views on a pilot METPEX passenger survey questionnaire (developed from the initial literature review, Cats et al, 2014) on important travel experience factors, target user groups, common practices and policy priority areas. Stakeholders discussed which variables were important from their perspectives, identified variables that may be missed or unique from city to city and offered suggestions concerning questionnaire format and survey design.

In the UK, the interview was delivered in a semi-structured manner in the stake-holder's office. Respondents were encouraged to read the questionnaire beforehand and provide supporting material (e.g. data collection tools).

Results

EU stakeholder perspective

A total of 45 stakeholders interviews were carried out, of which 12 were planning authorities from municipal, regional and national levels, 17 public transport agencies and operators, nine non-governmental special interest groups and seven miscellaneous (e.g. national research institutes). An overview of the result is provided in Table 1. Most the stakeholders appreciated and saw the benefit of the METPEX tool. They believed that the identification of passengers' travel needs could become essential for the future action plans and contract design for transport service modernisation. Further, they identified that one of the most useful outcomes of the project would be to quantify and rank the contribution of each quality factor to the overall experience and use this to guide investment in service improvement measures. One of the stakeholders highlighted the potential of this project to provide a better and more (economically) efficient service during the period of economic crisis.

As expected, different questions were valued differently by different classes of stakeholders. Operators were mostly interested and concerned about the impact of detailed level-of-service related variables on passenger experience (e.g. the use of travel information, time utilisation whilst on-board, more detailed impacts of disruptions, detailed trip pattern, etc.), whilst they expressed less interest in the overall satisfaction of whole journey and questioning the value of quantifying the impact of past poor experience. Furthermore, they also showed limited interest in those variables that could neither be used in understanding their customers' behaviour nor in detailed planning processes (e.g. the trip satisfaction can be improved in detailed level).

Table 1.   Variables most valued by different stakeholder groups.

Operator	Authorities	Specific Needs Groups	Other
Subjective well being			Subjective well being
	Attitudes and opinions towards model specific preferences, social norm, transfer preference, traffic congestion and pollution, safety and security while travelling		
	The main purpose of the trip		
	Carrying heavy or bulky items when travelling		
	Familiarity with the trip		
	Trip arrival constraint		
	Access to public transport card		
The use of pre trip information			
			Satisfaction level of current choice
		The occurrence and impact of disruptions	
Detailed trip stages, including waiting and on-vehicle time and speed, travel time and punctuality		Detailed information on perception of time reliability	Detailed trip stages, including waiting and on-vehicle time and speed, travel time and punctuality
Information acquisition		Information acquisition	
Time utilization on board and at stops		Time utilization on board and at stops	
	Overall satisfaction in general, compared to expectations, towards other mode choices and travel modifications		
Parking price and easiness to find parking spot	Travel experiences among car travellers, which include the reliability, travel time, speed and information provision, parking provisions and fees		
	Travel experiences among cyclists, which include the feeling of safe and being prioritised on the road, availability of the relevant information, route connectivity and the availability of bicycle parks at the destinations		
	Travel experiences among pedestrians, which include the quality and design of the pedestrian paths, feeling secure and safe while walking and the availability of relevant information		

(*Continued*)

Table 1.   Continued.

Operator	Authorities	Specific Needs Groups	Other
Open Suggestions to improve travel experience			
			Gender, age, disability, household composition, income and education information
Special group needs including way-finding, accessibility, stress, travel information and lighting			
Passenger satisfaction on: service availability (frequency and stop location), travel speed (both subjective and relative speeds), information at stations and on-board, information about ticketing, comfort (quality on on-board, fellow travellers, seat availability, seat comfort, easiness to buy ticket, crowding both at stops and on-board, station facilities), appeal (physical environment, vehicle quality, cleanliness both at stations and on-board), safety (at stops and on-board), overall reliability (including regularity and punctuality), personnel availability at stops and on-board, price (value-for-money and fairness), connectivity (network-wise and easy transfer), travel sickness, and environment issue.			

In contrast, the planning authorities were more interested with wider general urban and public transport planning issues and the multi-modal travel patterns (e.g. different impacts of level-of-service for different travel modes and trip purposes). This is expected given their responsibility is to improve the transport service at the network and city (or even bigger) level. They were also interested in the impact of congestions and pollution in general and what particular conditions and locations public transport is needed from a planning perspective.

The special interest groups were understandably more interested with their constituents' interests. Some argued that the questionnaire was not detailed enough to explore the disabled travellers' needs. For example, instead of asking 'whether the vehicle design has fulfilled their needs', detailed information should be collected about the quality of the space, ramps etc. Similar reactions were also received from car interest groups. Further, they also found personal related information such as asking age, education and disability level of the travellers to be sensitive.

The 'other' group of stakeholders was mainly government research institutes, many of whom were interested with more detailed trip patterns and behavioural variables that underlie the travellers' decision-making processes in order to inform policy decisions. They were also interested in multidisciplinary issues such as the role of subjective well-being, stress and the impact of travellers' time constraints.

Local stakeholders highlighted the benefits of having similar questions that are comparable with ones that they already use.

UK based results

Public transport users in the UK have the opportunity to comment on the quality of public transport regularly through market research and evaluation conducted by transport authorities (such as CENTRO and Transport for Greater Manchester), Passenger Focus (the independent passenger watchdog which conducts national surveys on buses, trains and trams), surveys conducted by transport operators such as Virgin, Arriva and National Express amongst others and major regional employers (such as Birmingham International Airport).

Although much information is collected on the passenger experience by different stakeholders, there is still a shortage of information about why people do not use public transport and the quality of the whole journey itself. It is argued that the wider traveller experience, including intermodal travel, the experience of travelling to the transport node, and the journey from the transport to the final destination, or the next stage of the journey may influence and restrict use of public transport. Currently information tends not to be collected by the transport operator if it is about issues that are beyond their control about, for example interchanges, stations or bus stops. They may include questions on it (for completeness) and pass information on. Indeed, most transport operators were only interested in the perceived quality of their service. The municipalities and regional transport authorities were more concerned with the overall quality of experience and the traveller's journey (for example the provision of way finding information, tactile pavements and dropped kerbs. Many respondents were starting to consider social media or apps as a way of collecting travel information, but were not sure how to use the information or how representative it was. Additionally, certain groups may be underrepresented in the surveys such as women with young children and those who did not use public transport.

Therefore, there is a clear role for METPEX in investigating the quality of the whole journey experience as experienced by an individual traveller, paying particular attention to intermodality and those parts of the journey conducted by more active forms of transport. For the municipality and local transport authority, detailed local information is important in addressing need and planning new transport provision. Current surveys may not be able to capture the level of information needed to start planning for new models of mobility.

Discussion

Whilst the stakeholders may fully aim to provide the best service for the users, the interviews highlighted that the factors they consider are much narrower than those that were empirically found to be significant for travellers. For example, it seems that the operators were more interested and concerned with the impacts of detailed level-of-service related variables (e.g. the use of travel information, more detailed impacts of disruptions), whilst less interested with the overall satisfactions of whole journey and questioning the usefulness of measuring the quantity of poor experience in the past. Furthermore, they also appreciate less the variables that are too general and neither can be used in understanding their customers' behaviour nor in detailed planning processes (e.g. the trip satisfaction can be improved in detailed level).

Acknowledgements

METPEX is funded under FP7-SST-2012-RTD-1, under the SST.2012.3.1-1. Research actions regarding the accessibility of transport systems work programme. The authors would like to acknowledge the contribution of all partners from Interactions (Ireland), SIGNOSIS (Belgium), ITENE (Spain), ZHAW (Switzerland), Euokleis s.r.l. (Italy), POLITO (Italy), ANGRE (Greece), KTH (Sweden), INTECO (Romania), FIA (Belgium), VTM (Portugal), SmartContinent (Lithuania), SBOING (Greece), TERO (Greece), RSM (Italy).

References

Cats, O., Susilo Y., Hrin, R., Woodcock, A., Diana, M., Egle Speicyte, E., O'Connell, E., Dimajo, C., Tolio, V., Bellver, P., and Hoppe, M. (2014), An Integrated Approach to Measuring the Whole Journey Passenger Experience, *Transportation Research Arena*, Paris.

Friman M., Fujii S., Ettema D., Gärling T. & Olsson L.E. (2011). Psychometric analysis of the satisfaction with travel scale. In: *Presentation at the 9th Biannual conference of Environmental Psychology in Eindhoven*, pp. 26–28, The Netherlands.

Golledge, R. & Gärling T. (2004). Cognitive Maps and Urban Travel. In D.A. Hensher, K.J. Button, K.J. Haynes and P.R. Stopher (Eds.), *Handbook of Transport Geography and Spatial Systems* (Volume 5, pp. 501–512), (Elsevier, Oxford).

TRANSPORT:
OPERATIONS

DESIGNING VISUAL ANALYTICS FOR COLLABORATIVE ACTIVITY: A CASE STUDY OF RAIL INCIDENT HANDLING

Tom Male & Chris Baber

Information Management/Group Business Services, Network Rail, Rugby,
Electronic, Electrical and Computer Engineering, University of Birmingham

Our goal in writing this paper is to consider how user interfaces for decision support could be developed using the concept of Influence Diagrams. Working from the question of how to support Visual Analytics in the rail domain, we observed personnel in Incident Control Centres in order to appreciate the challenges that they face and the information to which they currently have access. From these observations, we developed a set of decision rules which were captured in the form of Influence Diagrams and these were used to inform user interface designs. It is proposed that the approach provides a complimentary approach to Ecological Interface Design by reflecting key elements of decision tasks in the interface design.

Introduction

Visual analytics involves the combination of automated data analysis techniques with interactive visualisation. The role of the 'analytics' part of this process relates to the automatic processing of the data in order to spot trends and patterns in the mass of data and to provide recommendations on which course of action might be most appropriate. The role of the human, then, becomes potentially a consumer of the output of the complex analytics processes. This raises the question of how best to design visualisation for human decision making in complex, dynamic domains. This is, of course, a perennial challenge for ergonomics and much of this work has echoes of interface design for complex, dynamic systems, such as control rooms. Data are collected from a variety of sources some of which can be imprecise or with variation in the quantity of data to be analysed.

In this paper, we report a study which uses a combination of task analysis with decision modelling to develop a novel user interfaces. The study is based on incident handling in rail networks which, while not currently a domain in which visual analytics is applied, provides a challenging, complex, dynamic system with data arriving from many sources. The role of the operators in this system is to assimilate these data and then make appropriate routing decisions. Our focus lies in the ways in which this decision can be supported through the visualisation of the problem domain.

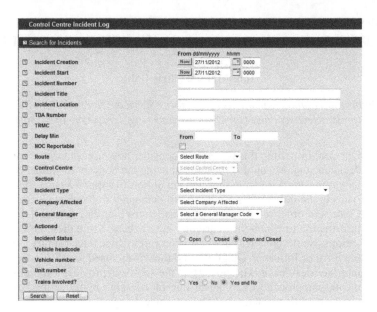

Figure 1.   User Interface to Control Centre Incident Log.

Figure 1 gives an example of the user interface, which allows operators to record details about incidents. Several of the operators interviewed commented negatively on the usability of this interface (and the associated system). We are cognisant of the ongoing developments in the rail industry to review and improve the user interfaces in many of its control operations (Wilson and Norris, 2005). The aim of this paper is not to conduct an overview of contemporary user interfaces in the railway industry, but rather to explore the potential for using systems dynamics models to develop novel designs of user interface. Our focus on incident handling is to explore a domain of activity, which involves a degree of uncertainty and decision making which could be supported through the design of user interfaces, which support not only information retrieval and display but also decision making. In part, our approach can be extended to other forms of control room and is intended to complement cognitive work analysis and ecological interface design (Vicente and Rasmussen, 1992). This latter point will be revisited in the discussion section of this paper.

Incident handling in the (UK) rail domain

Within each of the Integrated Electronic Control Centres (IECC) in the UK, information is received by the incident controller (Figure 2). The coordination of network traffic is managed by the train running controller, who will oversee the traffic

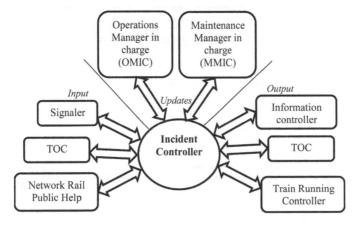

Figure 2.   Communications Links in LNW IECC.

conditions of the region and make decisions regarding the 'pathing' and regulation of trains. Information is generally received from a signaller, a public help line or Train Operating Company (TOC) staff. For minor incidents, information is processed and logged on the Control Centre Incident Log (CCIL) and Fault Management System (FMS) by the incident controller. In larger incidents, responsibility is passed to an Operations Manager In Charge (OMIC) or a maintenance manager in charge (MMIC). In addition to CCIL, staff use an application called *Control Centre of the Future (CCF)*, which visualises the real-time state of traffic on the network.

Field study

Two visits were made to the Network Rail London North Western Integrated Electronic Control Centre at The Mailbox, Birmingham in January 2013 to conduct interviews (with five participants). The participants were selected from a pool of ten members of staff who were working in an Incident Controller, Train Running Controller or Management capacity and were working a shift at the time of the visit to the IECC. Selection from the pool was based on the availability of staff, which in turn was determined by their workload at the time of the visit. The interview data were subsequently processed and analysed to produce two separate hierarchical task analyses: for the train running controller and for the incident controller. Following the interviews and HTA, the next step was to create influence diagrams to reflect the decision making of the operators. An influence diagram graphically represents a decision in the form of a directed graph of the decision maker's beliefs and preferences about decisions (Howard & Matheson, 1984). The web-based modelling tool 'Insightmaker' (http://insightmaker.com) was used to model decision scenarios faced by an incident controller.

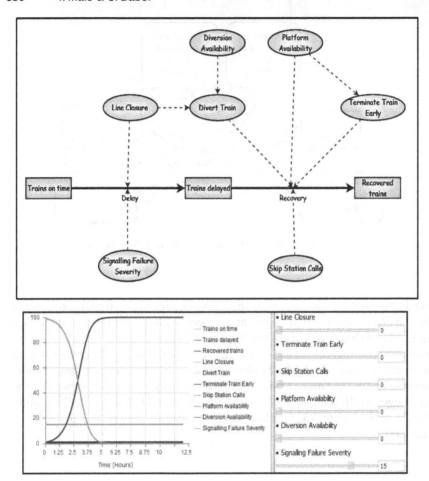

Figure 3.   Example of Influence Diagram and resulting model.

User interface design and initial evaluation

Interface design was carried out using the wire-framing and rapid prototyping package, Balsamiq. Three designs were made, each with the aim of assisting Incident Controllers with a specific task or identified problem.

Design 1: The upper section of the interface displays the track controlled by the IECC and landmark features such as stations and junctions. While this design is similar to the existing CCF, boundaries of each region and the locations of incidents have been added. The lower section indicates types and severity of incidents in own and neighbouring regions. For each incident displayed within this section it is possible to view recent events relating to the incident in the form of an incorporated incident log.

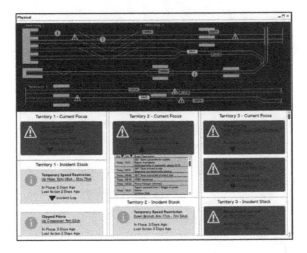

Design 2: The user builds an incident model, shown in the upper half of the interface. The controls to the left allow the user to input constraints which they are facing. The constraints are used to model the rate of service decay. The controls to the right show the decision options available to the user. The lower half of the interface shows the real-time output of the model. The incident controller can attempt to maximise the rate of recovery and minimise the rate of delay by altering the values of the controls and seeing the resultant effect.

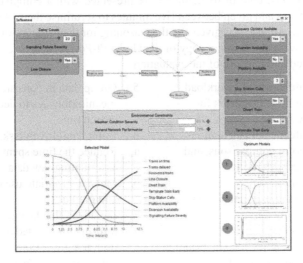

Design 3: The constraints for an incident model are displayed. In the lower right is the incident model. To the left of the incident model, a set of constraints are displayed. In contrast to design 2, the constraints in this design are environmental, such as weather, which the incident controller has no control over. The rationale for displaying two sets of constraints on different interfaces relates to the fact that the magnitude of all constraints displayed in design 2 can be varied by the incident

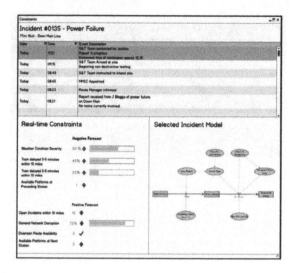

controller and subsequently displayed through the incident model output. The upper section of the interface displays the previous events associated with the incident in the form of an incident log.

Three experienced control room staff were presented with a complete set of the interface designs and given between five minutes to scrutinise them. A brief explanation of their purpose was given and the reasoning for their design was explained. Participants were also given a copy of the Gerhardt-Powals heuristics and briefed on the method and purpose of the evaluation.

1. Automate unwanted workload
a. Free cognitive resources for high-level tasks
b. Eliminate mental calculations, estimations, comparisons, and unnecessary thinking
2. Reduce uncertainty
a. Display data in a manner that is clear and obvious
3. Fuse data
a. Reduce cognitive load by bringing together lower level data into a higher level summation
4. Present new information with meaningful aids to interpretation
a. Use a familiar framework, making it easier to absorb
b. Use everyday terms, metaphors, etc.

5. Group data in consistently meaningfulways to decrease search time
6. Limit data-driven tasks
a. Reduce the time spent assimilating raw data
b. Make appropriate use of colour and graphics
7. Include in the displays only that information needed by the user at a given time
a. Allow users to remain focused on critical data
b. Exclude extraneous information that is not relevant to current tasks
8. Provide multiple coding of data when appropriate

Interface # 1: The evaluators liked to see all incidents for the whole route on a map and to have these incidents grouped by severity and region. The features can also be seen on CCF. The evaluators suggested that improvements could be made by allowing the user to add new events or to apply different ways of grouping events, e.g., in terms of type of event or weather conditions. They also felt that it would be useful to have a clearer means of indicating the criteria used to code incidents, and that it was not always necessary to expand the incident log.

Interface # 2: The evaluators were less keen on this design and suggested several ways in which it could be improved. It was not clear how 'optimum' values related to the sliders on the interface, and it would be useful to present more than 3 optimal solutions (at least on some occasions) to allow the user to consider a range of solutions. It would be useful to show an estimated recovery time and to show the knock-on effects of incidents. Finally, it would be useful to be able to add information or nodes on-the-fly and to be able to combine incident delay/recovery for some operations.

Interface # 3: The evaluators liked the idea of grouping real-time constraints and the manner in which the constraints were displayed. They felt, however, that it was not always easy to interpret these constraints and that this could be improved. As with interface #2, it would be useful to be able to add information or nodes on-the-fly and to be able to combine incident delay/recovery for some operations.

Discussion

Current user interface designs that we observed in the rail industry are based on the capture and display of data relating to network operations. Thus, for example, CCF shows a schematic of track with trains and delays marked on it. While these types of display can support some forms of decision making (particularly in terms of determining the current status of the network) they can be less useful for making decisions that involve either forecasting of future states or dealing with uncertainty. Indeed, the staff we interviewed commented to this effect. Thus, this work sought to develop user interface designs that were directed towards supporting predictions and dealing with system constraints.

Heath and Luff (1992) discuss the merits of using a visual display or interface in the context of a collaborative working environment. It is argued that the use of a large display can enable co-participation between colleagues and allows information to be shared without interruption thus allowing individuals to continue working on their respective tasks and enabling them to consult the large display for updates when convenient. This is in agreement with the findings of an investigation into the use of large off-workstation displays in a railway environment within Network Rail (Carey, 2008). From the study reported in this paper, we propose that there is merit in exploring new approaches to interface design that provide operators with oversight of the constraints under which they are working and the inter-relationships between these constraints.

The approach taken in this work could be compared with other approaches to user interface design that are based on revealing the constraints for a given domain, such as ecological interface design (Vicente and Rasmussen, 1992). Our aim is not to compete with such approaches but rather to indicate how the model of decision making that is reflected in influence diagrams can provided a useful foundation for considering the design of decision support. What we believe is particularly beneficial in the approach offered in this paper is the focus on decision making as a form of optimisation (using systems dynamics modelling) which presents to users the key parameters to which they should attend, and the constraints that can affect these parameters. From the ecological interface design approach, this is akin to taking the 'values and priorities measures' from the Abstraction Hierarchy (from cognitive work analysis, CWA) and visualising the relationship between these in the form of a decision model. The decision model can be contrasted with the control task analysis ('step ladder') model that CWA uses to describe the manner in which operators might make decisions. We suggest that, to date, the EID approach provides little guidance on *how* to display information to support operators in their decision-making, although it provides a structured approach to determining *what* to display. This paper provides a contribution to this literature by offering a template for considering decision making in terms of visual representation of key parameters and constraints.

References

Carey, M. 2008, Use of off-workstation displays in rail control applications, In P.D. Bust (ed.) *Proceedings of the International Conference on Contemporary Ergonomics*, (Taylor & Francis, London).

Gerhardt-Powals, J. 1996, Cognitive engineering principles for enhancing human-computer performance, *International Journal of Human-Computer Interaction* 8, 189–211.

Heath, C. and Luff, P. 1992, Collaboration and control: crisis management and multimedia technology in london underground line control rooms, *Journal of Computer Supported Cooperative Work* 1, 24–48.

Howard, R.A. and Matheson, J.E. 1984, Influence diagrams. In R.A. Howard & J.E. Matheson (eds.) *The Principles and Applications of Decision Analysis* (Strategic Decisions Group, Menlo Park).

Vicente, K.J. and Rasmussen, J. 1992, Ecological interface design: Theoretical Foundations, *IEEE Transactions on Systems, Man and Cybernetics, 22*, 589–606.

Wilson, J.R. and Norris, B.J. 2005, Rail human factors: past, present and future, *Applied Ergonomics* 36, 649–660.

THE EFFECT OF DRIVER ADVISORY SYSTEMS ON TRAIN DRIVER WORKLOAD AND PERFORMANCE

David R. Large, David Golightly & Emma L. Taylor

Human Factors Research Group, University of Nottingham, UK

A simulator study investigated the impact of using two designs of capacity-based driver advisory systems (DAS) on train driver workload and performance. Results indicate that using DAS in trains imposes higher workload on drivers compared to a control (no DAS) condition, though this varies by the type of support offered. Any driver performance benefits were inconclusive. The results are indicative of potential workload costs associated with using networked DAS in trains.

Introduction

There is increasing interest in the use of driver advisory systems (DAS) within trains, precipitated by recent trends towards in-cab rather than trackside signalling and the proliferation of other in-cab systems and technology (Fenner, 2002), typically designed to mitigate Signal Passed at Danger (SPAD) risk. In-cab technology may also be employed to improve train driving and rail traffic management on a strategic level, for example by maximising available capacity or increasing the energy efficiency of railway operations.

In contrast to driving a car, navigating railways requires highly specialised skills and considerable expertise (Naweed, 2013). On conventional rail networks (with trackside signalling), train drivers must acquire a detailed knowledge of their routes, and know all track speeds and signal positions in order to navigate (Branton, 1979). In addition to this they are responsible for the safe transport of people and freight in accordance with a strict timetable, whilst also maximising energy efficiency and available capacity (for example, by avoiding stops in front of red signals and improving the departure process in stations); passengers also expect to travel in comfort. Nonetheless, task-related characteristics such as monotony and low demand are an inherent part of train driving (Cabon et al., 1993). The repetitive nature of the train-driving environment and reduced external visual stimuli when driving underground, in a tunnel or at night, may add to the experience of monotony. These factors alone have been shown to contribute to performance decrements over time (Williamson et al., 2011) such as driver related SPADs. Some of these concerns may be obviated by *increasing* driver workload. For instance, by imposing a relatively minor increase in cognitive demand, drivers are more likely to remain alert and *in-the-loop*; this can mitigate the adverse monotony-related effects on their performance (Dunn and Williamson, 2012). However, research in car driving suggests that while driver support technology can reduce driving workload, inappropriate implementation can have adverse effects (Doi, 2006). Train driver workload and performance when using in-cab DAS is therefore of notable interest; a marginal increase in workload may have the serendipitous effect of keeping drivers *in-the-loop*. However, excessively high workload may distract drivers from the primary task of driving, consequently impairing their control of the train.

There are myriad DAS systems under discussion. These typically differ depending on the type of information presented to the driver and the degree of freedom it provides for them to act on it. For instance, DAS may present explicit driving instructions, such as a speed target or speed profile over time, or specific advice to speed up, slow down, or coast. Alternatively, drivers may be provided with a time target or an indication of whether the train is running early or late with regard to the optimum speed profile to realise the timetable (temporal advice). DAS may also provide decision-support information, such as the gradient profile, energy usage, or position of other trains in the vicinity to drivers. It is then the responsibility of the driver to formulate the correct course of action in response to this. In practice, systems may present a single piece of information or may combine several types.

Speed and *Timetable* DAS have been proposed and reached some degree of implementation in parts of the EU rail network, though little data about the success of such systems has been published in the literature. It is suggested that by providing advice in the form of a timetable, drivers are able to judge current and future performance against the plan. This allows them to respond proactively to the advice, as dictated by the current conditions and their expert knowledge. In contrast, responses to speed advice may be more reactive. Nonetheless, explicit speed advice can encourage drivers to adhere to specific profiles when optimisation is occurring across the network within tight margins.

A simulator study was conducted to examine the impact of using speed and timetable DAS on workload and driver performance. Early data from similar studies, suggests that trained novices may exhibit many similar patterns to experienced train drivers (Dunn and Williamson, 2012). Given this, and the difficulty in attracting large numbers of real drivers, the study used both experts and novices to provide a degree of comparison.

Method

Participants

20 people (12 male; 8 female) volunteered for the study. Participants included 13 students and staff members from the University of Nottingham, all of whom had never previously driven a train. In addition, 7 experienced train drivers were recruited from UK train operating companies and a local train heritage group. Participants provided written consent before taking part and were reimbursed with

vouchers for their time. The study procedure had been approved by the University of Nottingham, Faculty of Engineering Ethics Committee.

Design

A repeated measures design was used, with three driver advisory systems (DAS), a control condition with no DAS (C), speed advice (S) and timetable advice (T); and two driving workload levels, low (L) and high (H), as the independent variables (IV), resulting in six conditions. Driver experience (expert/novice) was also included as an IV. Workload was measured using the NASA Task Load Index (TLX) (Hart and Staveland, 1988), administered after each condition. Participants' ratings for the six subscales were aggregated to provide an overall workload rating (raw TLX) in line with common practice. Performance was measured objectively using the accumulated value of overspeeds. Speed data was captured directly from the simulation software and compared to the optimal line speed. For any infringement (i.e. where line speed was exceeded), the overspeed was recorded. An aggregate score was calculated for each condition.

Apparatus

The study took place in the Human Factors Research Group train simulator at the University of Nottingham. The train simulator comprises a medium fidelity, fixed-base train cab (based on a 319 commuter class train) situated in front of a large, single screen onto which the scenario is projected. The cab includes interactive power and brake handles, a horn and an Automatic Warning System (AWS), which are connected via a keyboard encoder to a PC running RailSimulator.com's Train Simulator 2013 software. A bespoke scenario was created. The route was geotypical and included points of interest such as stations, speed limit changes, road crossings, a tunnel and a bridge. These provided points of reference for route learning and were also used as timing points. The route was designed to last for approximately ten minutes in the low workload condition (L). During the high workload condition (H), where participants were required to stop at all stations and could be confronted by yellow and red signals, the route took approximately fifteen minutes to complete.

Participants were provided with two prototype DAS using MS PowerPoint. Advice was displayed on a screen situated in the centre console of the cab and updated at predetermined locations along the route by the experimenter using a Wizard-of-Oz approach. Driver advice comprised either an explicit speed target (see Figure 1) or a timetable, which highlighted the approaching and future timing points (see Figure 2).

Procedure

The study consisted of two phases – a training phase and the main study phase – which both took place using the same simulated environment. The training phase

Figure 1. Example of Speed	Figure 2. Example of Timetable
DAS.	DAS.

typically took an hour and provided the opportunity for participants to learn the controls of the simulator and the rules of train driving required for the study, including basic signalling and signage, and familiarise themselves with the route. Each participant was required to reach an acceptable standard of competence during training before progressing to the main phase of the study, judged as the ability to complete the high workload condition without major overspeeds, missed stations or SPADs. No driver advisory systems were used during training.

For the main phase of the study, participants were required to drive the same route six times, made up of a combination of the two different levels of workload (L and H) and the three different DAS levels (C, S and T). Participants completed the NASA-TLX immediately after each of the six conditions. All three conditions for each demand level were completed consecutively to avoid confusion and the control condition for each demand level was always undertaken first. To reduce learning effects, the order of the speed and timetable DAS conditions were varied between participants but remained consistent for the low and high workload conditions for each participant. The main study phase typically took 2.5 hours.

Results and analysis

Repeated measures ANOVAs compared the three DAS types across the two different workload levels. Significant interactions were investigated using simple main effects analysis and pairwise comparisons.

Mean TLX ratings and speed performance data are summarised in Table 1. For *TLX overall workload* there was a near significant main effect of workload, $F(1, 19) = 3.252$; $p = 0.087$, with means suggesting that ratings for *overall workload* were higher for the high workload condition than the low workload condition (see Figure 3). Analysed independently, this affect was significant for the control condition, $F(1, 19) = 4.905$; $p < 0.05$, with higher *overall workload* associated with the high workload condition. For the speed DAS, differences in *overall*

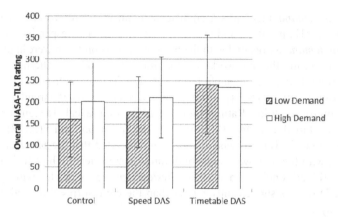

Figure 3.   Overall NASA-TLX Ratings (error bars indicate ± 1 SD).

Table 1.   Mean (SD) TLX ratings and speed performance data.

Variable	Low Workload (L)			High Workload (H)			Significant Results
	C	S	T	C	S	T	
TLX Overall	159.8	177.0	241.3	201.8	211.3	235.3	H>L
Workload	(87.1)	(82.6)	(115.2)	(88.5)	(93.5)	(119.0)	T>S>C
Mental	35.0	40.0	46.5	47.0	51.1	57.3	H>L
Demand	(22.0)	(21.1)	(20.0)	(23.8)	(24.3)	(26.9)	T>{C, S}
Physical	12.3	16.3	19.0	16.8	23.8	23.3	H>L
Demand	(9.0)	(12.8)	(13.8)	(12.4)	(19.6)	(19.3)	{T, S}>C
Temporal	25.3	30.3	45.8	37.0	37.0	38.3	T>{C, S}
Demand	(20.0)	(22.0)	(26.2)	(21.1)	(22.1)	(25.0)	
Performance	32.3	27.3	45.5	30.0	29.75	33.3	T>S
	(22.0)	(13.6)	(28.0)	(18.8)	(14.8)	(20.1)	
Effort	36.8	40.8	51.0	49.8	46.5	53.8	T>{C, S}
	(23.9)	(22.3)	(23.4)	(26.0)	(20.3)	(28.9)	
Frustration	18.3	22.5	33.5	21.3	22.8	29.6	T>{C, S}
	(16.2)	(17.5)	(29.7)	(15.8)	(15.9)	(24.5)	
Speed	11.7	2.5	14.5	3.9	2.9	3.0	L>H
Performance	(15.8)	(2.5)	(15.5)	(5.89)	(3.6)	(4.1)	{C, T}>S

workload between the low and high workload drives were approaching significance, $F(1, 19) = 3.603$; $p = 0.073$. Participants did not indicate any difference in *overall workload* between high and low workload conditions when using the timetable DAS. Furthermore, in contrast to the control and speed DAS conditions there was a trend towards participants indicating lower *overall workload* associated with the high workload condition when using the timetable DAS. There was a significant main effect of DAS, $F(2, 18) = 7.169$; $p < 0.01$. Participants associated the highest *overall workload* with the timetable DAS. This was significantly higher than the speed DAS and control condition (both $p < 0.01$). Speed DAS was also significantly higher than the control condition ($p < 0.05$) (see Figure 3).

For *mental demand* there was a significant main effect of workload and DAS ($F(1, 19) = 8.712$; $p < 0.01$, and $F(2, 18) = 7.514$; $p < 0.01$, respectively). Ratings for *mental demand* associated with the high workload condition were significantly higher than during the low workload condition ($p < 0.01$). Furthermore, *mental demand* when using the timetable DAS was significantly higher than the control condition and the speed DAS ($p < 0.01$). The speed DAS showed a trend to be higher than the control condition. Ratings for *physical demand* revealed a significant main effect of workload and DAS ($F(1, 19) = 5.736$; $p < 0.05$, and $F(2, 18) = 4.324$; $p < 0.05$, respectively). Pairwise comparisons revealed that the *physical demand* associated with the high workload condition was significantly higher than during the low workload condition ($p < 0.05$). *Physical demand* associated with speed and timetable DAS was significantly higher than the control ($p < 0.05$ and $p < 0.01$, respectively).

For *temporal demand* there was a significant main effect of DAS, $F(2,18) = 4.979$; $p = 0.019$, and a significant interaction between DAS and workload, $F(2, 18) = 5.647$; $p = 0.012$. Participants felt significantly more hurried or rushed when using the timetable DAS, compared to the speed DAS and control conditions ($p = 0.023$, $p = 0.006$, respectively). The difference between the speed DAS and the control condition was approaching significance ($p = 0.099$).

Ratings for *performance* were significantly different depending on DAS, $F(2, 18) = 4.640$; $p = 0.024$. There was also a near significant interaction between DAS and demand, $F(2, 18) = 3.188$; $p = 0.065$. Participants felt that they were most successful in accomplishing what they were asked to do when using the timetable DAS compared to the speed DAS ($p = 0.009$).

For *effort* there was a significant main effect of DAS, $F(2, 18) = 4.775$; $p = 0.022$, and the effect of workload was approaching significance, $F(1, 19) = 3.065$; $p = 0.096$. Participants felt that they had to work hardest when using the timetable DAS compared to both the control and speed DAS ($p = 0.007$ and $p = 0.011$, respectively). Ratings for *frustration* were significantly different depending on DAS, $F(2, 18) = 4.027$; $p = 0.036$. Participants were most frustrated when using the timetable DAS compared to both the control and speed DAS ($p < 0.05$).

Driving performance data revealed that there was a significant main effect of DAS and workload ($F(2, 18) = 8.141$; $p = 0.003$, and $F(1, 19) = 10.746$; $p = 0.004$, respectively). There was also a significant interaction between demand and DAS, $F(2, 18) = 7.682$; $p = 0.004$. *Overspeeds* were significantly higher when using both the control and timetable DAS compared to the speed DAS ($p = 0.014$ and $p = 0.001$, respectively). There were also more *overspeeds* during the low workload condition than high workload condition ($p = 0.004$).

A between subjects analysis was conducted to compare drivers and non-drivers in terms of their overall workload ratings. This revealed a significant interaction between demand and experience (driver or non-driver) for the control condition ($F(1, 18) = 5.789$; $p < 0.05$). However, no other significant effects or interactions were found between drivers and non-drivers.

Discussion

It was evident from the study that using a speed or timetable DAS increased overall driver workload compared to a control (no DAS) condition. This effect was evident during both the low and high workload scenarios. In line with Dunn and Williamson (2012), the absence of any significant between-groups results for overall workload or performance when using either DAS suggests that the behaviour of non-drivers was equally as valid as that of experienced drivers.

Driver performance was significantly different between high and low workload conditions: more overspeeds occurred during the low workload condition than during the high workload condition. During the latter, drivers were required to stop at stations. This potentially engaged the drivers to a greater deal, leading to better performance. Stopping at stations also reduced drivers' mean speed. This may have reduced the likelihood of overspeeds occurring compared to the low workload conditions where there were no requirements to stop.

Participants reported that driving was more physically demanding when using the timetable and speed DAS compared to the control. The hardest work was associated with using the timetable DAS, which was also the most mentally demanding. By providing regular targets, drivers are required to perform more control actions to maintain the advised level of performance. This naturally increases both physical and mental demand. Furthermore, by providing advice in the form of a timetable with *multiple* timing points, drivers are able to predict future performance against the plan. In line with existing scheduling techniques, there was limited ability for drivers to catch up if they deviated from the plan. This may therefore account for the higher temporal demand and frustration they associated with using the timetable DAS, particularly if they fell behind plan. Nonetheless, participants felt that they achieved the highest level of success when using the timetable DAS. In contrast, participants were least hurried or rushed when using the speed DAS. Speed advice is a more transient measure. If drivers are provided with an explicit target speed, they may be less aware of the impact of this on future performance. Responses to speed advice may therefore be more reactive, resulting in an erratic speed profile, and poor driving style, generally. This may account for the greater perceived success indicated by participants when using the timetable DAS. A solution may be to combine the benefits of both types of DAS by contextualising speed advice, for example, by presenting an explicit speed target within a speed profile. In this manner, drivers may be encouraged to adhere to longer-term speed profiles, rather than responding to short-term transient targets.

During the study, the greatest performance (i.e. lowest cumulative overspeeds) was achieved with the speed DAS. For consistency, overspeeds were calculated based on the maximum line speed in all conditions. When using the speed DAS, drivers were naturally advised to drive at lower speeds. The likelihood of exceeding the line speed was therefore greatly reduced. Further study should utilise a broader palette of performance measures to address this.

Conclusions

Maintaining the correct line speed is a fundamental and critical requirement for train drivers in order to comply with the timetable. In contrast, the extent to which any capacity improvement or energy efficiency tactics are realised may not be drivers' primary aims. Consequently, there may be a conflict between the operational aims of DAS and the short-term goals of individual drivers. Furthermore, by its very nature, DAS potentially requires additional, possibly conflicting, control actions *in addition* to those required by speed and signals, and therefore must involve extra physical and cognitive effort. This appears to be an inevitable cost of using such systems. A key measure when designing in-cab DAS must be to maximise the operational benefits (e.g. improvements in capacity) while minimising the driver workload costs. That said, there may be additional benefits that are realised with increased workload, such as enhancing driver arousal and keeping them *in-the-loop*. These effects need further study for fuller quantification.

Acknowledgements

The research was funded by and undertaken as part of EU FP7 ONTIME: Optimal Networks for Train Integration Management across Europe project, Agreement no. 285243.

References

Branton, P. 1979, Investigations into the skills of train-driving. *Ergonomics* 22: 155–164.

Cabon, P., et al., 1993. Human vigilance in railway and long-haul flight operation. *Ergonomics* 36 (9), 1019–1033.

Doi, S. 2006, Technological Development of Driving Support Systems based in Human Behavioural Characteristics. *IATSS Research*. Vol. 30, No. 2

Dunn, N., Williamson, A. 2012, Driving monotonous routes in a train simulator: the effect of task demand on driving performance and subjective experience, *Ergonomics*, 55: 9, 997–1008.

Fenner, D. 2002, Train protection. *IEE Review* 48: 29–33.

Hart, S.G., Staveland, L.E. 1988, Development of a multi-dimensional workload rating scale: Results of empirical and theoretical research. In: P.A. Hancock and N. Meshkati (eds.), *Human Mental Workload*, (Elsevier, Amsterdam), pp. 139–183.

Naweed, A. 2013, Investigations into the skills of modern and traditional train driving. *Applied Ergonomics* 10.1016/j.apergo.2013.06.006.

Williamson, A., Lombardi, D.A., Folkard, S., Stutts, J., Courtney, T.K., Connor, J.L. 2011, The link between fatigue and safety. *Accident Analysis and Prevention* 43, 498–515.

ERTMS TRAIN DRIVING-INCAB VS. OUTSIDE:
AN EXPLORATIVE EYE-TRACKING FIELD STUDY

A. Naghiyev[1, 2]*, S. Sharples[1], M. Carey[2], A. Coplestone[2] & B. Ryan[1]

[1] *Human Factors Research Group, Faculty of Engineering,
University of Nottingham, UK*
[2] *Ergonomics Team, Network Rail, UK*
*née Arzoo Buksh

The European Rail Traffic Management System (ERTMS) uses an in-train automation and control system, which changes a number of aspects of the driving task. Questions have been raised about the effect of ERTMS driving on drivers' allocation of visual attention. This is an exploratory eye-tracking field study with ERTMS and conventional drivers on their normal routes. The first level of analysis, presented in this paper, examines the allocation of visual attention for predefined areas of interest (AOIs) inside and outside the cab. The findings indicate a shift of typical visual attentional strategy from monitoring outside to inside the cab. Understanding visual attention allocation may contribute to driver training and design of other technologies and tasks for train drivers.

Introduction

Since the introduction of the European Rail Traffic Management System (ERTMS) in the UK there has been a human factors necessity to understand the impact it has on train driving in the UK. This need has been further highlighted by the Llanbadarn automatic barrier crossing incident in 2011 (RAIB 2012), where an ERTMS fitted train ran onto the level crossing with the barriers raised. One of the casual factors implicated in this incident was the in-cab signalling, as the driver reported he was observing the driver-machine interface (DMI) within the cab when he was required to be observing the lineside indicator for the crossing. This suggests that the presence of ERTMS changed the driver's visual behaviour, leading to the motivation for understanding the effect ERTMS driving has on drivers' allocation of visual attention.

ERTMS uses a relatively new automation and control system fitted into trains called the ETCS (European Train Control System). There are different levels of complexity for ERTMS; however for the purpose of this paper the focus will be on 'level two' as it is being implemented across the UK. In comparison to conventional driving (or class B driving), ETCS provides drivers with all movement authorities (i.e. their primary signalling information and instructions regarding whether or not they should proceed on a line) on a planning area on a DMI inside the cab, as opposed to the conventional method of issuing movement authorities by using lineside signals outside on the track. The ETCS also acts as a supervisory automatic protection system that provides the driver with speed profiles which must be adhered to; if a driver over-speeds, the system produces over-speed alarms and eventually applies the brakes.

This increase in information displayed inside the cab has raised concerns that there is a risk of UK train driving shifting from 'heads up' to 'heads up, heads down' driving. Attention outside the cab is very important, as all train drivers in the UK are required to monitor the environment outside the cab. This is not only to monitor parts of the infrastructure e.g. approaching a station and level crossings; but also for any hazards, damage to the infrastructure and unusual events. Therefore there is a need to understand drivers' visual behaviours and allocation of attention when using ERTMS.

It has been suggested previously that the enhanced automation in ERTMS driving could cause the driver to have an increase in monitoring tasks (Stoop et al 2008). A risk analysis conducted on the implementation of ERTMS in Sweden identified information overload and divided attention as possible issues (Kecklund et al 2011). However, ETCS does provide elements of attention control in the form of auditory alerts that are designed to direct drivers' attention to important visual information on the DMI to reduce information overload (Metzger et al 2012). There is also concern that if drivers have more monitoring tasks and less anticipatory control, they will shift from being proactive drivers to more reactive drivers (Jansson et al 2004).

An interview study conducted with ERTMS drivers in the UK (Buksh et al 2013) revealed that drivers perceived that their focus of attention had shifted from monitoring outside the cab to inside the cab at the DMI, reporting varying amounts of time focusing on the DMI for information on speeds and movement authorities. Drivers in this study also suggested that they experienced increased mental workload in particular scenarios where they had to closely monitor and control speed within narrow parameters of the ETCS.

An experimental eye-tracking study with train drivers using an ERTMS train simulator was conducted in Germany (Arenius et al 2012). Drivers were presented with an ETCS DMI that had the capability to activate and deactivate the 'planning area' (showing the upcoming route and movement authority) in both ETCS levels one and two. Their results showed a significant interaction effect between the planning area condition (activated and deactivated) and the ETCS level condition (level one and two). The average time spent visually attending to the middle view out of the window increased in ETCS level two when the planning area was deactivated, indicating that the display of the upcoming route is something that particularly encourages drivers to focus their visual attention within the cab. However, their results may not be relevant in the UK context as there are factors that create cross-cultural issues with the generalisability of these results. These factors include differences in the layout of the cab, the simulation of level one ETCS which is not implemented in the UK (this was one of the comparator conditions used in the study), and the inclusion of an additional driver advisory system in the simulator that is not normally implemented in UK rail cabs.

This paper presents an exploratory eye-tracking study that was conducted in the field with ERTMS drivers and conventional drivers on their normal routes. The aim of the study was to understand the visual behaviour of ERTMS drivers, in comparison to conventional drivers. The first level of analysis is presented in this paper, examining the allocation of visual attention for outside the window and inside the cab, but also for predefined areas of interest (AOIs) inside the cab. The eye movement measure presented in this paper is the total fixation duration. A fixation is an eye movement measure where the eye actually remains still over a period of time and it is generally considered that a fixation also relates to attention to that position (Holmqvist et al 2011). This paper aims to address the research question: Is there a difference in allocation of visual attention for outside and inside the cab between ERTMS drivers and conventional drivers?

Methodology

Design

A between groups exploratory eye tracking study was conducted in the field. The independent variable was the type of train driving and consisted of two levels; ERTMS drivers and 'conventional' drivers. The dependent variables were the eye movement measures in the areas of interest used in the analysis (described in further detail below).

Participants

Fourteen ERTMS drivers and fourteen conventional drivers participated in the study. The ERTMS drivers were all male ranging in age from 30 to 53 years ($M = 44.1$ years; $SD = 6.6$ years) and had an average of 9.1 years experience train driving ($SD = 6.5$ years). Their experience driving with the ETCS system ranged from 0.25 to 2 years ($M = 1.76$; $SD = 0.73$). The conventional drivers were all male ranging in age from 34 to 54 years ($M = 42.2$; $SD = 7.7$) and had an average of 10.0 years experience train driving ($SD = 5.0$). Apart from the signalling system(s) that they were currently driving under and the classes of trains they were driving, all drivers had previous experience with a variety of different signalling systems and classes of trains. The drivers were recruited with the assistance of the train operating company (TOC) and participation of the drivers was voluntary. Train drivers who wore glasses were excluded from the study.

Apparatus

The SMI eye tracking glasses were used to collect the data, which allowed binocular, non-invasive tracking with automatic parallax compensation. It had a sampling rate of 30 Hz and gaze position accuracy of 0.5°. The eye tracking glasses were used in conjunction with a notebook computer, which used the software iView to collect and record the data.

The ERTMS drivers used one of the following three routes, which were fitted with ERTMS:

- Machynlleth to Pwlheli
- Machynlleth to Shrewsbury
- Machnylleth to Aberystwyth

The conventional drivers used one of the following three routes, which consisted of a combination of signalling systems (other than ERTMS):

- Shrewsbury to Birmingham New Street
- Shrewsbury to Crewe
- Shrewsbury to Chester

Procedure

Drivers were briefed prior to the study. Once in the cab the drivers were instructed to put on and secure the eye tracking glasses. After approximately one minute of wearing the eye tracking glasses a one-point calibration and validation was conducted. The train drivers were then instructed to drive the train and conduct their duties as they normally would and the software started to record the visual scene and collect the eye movement data. The recording was stopped and restarted at each station and the duration of the study was dependent on the route that was been driven and the how long the driver was comfortable wearing the eye tracking glasses. The duration of a trial typically lasted approximately one hour. Drivers were not prompted to provide verbal protocol as part of the study; however some drivers spontaneously provided verbal data. This data will be analysed separately and is outside the scope of this paper. At the end of the study the participants were asked to remove the eye tracking glasses and were debriefed.

Analysis

The Begaze software was used to analyse the eye movement data. Areas of Interest (AOIs) were specified using reference images of inside the cabs. Seven AOIs were constructed for the ERTMS drivers and six AOIs for the conventional drivers. These AOIs were defined with the input of rail experts. Both sets of drivers had AOIs for outside (window), DRA (driver reminder appliance), brake pressure, buzzer and speed. The ERTMS drivers also had AOIs for the planning area and message box on the DMI. The conventional drivers had an additional AOI for the AWS (advance warning system) indicator. An example of AOIs on a reference image is shown in figure 1.

The data for each fixation was coded to either a corresponding AOI or to 'white space' which fell outside these areas (other). This process produced detailed statistics for each AOI which included information for fixations, saccades, pupil

Figure 1.   Example of the ERTMS AOIs used in Begaze.

diameter and blinks. This paper presents the results for the total fixation duration measure. As the lengths of the trials were unequal the distribution of visual attention is characterised by the ratio of the total fixation duration within a defined AOI and the total time taken in the trial, expressed as a percentage.

Results

The means and standard deviations of the total fixation duration ratio as percentages were calculated for the AOI outside (i.e. the window) and for the sum of all the other AOIs which were located inside the cab. This was done for both groups and is shown in table 1 (NB the total fixation duration percentages do not add up to 100%, as they use only the time when the eyes were fixated on an AOI, not the total dwell time).

A MANOVA was conducted to avoid type I errors. Preliminary tests revealed that the data passed Box's test for equality of covariance matrices (Box $M = 3.745$, $p = 0.33$). The MANOVA revealed a significant main effect of driving type (ERTMS vs. conventional) for total fixation duration ratio of the two AOIs ($F(2, 25) = 11.324$, $p < 0.001$). Univariate analysis showed that there was no main effect of driving type on the total fixation duration ratio for outside the cab ($F(1, 26) = 1.303$, $p = 1.3$), but there was for inside the cab ($F(1, 26) = 23.54$, $p < 0.001$).

Table 1.   Total fixation duration ratio as %s
for inside vs. outside cab.

	ERTMS		Conventional	
AOI	Mean	SD	Mean	SD
outside	36.4	14.0	42.7	15.2
inside	10.2	4.36	3.51	2.80

Table 2.   Total fixation duration ratios as %s for AOIs.

	ERTMS		Conventional	
AOI	Mean	SD	Mean	SD
outside	36.4	14.0	42.7	15.2
Speed	7.42	3.89	2.74	2.74
brake pressure	0.0969	0.126	0.137	0.149
DRA	0.0421	0.0466	0.0337	0.0343
Buzzer	0.103	0.0646	0.0551	0.0410
planning area	1.82	0.911	–	–
message box	0.360	0.244	–	–
AWS indicator	–	–	0.0279	0.0467
Other	0.390	0.426	0.512	0.518

The means and standard deviations of the total fixation duration ratios were calculated for each of the AOIs for the two driving types (ERTMS and conventional) and is shown in table 2 below.

A MANOVA was conducted to avoid type I errors. Preliminary tests revealed that the data passed Box's test for equality of covariance matrices (Box $M = 18.505$, $p = 0.48$). The MANOVA revealed a significant main effect of driving type for total fixation duration ratio of the different AOIs ($F(5, 22) = 3.922$, $p < 0.005$). Univariate analyses showed that there was only a main effect of driving type on the total fixation duration ratio for the speed AOI ($F(1, 26) = 13.496$, $p < 0.01$).

Conclusions

The findings from this study demonstrate that ERTMS drivers look outside the cab significantly less than conventional drivers; indicating that there has been a shift of typical visual attentional strategy from monitoring outside on the tracks to inside the cab. However, the difference in total fixation duration ratio for the two groups was 6.3%. Interestingly, the data also questions the assumed notion of 'heads up' driving in the UK, as the results demonstrate that a small proportion of monitoring is occurring inside the cab, in particular speed, even in the conventional driving condition (albeit not to the same degree that ERTMS drivers are monitoring inside the cab).

The findings demonstrate that inside the cab the ERTMS drivers are looking significantly more at the speed information than in conventional driving. The difference in total fixation duration ratio for the two groups was 4.68%. This suggests that ERTMS drivers are spending more time monitoring their speed as suggested by Stroop et al (2008) which implies that ERTMS drivers may have less anticipatory control than conventional drivers and are more reactive drivers than conventional drivers (Jansson et al 2004).

A comparison cannot be made for the planning area or the message box on the DMI; however their total fixation duration ratios were 1.82% and 0.36% respectively. This suggests that drivers are spending more time looking inside the cab at the DMI, however the majority of this visual attention is directed at monitoring speed. The results also support the qualitative study conducted by Buksh et al (2013) where drivers reported that they perceived their focus of attention had shifted from monitoring outside the cab to inside the cab. However, the results demonstrate that a large proportion of the monitoring inside the cab is at the speed, as opposed to the planning area. The Buksh et al (2013) results provides an explanation for the increase in visual inspection in speed as the ERTMS drivers reported increased workload in particular scenarios where they had to closely monitor and control speeds within narrow parameters of the ETCS.

This study only presents the initial analysis of the data collected. Future work will show further analysis of the eye movement data, as well as the analysis of the verbal data. Furthermore, the events in the infrastructure (e.g. stations, level crossings) and the ETCS (e.g. auditory alerts and alarms) will be coded using the video analysis software Observer and triangulated with the eye movement data. This will help to develop a deeper understanding of the impact of ERTMS on train driver behaviour.

Acknowledgements

This work is funded by Network Rail and the ESPRC. The main author would like to thank Arriva Trains Wales for all their assistance; in particular Gareth Jones for his continued support and enthusiasm. The author is grateful to all the staff at the Machnylleth and Shrewsbury depots for their kind assistance and to the train drivers for their invaluable involvement in the study.

References

Arenius, M., Metzger, U., Athanassiou, G. & Strater, O. 2012 Planning on Track-introduction of a planning device in the railway domain. In: *PSM 11& ESREL 2012*. Finland: Helsinki, 25–29 June 2012

Buksh, A., Sharples, S., Wilson, J. R., Morrisroe, G. & Ryan, B. 2013 Train automation and control technology – ERTMS from users' perspectives. In: Anderson, M.

(ed). *Contemporary ergonomics and human factors 2013*, (Taylor and Francis, London), 168–175

Holmqvist, K. & Nyström, M. 2011 *Eye tracking a comprehensive guide to methods and measures.* (Oxford University Press, UK)

Jansson, A. Olsson, E. & Kecklund, L. 2004 Acting or reacting? A cognitive work analysis approach to the train driver task. In J. R. Wilson, B. Norris, T. Clarke & A. Mills (eds) *People and rail systems: human factors at the heart of the railway* (Ashgate, UK)

Kecklund, L., Mowitz, A. & Dimgard, M. 2011. Human factors engineering in train cab design-prospects and problems. In P. C. Cacciabue, M. Hjalmdahl, A. Ludtke & C. Riccioli (eds) *Human modeling in assisted transportation* (Springer, Milan, Italy)

Metzger, U. & Vorderegger, J. 2012. Railroad. In M. Stein & P. Sandi (eds) *Information ergonomics: a theoretical approach and practical experience in transportation* (Springer, Berlin, Germany)

RAIB. 2012, "Rail Accident Report: Incident at Llanbadarn Auomatic Barrier Crosssing (Locally Monitored), near Aberystwyth, 19 June 2011." Retrieved 23rd September, 2012, from: http://www.raib.gov.uk/cms_resources.cfm?file=/ 120627_R112012_Llanbadarn.pdf

Stoop, J. & Dekker, S. 2008. The ERTMS railway signalling system; deal on wheels? An inquiry into the safety architecture of the high speed train safety. In E. Hollnagel, F. Pieri, & E. Riguad (eds) *Proceeding of the third resilience engineering symposium* (Ecole de Mines Paris, Paris), 255–262

VIGILANCE EVIDENCE AND THE RAILWAY LOOKOUT

Laura Pickup[1], Emma Lowe[2] & Stuart Smith[2]

[1] The University of Nottingham, Nottingham, NG7 2RD
[2] Network Rail, The Quadrant, Milton Keynes, MK9 1EN

Vigilance tasks have long been recognised as difficult to sustain, at a constant level of performance, over a period of time. The last 10 years of research in this field has seen a volte face on the reasons attributed to the difficulty of these tasks. This paper summarises the literature to highlight the risks associated with safety roles, such as the railway lookout. We contend this approach and evidence is equally applicable for understanding other safety relevant roles that involve similar vigilance tasks.

Introduction

Vigilance is a term used to describe monitoring tasks and has been investigated by generations of researchers. In the last ten years there has been a shift in the evidence on how human behaviour influences the ability to complete vigilance tasks. This paper reflects the findings of a literature review completed for Network Rail, the company which owns, maintains, and operates the UK railway infrastructure. This aimed to understand why railway lookouts could, and do, occasionally fail to remain vigilant; sometimes with fatal consequences.

A railway lookout is the individual with the responsibility to identify and alert track workers of trains travelling near or into an area of maintenance. They may work close to or at a distance to the actual site of work. They rely on physical touch, flags, horns or whistles to communicate the need for a track team to remove themselves to a place of safety; in sufficient time to avoid injury. This role may last for many hours and in all weathers.

Those uninitiated with the UK railway could understandably assume that by timetabling, train arrival times are predictable; ergo how vigilant does one need to be? However, timetables cannot be relied upon as un-timetabled traffic and slow running freight both confound predictability. Therefore, a lookout has to stay alert and not become distracted during periods of high and low work activities, for as long as they are required. Parallels can be drawn with other jobs (e.g. military sentries, anaesthetists) which can involve open ended tasks and risk physical harm to either the person required to remain vigilant or another (Huey and Wickens 1993, Weinger and Englund 1990). The findings from this work have informed guidance delivered to Network Rail to modify existing lookout training programs and manage the risks associated with vigilance tasks.

Terminology

The term vigilance was first adopted in the 1920's, a contemporary definition is;

> *"Vigilance or sustained attention refers to the ability of organisms to maintain their focus of attention and remain alert to stimuli over prolonged periods of time." Warm et al (2008) p115.*

The 'vigilance decrement' is the term used to describe when people's performance in detecting and responding during a monitoring task deteriorates (Mackworth 1948). Studies suggest this can be observed after just 15 minutes with the greatest decrement occurring within the first 30 minutes of a task. This further deteriorates, at a lesser rate, over subsequent hours (Mackworth 1948, Wickens et al 1997).

Evidence relating to vigilance tasks

Over time, vigilance researchers came to recognise that experimental findings often differed significantly (Deaton and Parasuraman 1993). They attributed this to differences in the type of vigilance task used in experimental designs. Therefore, to move the understanding of human behaviour in vigilance tasks forward, a taxonomy was proposed. This classified tasks by their characteristics: task type, event rate (fast or slow), sensory modality (visual/auditory), cognitive (a cognitive operation e.g. calculation/logical argument is necessary to distinguish target), source complexity (single or multiple targets), and successive (absolute judgments made to compare a current target with one retained in working memory) or simultaneous (comparative judgments where all the information is present in the target itself to make a judgment). This taxonomy has been heavily relied upon to draw general conclusions relevant to human behaviour specific to the *type* of vigilance task. Categorisation of the type of vigilance task seems prudent to ensure information gathered can reflect, and fully inform, the risk factors to an operational setting. A classification of the railway lookout task is proposed in table 1.

Table 1.   Categorisation of the railway Lookout task.

Vigilance Taxonomy	Classification	Justification
Task Type	Simultaneous	The lookout is required to identify trains on specific tracks
Event Rate	Variable	Traffic will vary in frequency
Sensory Modality	Visual/Auditory	Identification of train requires auditory and visual information
Source Complexity	Single	The train is the only target
Task Complexity	Sensory	The task is perceive and detect, independent of calculations and logical arguments

Historically poor performance of vigilance tasks has been attributed to monotony, low levels of arousal and attention due to the task being tedious and under stimulating (Frankman and Adams 1962, Robertson et al 1997). These factors imply that the repetitive nature of the task causes the individual to actively reduce their attentional effort due to reduced wakefulness. Recent evidence contradicts these beliefs and suggests the sustained attention, necessary to remain vigilant, uses considerable mental effort and alertness to apply high levels of information processing to sustain our attention (Mathews et al 2000, Warm et al 2008). This has been demonstrated using brain imaging technology that records cerebral blood flow. When the brain becomes metabolically active, it releases carbon dioxide into the blood stream, which subsequently causes vasodilation of the blood vessels and an increase in blood flow. Warm et al (2009) reports blood flow most significantly increases in the right cerebral hemisphere during a vigilance task, and is directly related to task demand. These findings support the theory that vigilance requires mental resources. Therefore, when an individual's vigilance performance declines (vigilance decrement) exhaustion of the mental resource of attention would show a reduction in blood flow. Warm et al (2008) confirmed that this did occur but only in observers actively involved and tasked with a monitoring duty. Passive onlookers (with no responsibility for monitoring) did not show evidence of reduced blood flow, as would have been the case if reduced wakefulness was responsible for any vigilance decrement.

In summary, evidence now suggests prolonged simultaneous vigilance tasks, such as the task of the railway lookout, are resource-demanding and associated with perceived high workload, higher error rates and significant deterioration in perceptual sensitivity. The perception of high workload should not be considered as a consequence of fighting boredom, but rather the effort to sustain the necessary attention (Hitchcock et al 1999). A summary of factors influencing vigilance performance relevant to the railway lookout are outlined in table 2.

Associated risks

The research concludes that vigilance tasks, like that of the railway lookout, are inherently stressful and lead to a reduced ability for individuals to stay engaged in the task (Finomore et al 2009, Hancock and Warm 1989, Rose et al 2002, See et al 1995, Temple et al 2000, Warm et al 2008). Put simply, humans are not well designed to remain stationary, monitoring and alert for long periods. The evidence from the literature revealed a number of strategies relevant to the rail industry to address the risks inherent to the task of the lookout. Some of these are summarised below:

The person – evidence associated with an individual's personality type, intelligence, gender and age are inconclusive (Rose et al 2002). However, a strong correlation has been noted between poor vigilance performance and individuals prone to boredom and daytime sleepiness (Wallace et al 2003). A screening program could identify lookouts with these traits or conversely those more predisposed

Table 2.   Factor Influencing Vigilance Tasks.

Influencing Factor	Evidence	Summary of Findings
Temperature	Hancock 1984	Extreme temperatures, or inability to sustain a constant core temperature, are detrimental.
Noise	Becker et al 1995, Hancock 1984	High intensity intermittent noise can be detrimental to monitoring tasks. Some evidence suggests noise can reduce perceptual sensitivity, increase workload and impede information processing.
Individuals	Reinerman-Jones et al 2010, Rose et al 2002	Conscientiousness has been suggested as relevant to higher perceptual sensitivity, considering sensation seeking behaviour as the polar opposite implies this is a less desirable characteristic. Studies on introversion, extroversion, neuroticism, age, gender are either inconclusive or unreliable indicators.
Boredom	Oborne et al 1993, Smallwood and Schooler 2006, Thackary et al 1977	Individuals increase their level of activity during periods of low stimulation, either cognitively or physically. Mind wandering or Task Unrelated Thoughts (TUTs) are common and associated with error. Recalling the duration of a lapse in attention is also poorly estimated.
Fatigue/ sleep loss	Huey and Wickens 1993, Shaw et al 2010, Temple et al 2000	Fatigue and sleep loss reduces the availability of attention resources and performance of low load monitoring tasks. Shift patterns and time of day (early hours of the morning) can both negatively influence vigilance performance. Caffeine is recognised as impeding the vigilance decrement.
Task engagement	Helton et al 2005, Mathews et al 2002, Warm et al 2008	Task engagement is the motivation towards the task, and can positively influence vigilance performance and delay the onset of the vigilance decrement. It has an inverse relationship to perceived workload and has been suggested as an indicator of resource availability. Sensory tasks are more prone to loss of task engagement and increase stress.
Coping	Lazarus 1999, Shaw et al 2010	Coping styles include: problem focused, emotion focused and avoidance orientated. Individuals with low avoidance coping strategies perform better.

to remaining alert. Educating people on the difficulties associated with prolonged vigilance tasks and encouraging staff to risk assess their own levels of sleepiness and fatigue has also been found to improve vigilance performance (Adams-Guppy and Guppy 2003, Bolstad et al 2010). Fatigue is inevitable at some time; ensuring regular breaks and caffeine intake were both cited as compensating strategies that reduced the vigilance decrement (Temple et al 2000).

Job design – rest breaks and shorter shifts would assist task engagement. Evidence from the military and air traffic control work domains suggest breaks between 1–2 hours are most commonly reported (McBride 2007, Wickens et al 1997). Mackworth (1948) noted that interrupting a vigilance task after 30 minutes, by a telephone call, served to return the level of vigilance performance back to the original state. Radio communications are used by the military and could be by the rail industry if the risk of distraction is not assessed to be too great (McBride 2007).

Knowledge of results – rate of detections can be improved by informing operators of how successful their last detection was (Hitchcock et al 1999). Why this occurs remains unclear; as it may just be the interruption and motivational aspect that achieves the improvement (Wiener and Attwood 1968). Training on the use of visual scanning patterns can also promote active engagement; regular scanning towards the direction of potential cues can improve detection rates (Underwood 2007). Introducing lookout training to promote an 'automatic attention response' would promote automatic processing (skilled behaviours) compared to controlled processing (new and inconsistent information). This type of task is more likely to impose lower levels of workload and become resilient to environmental factors (Fisk and Scerbo 1987).

General wellbeing – a common cold is associated with a reduction in task engagement (Warm et al 2008). Petri et al (2006) also suggests impaired cognition, mental endurance and psychomotor reaction times will occur after 3 hours without fluids. Involvement in celebrations and certain festivals can alter the type or frequency of sleep and the intake of food and drink (BaHammam 2003). Railway lookouts should be educated on how changes in health, sleeping, eating and drinking patterns can impact wellbeing and vigilance tasks.

Cueing – the railway lookout relies on both visual and auditory cues to identify traffic. A task can be considered inherently more complex and have a higher workload where there is difficulty discriminating a target from the background environment (Warm et al 2009). Vigilance tasks that have cues with reduced salience and are temporally and spatially unpredictable are recognised as detrimental to vigilance performance (Warm and Jerison 1984). Weather or visual clutter can impact the clarity of the visual field for the railway lookout. The background noise in the operational railway environment may also mask the sound of a train, furthermore, trains are designed to limit noise emissions (Hardy 2004). A design to enhance the auditory salience of a train could improve detection by a lookout, however, localising and judging the distance may remain a challenge (Warm and Jerison 1984, Warm et al 2008, Scharine and Letowski 2005). Reliably cueing a lookout to focus their attention to the likely direction of a forthcoming train is the approach most likely to improve detection rates and reduce perceived workload (Hitchcock et al 1999).

Conclusion

Traditionally, vigilance tasks have been associated with boredom and monotony, but this can no longer be considered the case. Prolonged sensory vigilance tasks create high subjective workload and stress, and prove challenging for individuals to remain engaged hence a risk to task performance. This paper has summarised a large body of literature to illustrate the risks associated with the railway lookout, a safety relevant role. Future publications intend to expand on how the evidence has been applied by Network Rail.

References

Adams-Guppy, J. and Guppy, A., 2003. Truck driver fatigue risk assessment and management: a multinational survey. *Ergonomics* 46, 8, 763–779.

BaHammam, A. 2003. Sleep pattern, daytime sleepiness, and eating habits during the month of Ramadan. *Sleep and Hypnosis* 5, 4, 165–172.

Becker, A. B., Warm, J. S., Dember, W. N., and Hancock, P. A. 1995. Effects of jet engine noise and performance feedback on perceived workload in a monitoring task. *International Journal of Aviation Psychology* 5, 49–62.

Bolstad, C. A., Endsley, M. R. and Cuevas, H. M. 2010. In D. H. Andrews, R. P. Herz, M. B. Wolf (eds). *Factors Issues in Combat Identification*. (Ashgate Publishing Limited, Surrey), 147–160.

Deaton, J. E. and Parasuraman, R. 1993. Sensory and cognitive vigilance: Effects of age on performance and subjective workload. *Human Performance* 6, 1, 71–97.

Finomore, V., Matthews, G., Shaw, T. and Warm, J. 2009. Predicting vigilance: a fresh look at an old problem. *Ergonomics* 52, 7, 791–808.

Fisk, A. D. and Scerbo, M. W. 1987. Automatic and control processing approach to interpreting vigilance performance. *Human Factors* 29, 6, 653–660.

Frankman, J. P. and Adams, J. A 1962. Theories of vigilance. *Psychological Bulletin* 59, 257–272.

Hancock, P. A. 1984. Environmental stressors. In Warm, J. S'. *Sustained Attention in Human Performance*. J. Wiley and Sons Ltd. Chichester, UK. 103–142.

Hancock, P. A. and Warm, J. S. 1989. A dynamic model of stress and sustained attention. *Human Factors* 31, 5, 519–537.

Hardy, A.E.J. 2004. The implications of the physical agents directive (Noise) workstream 2 noise exposure of bystanders. RSSB Report AEATR-II-2004-010.

Helton, W. S., Hollander, T. D., Warm, J. S., Mathews, G., Dember., W. N., Wallart, M., Beauchamp, G., Parasuraman, R. and Hancock, P. A. 2005. Signal regularity and the mindlessness model of vigilance. *British Journal of Psychology* 96, 249.

Hitchcock, E. M., Dember, W. N., Warm, J. S., Maroney, B. W. and See, J. 1999. Effects of cueing and knowledge of results on workload and boredom in sustained attention. *Human Factors* 41, 365–372.

Huey, B. M. and Wickens, C. D. 1993. *Workload transitions: Implications for Individual and Team Performance*. Washington, DC: National Academy Press, 139–170.

Lazarus, R. S. 1999. *Stress and Emotion: A New Synthesis*. New York: Springer.

Mackworth, N. H. 1948. The breakdown of vigilance during prolonged visual search. *The Quarterly Journal of Experimental Psychology* 1, 6–21.

Matthews, G., Davies, D. R., Westerman, S. J, Stammers, R.B 2000 Vigilance and sustained attention. In *Human Performance: Cognition, Stress, and Individual Differences*. (Psychology Press, East Sussex), 107–124.

Matthews, G., Campbell, S. E., Falconer, S. E., Joyner, L. A., Huggins, J., Gilliand, K., Grier, R. and Warm, J. S., 2002. Fundamental dimensions of subjective state in performance settings: task engagement, distress, and worry. *Emotion* 2, 4, 315–340.

McBride, S. A., Merullo, D. J., Johnson, R. F., Banderet, L. E. and Robinson, R. T. 2007. Performance during a 3-hour simulated sentry duty task under varied work rates and secondary task demands. *Military Psychology* 19, 2, 103–117.

Oborne, D. J., Branton, R., Leal, F., Shipley, P. and Stewart, T. 1993. *Person-Centred Ergonomics,* (Taylor and Francis, London).

Petri, N, M, Dropuliæ, N, and Kardum, G. 2006. Effects of voluntary fluid intake deprivation on mental and psychomotor performance. *Croatian Medical Journal* 47, 855–861.

Reinerman-Jones, L. E., Matthews, G., Langheim, L. K. and Warm, J. S. 2011. Selection for vigilance assignments: a review and proposed new direction. *Theoretical Issues in Ergonomics Science* 12, 4, 1–24.

Robertson, I. H., Manly, T., Andrade, J., Baddeley, B. T. and Yiend, J. 1997. 'Oops' performance correlates of everyday attentional failures in traumatic brain injured and normal subjects. *Neuropsychologia* 24, 5, 636–647.

Rose. C. L., Murphy. L. B., Byard. L. and Nikzad, K. 2002. The role of the big five personality factors in vigilance performance and workload. *European Journal of Personality* 16, 185–200.

Scharine, A. A. and Letowski, T. R. 2005. Factors affecting auditory localization and situational awareness in the urban battlefield. Army Research Laboratory, *Report ARL-TR-3474,* 1–60.

See, J. E., Howe, S. R., Warm, J. S. and Dember, W. N. 1995. Meta-Analysis of the sensitivity decrement in vigilance. *Psychological Bulletin* 117, 2, 230–249

Shaw, T. A., Matthews, G., Warm, J. S., Finomore, V. S., Silverman, L. and Costa.P.T. 2010. Individual differences in vigilance: personality, ability and states of stress. *Journal of Research in Personality* 44, 297–308.

Smallwood, J. and Schooler, J.W. 2006. The Restless Mind. *Psychological Bulletin* 132, 946–958.

Temple, J. G., Warm, J. S., Dember, W. N., Jones, K. S., LaGrange, C. M. and Matthews.G. 2000. The effects of signal salience and caffeine on performance, workload, and stress in an abbreviated vigilance task. *Human Factors* 42, 2, 183–194.

Thackary, R. I., Bailey, J. P. and Touchstone, R. M. 1977. Physiological, subjective and performance correlates of reported boredom and monotony while performing a simulated radar control task. *Technical Report FAA-AM-75-8.* FAA Civil Aeromedical Institute, Oklahoma, USA. 1–12.

Thiffault, P. and Bergeron, J. 2003. Fatigue and individual differences in monotonous simulated driving. *Personality and Individual Differences* 34, 159–176.

Underwood, G. 2007. Visual attention and the transition from novice to advanced driver. *Ergonomics* 50, 8, 1235–1249.

Wallace, J. C., Vodanovichb, S. J. and Restino, B. M. 2003. Predicting cognitive failures from boredom proneness and daytime sleepiness scores. *Personality and Individual Differences* 34, 635–644.

Warm, J. S., and Jerison, H. J. 1984. The psychophysics of vigilance. In J. S. Warm (Ed.), *Sustained attention in human performance*, (Chichester, England: Wiley), 15–60.

Warm, J. S., Mathews, G. and Finomore, V. S. 2008. Vigilance, workload and stress. In Hancock, P. and James, S. (Ed) *Performance under Stress,* Ashgate Publishing Ltd, Oxon, 115–141.

Warm, J. S., Matthews, G. and Parasuraman, R. 2009). Cerebral haemodynamics and vigilance performance. *Military Psychology Report.* Air Force Research Laboratory, Wright Patterson AFB, OH. 1–27.

Wiener, E. L. and Attwood, D. A.1968. Training for vigilance: Combined Cueing and Knowledge of Results. *Journal of Applied Psychology* 52, 6, 474–479.

Weinger, M. B. and Englund, C. E. 1990. Ergonomic and human factors affecting anaesthetic vigilance and monitoring performance in the operating room environment. *Anesthiology* 73: 995–1021.

Wickens, C. D., McGee, J. P. and Mavor, A. S. 1997. *Flight to the Future: Human Factors in Air Traffic Control.* Washington DC, USA, National Academic Press. 125–134.

UNDERSTANDING HUMAN DECISION MAKING DURING CRITICAL INCIDENTS IN DYNAMIC POSITIONING

Linda J. Sorensen, Kjell I. Øvergård & Tone J.S. Martinsen

Department of Maritime Technology and Innovation, Maritime Human Factors Group, Vestfold University College, 3103 Tønsberg, Norway

Critical incidents in Dynamic Positioning (DP) have potentially disastrous consequences. DP operators have to make time-critical decisions in order to rapidly and effectively handle such unexpected incidents. The purpose of this study was to identify characteristics of critical incidents in DP operations and characteristics of decision-making. Semi-structured interviews using the Critical Decision Method were conducted with 13 experienced DP operators. It was found that decision-making in critical incidents in DP is naturalistic and recognition-primed. This study contributes to an increased understanding of critical incidents in DP operations and the key decision-making processes.

Introduction

Sophisticated automation intends to reduce operator error and enhance efficiency (Parasuraman, Mouloua & Molloy, 1996). Large-scale accidents, such as Chernobyl and Three Mile Island, have primarily been attributed to operator error (Meshkati, 1991), which has been described as an undesired consequence of human automation interaction. The maritime and offshore industry is increasingly becoming dependent on automated vessel station keeping for demanding operations at sea (Fossen, 1994). Critical incidents are events that are unplanned, non-routine but do not end tragically, yet have the potential to develop into large-scale accidents. Industries use the term "near misses" and have in common the fact that they are recovered throughout the sequence of events. Identifying specific characteristics of critical incidents could reveal important information about how large-scale incidents can be prevented.

Automation in the maritime domain

In the maritime fields, automation has been introduced as a technical aid, taking over the performance of tasks previously performed by people, with the intention of increasing performance and safety (Parasuraman, et al., 1996). Dynamic Positioning (DP) is an automated system for vessel station keeping. A computer control system automatically maintains a vessel's position and heading by controlling machinery power, propellers and thrusters. Position reference sensors, along with wind sensors, motion sensors and gyro compasses provide input to the computer in order to maintain the vessel's position, making allowances for the size and direction of environmental forces (Sørensen, 2011).

Automation affects humans in their work. The introduction of automated systems imposes new demands on the socio-technical systems, including the human operator (Sarter & Woods, 1995; Øvergård et al., 2008). When new automation is introduced into a system, or when there is an increase in the autonomy of automated systems, developers often assume that adding automation is a simple substitution of a machine activity for human activity (Woods & Sarter, 2000). Empirical data on the relationship of people and technology suggest that this is not the case and that traditional automation has several negative performance and safety consequences associated with the human out-of-the-loop performance problem (Kaber & Endsley, 2004). When a human operator is out of the loop, instances will occur when he or she cannot maintain control over the system (Norman, 1990). The need for the operator to be "in the loop" refers to the operator's Situation Awareness (SA). Jentsch et al. (1999) show how the loss of SA can lead to errors in assessments that could result in major accidents by describing how inadequate detection of changes in the position of a hostile aircraft. This may lead to an incorrect understanding of the situation which in turn leads to poor decisions in regards to placement of own aircraft. SA is therefore an important component of sound decision making.

Decision making in the maritime domain

Revisions to the International Convention on Standards of Training, Certification and Watchkeeping for Seafarers (the STCW Convention), and its associated Code were adopted in Manila in June 2010. New human factors-related elements included minimum training requirements for maritime officers' decision-making skills as well as resource management (IMO, 2011).

The Naturalistic Decision Making (NDM) theory has been proposed to explain real-life decision making, and Zsambok (1997) defines NDM as: "how experienced people, working as individuals or groups in dynamic, uncertain, and often fast paced environments, identify and assess their situation, make decisions and take actions whose consequences are meaningful to them and to the larger organization in which they operate" (p. 5). Klein et al. (1986) identified the following specific features about NMD decision-making. *First*, the fire ground commanders drew on their previous experience to recognize a typical action to take. *Second*, they did not have to find an optimal solution, merely a workable one. *Third*, they mentally simulated the solution to check that it would work. Klein et al. (1986) proposed the Recognition-Primed Decision-Making (RPDM) model which focuses on situation assessment and explains how an experienced professional can make rapid decisions. Situation assessment in the RPDM model considers understanding of plausible goals, recognition of important contextual cues, the forming of expectations and identification of courses of action as the four most vital aspects (Klein, 1993). Such a situation assessment, including mental simulation, explains how experienced decision makers can identify a reasonably good option as the first one they consider, rather than generating and evaluating a series of alternatives.

Expertise has been found to be essential in order to make decisions in uncertain contexts (Kahneman & Klein, 2009). Expertise is characterized by a high ability of skill and/or and knowledge within a domain (Salas et al., 2010). Expertise-based intuition, also called recognition-primed decision-making (Kahneman & Klein, 2009), is the rapid, automatic generation of single decision options, rooted in extensive domain-specific knowledge and the recognition of patterns from past events (Salas et al., 2010).

The study presented here investigated work situations where DP operations suddenly escalate from a routine situation to a critical incident with potentially disastrous outcomes. In all incidents, the human operator avoided accidents either by a recovery to normal mode or by aborting the operation. The research employed interviews with experienced DP operators with the aim of addressing one research question: *What characterises human operator decision-making in critical incidents in DP operations?* Increased understanding of what characterises critical situations that end well could reveal important information about how large-scale incidents can be prevented.

Method

Sampling

Using a purposive sampling strategy to target experienced DP operators, informants were contacted through various channels in Norway (such as DP training centres, maritime educational institutions, drilling companies and shipping companies). Initial contacts with the informants were through e-mail, telephone or personal communication. The inclusion-criteria were a minimum of 5 years seagoing experience and 3 years experience as a fully trained DP operator. A total of 42 candidates were approached of which 13 qualified informants agreed to take part.

Critical incidents

The informants had to have been on board the vessel at the time of the incident and been actively involved in the incident. Critical incidents were defined as events that were non-routine and that had a high damage potential. The industry uses the term "near misses". Incident recollections that were not personally experienced were excluded from the study. Each informant was interviewed about two critical incidents. Two of the informants provided information about only one incident. Hence the 13 informants provided 24 incident reports for further analysis.

Procedure

Prior to each interview each informant was informed of the purpose of the study, ethical considerations taken, safeguards for their confidentiality and their opportunity to withdraw from the study at any time. If the candidate agreed, an interview session was scheduled and signed consent was collected. A demographic questionnaire was administered before the interview began. The study was – according to Norwegian data protection regulations – reported to and approved by the Norwegian Science Data Services (project number 33042).

Data analysis and reduction

A semi-structured interview technique was applied using the Critical Decision Method (Klein et al., 1989). These were translated to Norwegian and used in this study. The interviews were analyzed using thematic analysis to find patterns of meaning within the qualitative data. The procedure involved five phases (Braun & Clarke, 2006), familiarizing with the data; initial generation of codes; searching for themes; evaluation of themes and final definition of themes.

Discussion of results

Demographics

All informants had a nautical education and unlimited DP certificates. Their age ranged from 29 to 69 ($\bar{x} = 44.3$; $\sigma = 12.1$). Seagoing experience ranged from 5 to 40 years ($\bar{x} = 20.2$; $\sigma = 11.4$). Experience as DP operators ranged from 4.5 to 33 years ($\bar{x} = 12.9$; $\sigma = 8.1$). Three informants had experience from one DP vessel type only, while one informant had experience from 8 different DP vessel types. On average, operators had experience of 4.3 DP vessel types ($\sigma = 2.3$).

Characteristics of critical incidents in dynamic positioning

Four themes occurred in all 24 incidents and are considered the main results of the thematic analysis. These categories have been labelled as "Experience and Recognition," "Situation Awareness," "Decision Strategy," and "Human and Automation". A total of 16 sub-themes contributed to the description of critical incidents.

Experience and recognition

The incident accounts presented a picture of the DP operator as action takers during incidents, assessing the event based on prior experience, recognition and planning within operational limitations in order to avoid serious consequences. In 19 out of 24 incidents, the DP operator stated that he used experiences from similar past decisions, thus being indicative of a type of RPDM (Klein, 1993). In the remaining five incidents, the operators all stated that they were inexperienced with regards to the operation, position on board or vessel at the time of the incident. The informants explained how prior experiences affected incident decision-making and how they collected experiences. The experience-collection can be compared to a mental database of patterns utilised for immediately knowing how to respond to various situations (Lipshitz and Shaul, 1997). Specific patterns were often developed for vessel or operation characteristics and originated not just from real-life experiences, but also from training sessions or mental simulation. The DP operators referred to situations where they sat and imagined "what if" incidents and reflected on how to solve and prevent such situations – thus showing that mental simulation *pre factum* can be used as a proactive measure to improve operator response to critical incidents. In all incidents, the informants referred to work procedures and emergency procedures as the baseline pattern for performing operations.

Situation awareness

In critical incidents DP operators are directed by an overarching situation awareness related to the risks involved, and the level of awareness was determined through an assessment process (see e.g. Klein et al., 1986). The findings implied that the situation assessment process was affected by cues, expectancy, problem and goal identification, time limitation, uncertainty and the identification of base events.

Further, sudden changes and continuous updating characterised SA in critical incidents. The DP operators strove to reach an optimal level of SA through an assessment of the situation. The assessment of the situation involved an overarching evaluation of perceived potential risk and the problem awareness was triggered by a cue, e.g. a visual or auditory signal, in the external environment. The findings revealed that all 24 informants defined a goal for their further actions, although they did not fully understand the problem to be acted on.

The DP operators' sense of time in the incidents was also affected. In 19 of the incidents, the DP operator did not feel he had adequate time to think. Uncertainty was described as an issue affecting the DP operator in 17 of the incidents. In all of the 24 recollections, the incident brought with it a sudden shift in SA. In situations where the automated system no longer projects the next correct action, the human operator had to take over, yet in order to do so successfully the operator needed a good understanding of the situation. Consequently, the DP operators immediately engaged in an intense evaluation of the situation, producing a recipe for problem solving. In other words, the DP operator's SA was determined by the availability of information and the ability to undergo a process of situation assessment quickly enough to make a sound decision.

Decision strategy

Informants described how decision strategies were formed based on experiences and comprehension of the situation as well as by interacting with automated technical artifacts. In critical incidents, DP operators seek compatibility between experience and the actual situation to develop decision strategies. Furthermore, DP operators recognised a limited number of options in decision-making scenarios and therefore employed different decision strategies. Three types of decision strategies were identified from the data analysis: 1) *prescriptive use of procedure following*, 2) *flexible adaptation of a procedure* or 3) *purposeful violation of procedure*. A process of matching experiences with the ongoing situation usually did not produce a large number of alternative options to the DP operators.

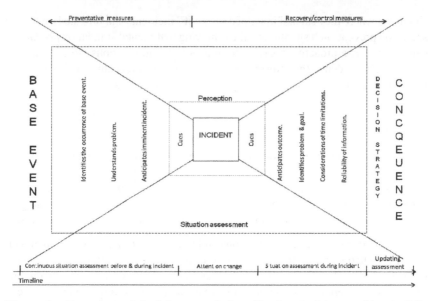

Figure 1.   Bow-tie model of characteristics affecting decision-making in DP.

Human & automation

The findings suggested that the DP operators' role transforms from monitoring to intervening during incidents. The DP operator's intervention involved reducing the level of automation during incidents. Furthermore, understanding and knowledge about the DP system affected the DP operators' actions in critical incidents. The DP operator was involved in the recovery of all 24 critical incidents. The DP system was not operational in seven of the incidents and the DP operator was forced to take over. In 17 of the cases, the DP operator chose to take over control of DP system. Whether or not the DP operators are forced to take over, or choose to, they all do so by manually controlling all or parts of the technical system. One reason for choosing to control the DP system manually was uncertainty and lack of knowledge about how the system would act.

A model of decision making in critical incidents

A bow tie model that represents the elements that affect the DP operators' situation assessment during critical incidents is shown in Figure 1. The model links base events (e.g. initiating factors of the incident) to consequences through a sequence of factors that affected DP operators' situation assessment and decision-making. DP operators were able to reason using facts, specific cues and general knowledge to identify base events and predict imminent events and final consequences.

The model shows that during an incident the human operator engaged in a cause-consequence assessment. On the left side of the model, the operator tried to identify the base event. Being able to identify the base event was helpful for imminent event prediction, assisting the human operator in identification of possible preventive actions, as would be expected in line with the NDM literature (Klein et al., 1989; Klein, 1993). On the right side of the bow-tie model, the DP operator predicted the outcome of the situation through an assessment that was affected by cues, anticipation, identification of problem and goal, consideration of time limitations and the reliability of the information. The assessment assisted the DP operator in finding potential control or recovery actions and formulating a decision strategy.

The model represents all four themes from the thematic analysis that were found to characterise decision-making in critical incidents in DP operations. Experience and Recognition was fundamental in development and maintenance of SA and relevant for the whole cause-consequence assessment. SA in turn influenced the choice of decision strategy. The model shows how SA, Experience and Recognition and Decision Strategy affected how the DP operator handled unexpected incidents involving the automated DP system.

Conclusion

A majority of operators' decision making during critical incidents in DP operations can be characterised by matching information to experience, recognition of salient cues and the creation of a few alternative courses of action – in accordance with the recognition-primed decision making framework (e.g. Klein, 1993). Further, experience and recognition affected the operator's SA that in turn influences decision strategies. The lack of time to react and recover during critical incidents in DP leads, to a large extent, to highly procedural decision strategies. As an adaptation to the fact that critical incidents impose strict time limitations, DP operators often perform mental simulations of imaginary, but potential, future incidents. This mental simulation might allow operators to react faster and more appropriately to critical incidents.

Acknowledgements

This research was funded by the Norwegian Research Council's MAROFF program (SITUMAR project, NFR Project number 217503).

References

Braun, V., and Clarke, V. 2006. "Using thematic analysis in psychology." *Qualitative research in psychology*, 3(2): 77–101.

Fossen, T. I. 1994. *Guidance and control of ocean vehicles* (John Wiley & Sons, West Sussex, England).

IMO. 2011. *STCW Including 2010 Manila Amendments, 2011 edition* (IMO Publishing, London, UK).

Jentsch, F., Barnett, J., Bowers, C. A., and Salas, E. 1999. "Who is flying this plane anyway? What mishaps tell us about crew member role assignment and air crew situation awareness." *Human Factors: The Journal of the Human Factors and Ergonomics Society*, 41(1): 1–14.

Kaber, D. B., and Endsley, M. R. 2004. "The effects of level of automation and adaptive automation on human performance, situation awareness and workload in a dynamic control task." *Theoretical Issues in Ergonomics Science*, 5(2): 113–153.

Kahneman, D., and Klein, G. 2009. "Conditions for intuitive expertise: a failure to disagree." *American Psychologist*, 64(6): 515.

Klein, G. A. 1993. "A recognition-primed decision (RPD) model of rapid decision making." *Decision making in action: Models and methods*, 5(4): 138–147.

Klein, G. A., Calderwood, R., and Macgregor, D. 1989. "Critical decision method for eliciting knowledge." *IEEE Transactions on Systems, Man and Cybernetics*, 19(3): 462–472.

Meshkati, N. 1991. "Human factors in large-scale technological systems' accidents: Three Mile Island, Bhopal, Chernobyl." *Organization & Environment*, 5(2): 133–154.

Norman D. A. 1990. "The 'problem' with automation: inappropriate feedback and interaction, not over-automation." *Philosophical Transactions of the Royal Society of London B*, 327: 585–593.

Øvergård, K. I., Bjørkli, C. A. and Hoff, T. 2008. "The bodily basis of control in technically aided movement." In S. Bergmann, T. Hoff and T. Sager (Eds.) *Spaces of Mobility: The Planning, Ethics, Engineering and Religion of Human Motion*, pp. 101–123 (Equinox Publishing., London, UK).

Parasuraman, R., Mouloua, M., and Molloy, R. 1996. "Effects of adaptive task allocation on monitoring of automated systems." *Human Factors: The Journal of the Human Factors and Ergonomics Society*, 38(4): 665–679.

Salas, E., Rosen, M. A., and DiazGranados, D. 2010. "Expertise-based intuition and decision making in organizations." *Journal of Management*, 36(4): 941–973.

Sarter, N. B., and Woods, D. D. 1995. "How in the world did we ever get into that mode? Mode error and awareness in supervisory control." *Human Factors: The Journal of the Human Factors and Ergonomics Society*, 37(1): 5–19.

Sørensen, A. J. 2011. "A survey of dynamic positioning control systems." *Annual Reviews in Control*, 35(1): 123–136.

Woods, D. D. and Sarter, N. B. 2000. "Learning from automation surprises and 'going sour' accidents." In N. Sarter & R. Amalberti (eds.) *Cognitive Engineering in the Aviation Domain*, pp. 327–354 (CRC Press, Boca Raton, FL).

Zsambok, C. E. 1997. "Naturalistic decision making: where are we now?" In Zsambok, C. E. & Klein, G. (eds.) *Naturalistic decision making*. pp. 3–16 (Lawrence Erlbaum & Associates, Mahwah, NJ).

OPERATORS IN TUNNEL CONTROL – ANALYZING DECISIONS, DESIGNING DECISION SUPPORT

Sebastian Spundflasch & Heidi Krömker

Department of Media Production, Ilmenau University of Technology, Ilmenau, Germany

To support the decision making process of tunnel control operators, specifically in view of the incoming flood of data, a Decision Support System should be designed within the research project "Real Time Safety Management for Road Tunnels". According to the requirements of tunnel control operators the system should reduce the amount of information and improve the information base for decision making. To determine the requirements for such a system, studies were performed in 12 tunnel control centers in Germany, Austria and Luxembourg. The aim was to point out the individual action strategies of the operators, in order to derive requirements for the system to be designed.

Introduction

Operators in tunnel control are monitoring the activities in road tunnels to ensure the safety of road tunnel users. In the case of incidents and risks, the operators must make decisions about intervention strategies to be initiated. The aim of these decisions is to avert potential danger from tunnel users and to establish a defined normal state again. As a basis for these decisions, information captured from the sensors installed in the tunnels are used e.g. the measuring of air quality or traffic density. This information is displayed to the operator at the human computer interface in the control centres. Starting from that point, the operator tries to understand the particular situation in order to decide on an appropriate intervention strategy. Depending on the order of the control center, some operators have to monitor up to 20 tunnels, which is around 30 miles of roadway. Due to the resulting flood of information, it is almost impossible for the operators to know the current situation in all tunnels. This increases the risk of overlooking details. While new detection technologies and sensors in the tunnels increase the amount of data, strategies are needed to improve the handling of the data for the operator. As part of the research project "Real Time Management System for Road Tunnels"[1] a Decision Support System should be designed that supports the operators by increasing the quality of data and provides a comprehensible basis for decision-making processes.

---

[1]ESIMAS joint project (Real Time Management System for Road Tunnels) is funded by the Federal Ministry of Economics and Technology. For more information visit www.esimas.de

Methodology

For us the fundamental question that arises is: how can we support the operator's decisions in management of incidents and risks. An analysis of the activities of operators should highlight situations in which the operators make decisions and reveal on what information base these decisions are made. In detail, this means to find out how decisions are made with a maximum of reliability and which decisions are afflicted with risk and uncertainty.

In decision theory, the decision process is generally divided into five phases (Betsch, 2005):

- Identification of a decision problem
- Generating of alternative behaviour
- Information gathering
- Evaluation and decision making
- Implementation of the selected behaviour

The Decision Support System may especially support the first and third phases, since the extraction of large amounts of information from the control system has the highest relevance here.

In their studies, Stanners & French (2005) were able to show that a direct correlation between the quality of decisions and a strong situation awareness exists. The concept of situation awareness describes "a person's perception of the elements of the environment within a volume of time and space, the comprehension of their meaning and the projection of their status in the near future." (Endsley, 1995). Endsley identified three stages of situation awareness:

- Level 1: perception of elements in the environment
- Level 2: comprehension of the current situation
- Level 3: projection of future status

Because the understanding of the situation is a prerequisite for making decisions, an analysis of operators in tunnel control should provide essential insights, especially with regard to the first two stages of the situation awareness. The findings should serve for the deriving of requirements for the design of a Decision Support System.

The central research question "How will a Decision Support System be able to support the decisions of operators in tunnel control" is broken down to the following sub-questions, inspired by the model of Endsley:

- What challenges are operators faced with when deciding?
- What are the points at which operators make their decisions?
- What information do operators need in different situations to make decisions? (Level 1)
- How are operators developing a comprehension of the overall situation starting from the perceived information? (Level 2)

Table 1.   The combination of methods used in the study.

Method Participants	Interview	Observation	Walkthrough
Operators (n = 18)	X	X	X
Tunnel-Managers (n = 12)	X		

- How can operators predict the future status on the basis of this understanding? (Level 3)

To answer these questions, we analyzed 12 tunnel control centres in Germany, Austria and Luxembourg and conducted interviews with tunnel managers and operators. In addition, operators were observed during their work and a cognitive walkthrough (Lewis & Wharton, 1997) through various situations of incidents and risks was conducted.

Results

The following presentation of the results is based on the sequence of the research questions.

Challenges that operators are faced with in the process of decision-making

- In event situations quick decisions must be made that require a quick assessment of the current situation
- A high parallelism of information perception is required. The monitoring of up to 20 tunnels makes it almost impossible to keep the situation in all tunnels in mind simultaneously.
- To assess the situation holistically, information is needed from various sources. The process of collecting this information is very time consuming and contradicts the requirement of quick decisions. This leads to decisions that are fraught with risk.
- In complex situations, the interplay of the key factors is to be understood.
- In infrequent situations, a decision uncertainty exists. When initiating intervention strategies this will be caused by a fear overlooked or forgotten something.

Points where decisions are made and the information that is needed

With regard to the use of a Decision Support System, four tasks could be identified which can be supported in an appropriate way. These are tasks of *monitoring, diagnosis, management of incidents* and *management of risks*. This is because decisions here are made on a broad information basis and in context of the challenges described above.

Monitoring
In the monitoring phase, the operators are pursuing two essential goals:

- Getting an overview of the current situation in the individual tunnels.
- Detecting deviations from the normal state.

The primary information resources are the video images from inside the tunnels. The video images are enriched with additional information from the process control system by the operators. Decisions to be made at this stage are:

- on which areas in the tunnels does the operator direct his attention and
- which additional information is needed from the control system for the assessment of the situation?

For this, different strategies were observed which heavily depend on the experience of each operator. This manifests itself in a targeted control of known critical areas or in the perception of indicators that point to certain events. For example, one operator perceived a break light on one of the many monitors, which may be an indication of a hazard to the operator that triggers the decision for a closer look. It could be observed that with an increasing number of tunnels to be monitored it becomes more difficult for operators to use these strategies for all monitoring objects.

Diagnosis
The aim of the diagnosis phase is the assessment of identified incidents or risks.

This assessment is carried out by:

- the analysis of the incident and its environment or the cause of a risk
- setting the results of the analysis in connection to the security level of the corresponding tunnel.

The operator has to make a decision whether an intervention is necessary or not. If so, an assessment and classification of the severity of the incident or risk must be done by the operator. For this assessment, a holistic understanding of the situation in the tunnel is required.

In this phase, the decision of the operator is to:

- select the relevant information from the control system
- combine the selected information elements in a way to get a holistic understanding of the situation, depending on his experience and knowledge.

This holistic understanding of the situation will be referred to the term security level. The security level influences the diagnosis of events, since it determines the boundary conditions. It is essential for a holistic understanding of an event and the event environment as well. For example, a car breakdown at night in an empty

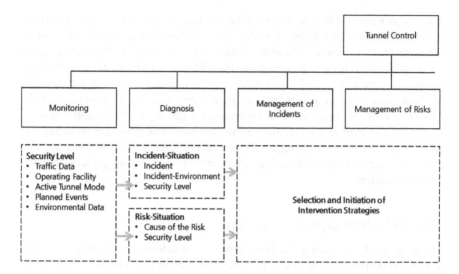

Figure 1.   Hierarchical Task Analysis of tasks to be supported and the relevance of the security level.

tunnel requires a different intervention strategy and thus a different decision than a car breakdown during the rush hour with a potentially high truck proportion.

The identified information elements the operators use to access the security level can be classified into five categories: traffic data, operating facility, active tunnel mode, planned events and environmental data. Figure 1 illustrates the influence of the perceived security level for the diagnosis of incidents and risks and its influence on the decision for the selection of intervention strategies.

Management of incidents and risks
In this phase, intervention strategies will be selected and initiated by the operator. The selected strategies aim to:

- improve the security level in the tunnel
- recover to the defined normal state.

The decisions taken by the operator heavily depend on his assessment of the current situation.

Operator's comprehension of the current situation

The strategies for linking the perceived information elements to an understanding of the current overall situation are only to some extent aware to the operators. The fact that it takes several years of experience in order to understand the current situation quickly leads to the assumption that the information elements are associated with information stored in memory (see Fracker, 1991).

The analysis showed that an understanding of the current security level is mainly developed through the combination of experiential knowledge with the information elements of the categories. These are: traffic data, operating facility, active tunnel mode, planned events and environmental data. This obviously creates the operators' model of the security level. The experiential knowledge of the operator, for example, is: knowledge about tunnel infrastructure and environment, experience with the behaviour of traffic participants or experiences in dealing with the control system. Because the overall understanding is generated individually, the Decision Support System can provide only the required information. But it should do it in a way that allows the operators to use their individual strategies for the comprehension of the current situation.

Operator's ability to predict future status

The strategies for predicting the development of situations are based on the tunnel-specific experience of the operators. They can predict the expected dynamics of information elements and their influence on the situation. For example, in a certain tunnel the operators know that in certain weather conditions fog will arise at the exit of the tunnel, which increase the traffic density and thus the risk of accidents. It could be observed that the ability to predict improves with the increasing experience of the tunnel operators. Because of that, the support of this capability is very limited and would require further investigation.

Conclusion

Methods

The fit of Decision Support Systems in highly complex work surroundings requires a methodical holistic approach to the tasks of the operators. The method combination of Hierarchical Task Analysis (Shepherd, 2001), interviews, and observation showed the various phases of action and the respective goals of the operators. From the partial actions, decisions could be extracted that can be supported by a Decision Support System. The method combination proved useful, as the Hierarchical Task Analysis and the interviews gave the task context for the Decision Support System. The observation and the walkthrough were suitable to define the functional and non-functional requirements on the system.

Findings

The findings of the analysis showed, that the decisions of the operator in all phases are accompanied by a certain degree of uncertainty:

- In the phase of *monitoring* the operator is aware of the fact that he cannot keep all the information in mind, to assess the situation in all tunnels.

- In the phase of *diagnosis* the operator has to struggle with the time pressure while the situation has to be understood holistically.
- In the phase of *incident- and risk management* it is their fear to have overlooked or forgotten something.

With regard to the support of a Decision Support System it became obvious that it is not the main decision that has to be supported, but the decisions in the previous phases. This is because in these phases the operator has to develop a holistic understanding of the situation, which serves as an information base for the main decision. In the end the main decision will only be as good as the information base on which it is made.

Requirements

Within the three phases a Decision Support System may support the operators as follows:

- In the phase of *monitoring* the Decision Support System must draw the attention of the operator to critical sections or detected events in the tunnels which require an assessment. This is important because all data cannot be evaluated by the operator in the quantity and at the required speed. In addition, the operator receives a summary of all monitoring objects, by which he or she can estimate the current security level in the various tunnels. The key factor here is the grouping of information, which is suitable for the task of the operator. In our case this could be realized through the information categories worked out. If all the information elements contained in an information category are in order, it is sufficient for the operator to check the category instead of checking the contained individual items.
- In the phase of *diagnosis*, the system has to offer possibilities for increasing the depth of information. The operators must be able to combine their knowledge and experience with the individual information elements. In event situations, the starting points for the information gathering are the video images from inside the tunnel. The system has to support the operators, by enriching the video images with further information of the event and the event environment. Important in assessing an event is the context in which the event takes place. An important task of the Decision Support System will be to show the relevant information that allows a holistic assessment of the environment. The attention of the operator must be directed precisely on factors that adversely affect the security level in the selected tunnel. A grouping of information, suitable for the task of the operator, plays an important role here too.
- In the phase of *incident- and risk management* the selection of the intervention strategy depends on the information base created in the phase of monitoring and diagnosis. The Decision Support System can only show changes of the state in this phase.

References

Betsch, T. 2005. *Wie beeinflussen Routinen das Entscheidungsverhalten?* Psychologische Rundschau, 56, 261–270.

Endsley, M. 1995. *Measurement of Situation Awareness in Dynamic Systems.* Human Factors 37, 1, 65–84.

Fracker, M. 1991. *Measures of Situation Awareness: Review and Future Directions.* (Armstrong Laboratories, Ohio).

Helander, M. G., Landauer, T. K. and Prasad, V. P., Eds. 1997. *Handbook of Human-Computer Interaction.* (Elsevier Science B.V., Amsterdam, Netherlands).

Lewis, C. and Wharton, C. 1997. Cognitive walkthroughs. In Helander, M., Landauer, T. H. and Prabhu, P. (eds). *Handbook of Human-Computer Interaction 1997* (Elsevier, New York), 717–732.

Shepherd, A. 2001, *Hierarchical Task Analysis*, (Taylor & Francis, London).

Stanners, M. and French, H. 2005. *An Empirical Study of the Relationship Between Situation Awareness and Decision Making.* Defence Science and Technology Organisation. (Edinburgh).

USER SYSTEMS

THE ANALYSIS OF INCIDENTS RELATED TO AUTOMATION FAILURES AND HUMAN FACTORS IN FLIGHT OPERATIONS

Wen-Chin Li[1], Davide Durando[1] & Li-Yuan Ting[2]

[1] *Safety and Accident Investigation Centre, Cranfield University, UK*
[2] *Institute of Transportation and Communication Management Science, National Cheng Kung University, Taiwan*

This research investigated the human role in human-automation inter-actions in landing incidents. A total of 257 incidents obtained from Part 121 of the FAA Aviation Safety Report System (ASRS) were related to automation during the landing phase. The aim was to clarify if the actions of the pilots after an inflight technical failure caused or prevented an incident/accident. The reports were analysed by Human Factors Analysis and Classification System (HFACS). Occurrences were subdivided into three automation failure categories. Whilst pilots are the last line of defence for aviation safety during automation failures, they sometimes commit active failures. Pilots' action and decision during automation malfunction made the aircraft land safely.

Introduction

Several decades have now passed since the introduction of advanced automation systems in aircraft cockpits and, despite the considerable efforts spent in dedicated research, human-computer interaction breakdowns are still a key safety concern (Billings, 1997). The literature has consistently shown that in aviation accidents the weak link in the chain is often the human component. Human error, as a con-tributory factor, accounts for 70–80% in incidents and accidents, and this figure has remained constant over the last decade, with no sign of decreasing (O'Hare & Chalmers, 1999). Tenney, Rogers and Pew (1998) suggested that failures of automa-tion or malfunctions could generate unexpected crew behaviour that could lead to adverse consequences. Commonly known as "automation surprises", these human-machine breakdowns are defined as "situations where crews are surprised by actions taken (or not taken) by the auto-flight system" (Woods & Sarter, 2000). The term "surprise" is adopted since crews are not fully aware of the automation's status or the aircraft's status, until some cue or event eventually trigger the operators' attention, and contradicts the current shared mental model between crews. These situations may occur as a consequence of undetected malfunctions or in a fully operative system affected by faulty inputs or "autonomous" system operations.

Wiener and Curry (1980) proposed that flight automation systems have superior perception and are able to deal with large amounts of information whilst simultaneously maintaining computational performances. Computers moreover are reliable and consistent in applying repetitive actions and monitoring small status changes, not easily detectable by crews. An important feature, which differentiates between automation and human operator is the capability to maintain constant performance over time without suffering fatigue. Through the mixing of the best features of human and computer, automated systems provide an incredible service; for example auto-throttles and flight management computers are more efficient in terms of fuel consumption and energy management which enables increased productivity. Integrated navigational systems are able to autonomously fly the aircraft with remarkable precision in both vertical and horizontal, and can enable landing in almost zero visibility. Automation has also contributed greatly to the current level of safety (Amalberti, 1999). A recent EASA 'cockpit automation survey' confirmed that the most common comments relating to the peaks of stress and workload generated by unanticipated situations requiring manual override of automation, are those which are difficult to understand and manage, this could subsequently create a surprise or startle effect (EASA, 2012). The evolution of the advanced technology in the cockpit, such as auto-thrust (AT), enhanced ground proximity warning system (EGPWS), flight management system (FMS) or flight data recorders (FDR) are only a part of the solution to overall safety improvement for the flight operation. Aircrew must also evolve in order to keep up with the changing environment. In a worldwide survey of causal factors in commercial aviation accidents, it was found that in 88% of the cases the crew was identified as a causal factor; in 76% of instances the crew were implicated as the primary causal factor (CAA, 2004). Therefore, the study of the human errors is of key importance in order to understand how incidents/accidents happen and how they can be prevented.

Human factors analysis and classification system (HFACS) is a human error taxonomy which based its theoretical framework on Reason (1990). In this framework, accidents and incidents include two kinds of failures: active failures, with direct impact on front-line operators, in this case flight crews; latent failures, dormant problems with indirect impact on operators, often found in upper levels of the system, including management and supervision (Wiegmann & Shappell, 2003). The incidental outcome is a 'trajectory' path, which crosses a series of levels of failures. This path associates the active failures performed by front-line operators with the latent failures generated in upper levels and which are normally dormant in the system. Combined with other contributory factors, these failures are able to break the system's defense and leading to adverse outcomes. Reason (1997) explained how the human component is commonly a contributory factor in an incident, given the fact the systems are designed, maintained, operated and managed by humans. Is it indeed a logical consequence that pilot's actions and decisions are able to affect system's safety, or pilots are the last line of defense for safety during automation system failure. The aim of the research was to clarify if the actions of the pilots taken after an automation failure either caused or prevented an incident/accident.

Methodology

Data

The data were obtained through a detailed search of the aviation safety report system (ASRS), taking into account only commercial flights (FAA part 121) that involved automation-related incidents in the landing phase. The ASRS database is coded by NASA analysts with a specific taxonomy used to classify type of incident; the following criteria were inputted into the research field on the database form ASRS:

* Data: 257 reported incidents between January 2000 and June-2013.
* Federal Aviation Regulation: FAR Part 121 – Operating Requirements of Commercial Aviation.
* Flight phases: Descent, initial approach, final approach, landing.
* Automation: Failed, critical malfunction, minor malfunction.
* Component Breakdown: approach coupler, autoflight system, autothrottle/speed control, ILS/VOR, autoland, flight control computer, radio altimeter, EGPWS, ILS facility, Alt hold/capture, database, INS/NAV system, PFD, MCP, air data computer, flight director, HUD.

Analysis framework

Wiegmann and Shappell (2003) developed HFACS to link the theoretical framework of Reason's model with an operational tool for accident analysis and classification. This research applied HFACS as an analysis framework. Level 1 of HFACS, underlying the majority of events, focuses on errors and violations. Errors are further divided into; skill-based errors, decision errors, perceptual errors, violations in routine and exceptional errors. In level 2 of HFACS, the focus is on the preconditions of unsafe acts. It is necessary to examine more deeply and explore what caused the unsafe acts to occur in the first place. Wiegmann and Shappell (2003) classified preconditions into the following seven subcategories: adverse mental state, adverse physiological states, physical/mental limitations, crew resource management, personal readiness, physical environment, and technological environment. As a result of the format of the anonymous reporting narratives, it was not possible to gather information regarding the management levels of airlines involved in the incidents and no extra information was available regarding organizational and supervision levels. Although HFACS has four levels, the level 4 and level 3 have not been included in this study, it is due to the lack of information as a result of the confidential nature of the safety reports used in this research.

Coding process

Content analysis uses a set of procedures to make valid inferences from text. It systematically identifies properties, such as the frequencies of terminology. The data were coded separately using content analysis into the HFACS framework by level 1 (unsafe acts of the operators) and level 2 (preconditions for unsafe acts), by an aviation human factors specialist and a flight engineer. Each incident was classified into a dedicated MS Excel framework structured to accommodate information provided by ASRS query and the HFACS taxonomy. The presence or absence of any HFACS category was assessed in each narrative respectively with 1 or 0. To avoid over representation, each category was counted a maximum of once per accident and the count acted as indicator of presence or absence of each HFACS category. Li, Harris and Yu (2008) have demonstrated inter-rater reliabilities of HFACS categories between 72.3% and 96.4%. The average of the inter-rater reliabilities of the data gathered from Indian accidents, calculated by percentage of agreement, was 87% (Gaur, 2005); the US data showed an average inter-rater reliability of 76% (Shappell & Wiegmann, 2001).

Results and discussions

A total of 257 Commercial Aviation ASRS reports associated with automation-related incidents which occurred during approach and landing were analysed. The sample consisted of seven aircraft manufacturers with 99 Boeing (38%), 52 Airbus (20%), 24 McDonnell Douglas (9%), 20 Bombardier (8%), 15 Embraer (6%), 3 Dash 8 (1%), 1 Citation X (0.4%). Because of the voluntary nature of reporting, the type of aircraft was not available for 43 cases (16%). The ASRS taxonomy included eleven types of incident (see Table 1).

By applying HFACS to incidents data, it can be found that the highest frequency occurrence at Level 1 were (in order, from highest to lowest): 1) decision errors included inappropriate procedures, inappropriate knowledge of the systems and failure to prioritise attention; 2) skill-based errors, including inadvertent operation, failure to set FMC, failure to set ILS, failure to set MCP or failure to set A/T; 3) violations, including applied inappropriate SOPs, and violation of SOP; 4) perceptual errors, including spatial disorientation and instrumentation misreads. At Level-2 of the HFACS framework the most frequent category of error occurrence

Table 1. Incident type.

Incident Type	Frequency	Percentage
Altitude Cross Restriction not met	52	20.23%
Altitude Overshot	17	6.61%
Excursion from assigned Altitude	21	8.17%
Track/Heading	57	22.18%
Speed	12	4.67%
Unstable Approach	22	8.56%
LOC-I	13	5.06%
Encounter CFIT	10	3.89%
Uneventful	46	17.90%
Procedure Deviation	5	1.95%
Airborne Conflict	2	0.78%

related to crew resource management, which included lack of teamwork, failure to cross-check crew actions and lack of leadership. The second highest frequencies related to technological environment, followed by adverse mental state, which encompasses task saturation, over confidence, tunnel vision and distractions. Personal readiness took into account mainly inadequate training and fatigue, followed by physical environment exclusively bad weather conditions and terrain. There were only a very small number of physical/mental limitations and adverse physiological state reported.

The frequencies and percentage of HFACS categories and reliability of Cohen's Kappa and percentage agreement are shown in Table 2. Decision errors were related to active failures by pilots. Perception error and skill-based errors were related to automation malfunction. Pilots' failures to prioritise attention encompassed task-related issues were connected to automation failure and over-reliance on automation in flight operations. With regards to decision error, the majority were concerned with 'improper use of automation', while inadequate knowledge of system and inappropriate procedure were connected to task-related issues. Perception and violation accounted for relatively low frequencies of unsafe acts, in turn spatial disorientation and violation of SOP during landing.

For pilot-automation breakdowns, the results demonstrated that skill-based error was involved in incidents less than decision errors, as pilots still can manually control the aircraft using problem-solving skill by manipulating alternatively backup systems during automation malfunction. It is the reason that pilots are the last defence for flight safety. There were some incidents involved in failure of monitoring automation, which related to failure to 'monitor' the status of FMC, ILS, auto-throttle or auto-flight system. According to Reason's Model (1990), human

Table 2. Frequencies, Percentage and Inter-raters Reliability of HFACS Categories for Level-1 and Level-2.

HFACS Category		Frequency (wen chin)	Percentage (wen chin)	Cohen's Kappa	Agreement of percentage
Level-2 Preconditions for Unsafe Acts	Technology environment	132	51.3 %	.812***	95.0%
	Physical environment	34	13.2 %	.548***	92.2%
	Personal readiness	58	22.6 %	.843***	94.9%
	Crew resource management	156	60.7 %	.910***	95.7%
	Physical/mental limitation	13	5.1%	NC	94.9%
	Adverse physiological states	10	3.9%	NC	96.1%
	Adverse mental states	122	47.5 %	.707***	85.2%
Level-1 Unsafe Acts of Operators	Violations	32	12.5 %	.329***	90.3%
	Perceptual errors	26	10.1 %	.444***	93.0%
	Skilled-based errors	101	39.3 %	.749***	87.5%
	Decision errors	127	49.4%	.222***	61.5 %

NC = not countable.

Table 3.   Significant associations between HFACS level-1 and level-2.

Level 2 association with Level 1 categories	Chi-square (χ2)			Goodman & Kruskal τ		Odds ratio
	Value	df	P	Value	p	
Technology environment * Decision errors	20.611	1	.000	.080	.000	3.292
Adverse mental states * Decision errors	7.163	1	.007	.028	.008	1.964
Crew resource management * Decision errors	23.335	1	.000	.091	.000	3.613
Technology environment * Skill-based errors	53.927	1	.000	.210	.000	9.731
Adverse mental states * Skill-based errors	31.817	1	.000	.124	.000	4.516
Crew resource management * Skill-based errors	60.284	1	.000	.235	.000	12.740
Adverse mental states * Perceptual errors	5.493	1	.019	.021	.019	2.748
Crew resource management * Perceptual errors	4.884	1	.027	.019	.027	2.987
Technology environment * Violations	7.392	1	.007	.029	.007	3.405
Adverse mental states * Violations	6.637	1	.010	.026	.010	2.750
Crew resource management * Violations	8.588	1	.003	.033	.003	4.019
Personal readiness * Violations	4.664	1	.031	.018	.031	2.335

behavior is the result of the relationship between personal factors and environmental factors. The results showed solid evidence for these assumptions with strong associations between categories of HFACS level 2 and level 1. The landing phase in advanced automated cockpits represents a highly dynamic situation with complex operations to be executed in short periods of time. These operations make cognitive demands and require high rates of information processing, fast communication between crews and critical systems settings to be inputted (Parasuraman & Manzey, 2010). The exposure of flight crews in such a complicated and time-critical phase increases the likelihood of the occurrence of active failures. Analysis of the strength between HFACS level 2 and level 1 showed a further 12 pairs of significant associations summarized in Table 3. The strength of these associations is evaluated by means of a χ^2 analysis supported by a further calculation of proportional reduction in error (PRE) in function of Guttmann and Kruskal's tau, indicating the strength of directional relationship between categories of human factors in level 1 and level 2 of HFACS.

Crew resource management (CRM) showed an extremely high correlation with skill-based errors, with an odds ratio of 12.7 times more likely to occur when cases of poor CRM are involved, and providing a PRE of .235%. The odds ratio evaluated for decision errors gave an increased chance of occurrence of 3.6 times providing a 9% value of PRE. The significant differences in term of statistical analysis allow for developing accident prevention strategies in relation to the nature of the human-automation breakdown. In particular, the relationship between CRM and skill-based errors is the key issue for developing the effective human-automation interactions training, as CRM techniques seemed to be more prone to fail in the two-side control and cross-check of flight operations. The design of PF (pilot flight) and PM (pilot monitor) is to improve the safety of flight operations, however, it has potential risk of failure to set and failure to monitor the automation if CRM failure as well.

This data-driven result showing human-automation breakdown was widespread throughout the occurrences analysed, specifically affecting flight-path deviation and, high risk events such as unstable approaches and CFIT. Unstable approaches are often symptomatic of a lack of prior manoeuvre planning and represent a serious safety issue since they demand critical cognitive efforts in order avoid adverse outcomes, mainly concerning go-around decisions. Incidents related to landing were frequently related to poor flight path monitoring and improper settings of the FMC/MCP and could potentially lead to disaster.

Conclusion

This research investigated the role of the pilots in the breakdown of human-automation interactions involved in landing incidents. The aim of the research was to clarify if the actions of the pilots taken after an inflight technical failure either caused or prevented an incident/accident. The results show that pilots are the last line of defence for aviation safety during automation failures, though sometimes pilots commit active failures. Reason (1997) has suggested that there is a 'many to one' mapping of precondition of unsafe acts and the actual errors themselves, making it difficult to predict which actual errors will occur as a result of which pre-conditions. Those categories at HFACS level 2 exhibited Reason's classic 'many to one' mapping of psychological precursors to active failures of decision-errors, skill-based errors, perceptual errors and violations on the level 1. Incidents involved in failure of monitoring automation which related to failure to 'monitor' the status of FMC, ILS, auto-throttle or auto-flight system could potentially lead to disaster. For automation failures, particularly in loss of control situations, it is the pilots' action and decision-making that ensure the aircraft to land safely.

References

Amalberti, R. 1999, Automation in Aviation: A human factors perspective. In D. Garland, J. Wise & D. Hopkin (Eds.), *Aviation Human Factors*, pp. 173–192, (Lawrence Erlbaum Associates, Hillsdale, New Jersey).

Billings, C. E. 1997, *Aviation automation: The search for a human-centered approach*. (Lawrence Erlbaum Associates, Mahwah, NJ).

Civil Aviation Authority. 2004, *Flight Crew Reliance on Automation* (CAA report 2004/10, Retrieved from http://www.caa.co.uk/docs/33/2004_10.pdf.

European Aviation Safety Agency. 2012, *European Aviation Safety Plan 2012–2015*. Retrieved from http:// www.easa.europa.eu/sms

Gaur, D. 2005, Human factors analysis and classification system applied to civil aircraft accidents in India. *Aviation, Space, and Environmental Medicine* 76, 501–505.

Li, W. C., Harris, D. and Yu, C. S. 2008, Routes to failure: Analysis of 41 civil aviation accidents from the Republic of China using the human factors analysis and classification system. *Accident Analysis and Prevention* 40, 426–434.

O'Hare, D. and Chalmers, D. 1999, The incidence of incidents: A nationwide study of flight experience and exposure to accidents and incidents. *The International Journal of Aviation Psychology* 9, 1–18.

Parasuraman, R. and Manzey, D. H. 2010, Complacency and bias in human use of automation: An attentional integration. *Human Factors* 52, 381–410.

Reason, J. 1997, *Managing the risks of organizational accidents.* Aldershot, England: Ashgate.

Reason, J. 1990, *Human error.* Cambridge, England: Cambridge university press.

Tenney, Y. J., Rogers, W. H., & Pew, R. W. 1998, Pilot opinions on cockpit automation issues. *The International Journal of Aviation Psychology* 8, 103–120.

Shappell, S. A., & Wiegmann, D. A. 2001, Human error analysis of commerical aviation accidents: Application of the human factors analysis and classification system. *Aviation. Space, and Environmental Medicine* 72, 1006–1016.

Wiegmann, D. A., & Shappell, S. A. 2003, *A human error approach to aviation accident analysis: The human factors analysis and classification system.* Aldershot, England: Ashgate.

Wiener, E. L., & Curry, R. E. 1980, Flight-deck automation: Promises and problems. *Ergonomics* 23, 995–1011.

Woods, D. D., & Sarter, N. B. 2000, Learning from automation surprises and "going sour" accidents. In N. B. Sarter & R. Amalberti (Eds.), *Cognitive engineering in the aviation domain,* pp. 327–353, (Lawrence Erlbaum Associates Mahwah, New Jersey).

USING N SQUARED ANALYSIS TO DEVELOP THE FUNCTIONAL ARCHITECTURE DESIGN FOR THE HIGH SPEED 2 PROJECT

Peter Nock[1], Eddie Walters[1], Nigel Best[2] & Wilson Fung[2]

[1]*HS2*
[2]*Network Rail*

HS2 Ltd is promoting and delivering the new High Speed Rail network on behalf of the Department for Transport. As part of the development a process was needed to develop a functional architecture and then use this to identify all high level interfaces. In addition, the process would be used to align RAMS, safety, requirements, commissioning and overall systems demonstration and assurance. N Squared analysis was selected as the appropriate technique for this work. This paper describes the process used and the work carried out to date.

Introduction

The Operations and Engineering strategy for HS2 has been developed based upon considering a hierarchical definition of the railway as a complete system. The underlying philosophy being applied behind this strategy is a classical "middle-out" systems engineering approach, where the starting point is to consider the functions required of the system and then to connect these to both the inputs and the outputs required of the system. A simple hierarchy is presented in Figure 1.

The business development element of the diagram aligns functions to output capabilities and once established enables requirements established at railway level to flow into function level physical and geographical requirements.

Once an apportionment of railway level requirements down to functional requirements has been established, a functional to asset/physical allocation of requirements can then be completed and in this case aligned with RAM, safety and performance requirements through the system.

An initial functional architecture for the HS2 system has been developed by the HS2 Operations & Engineering team, with the functions defined based upon the functional classes identified in Table 1.

The function classes for HS2 have been aligned to Technical Standards for Interoperability (TSI) classifications, the formal standards that will apply to HS2. A key requirement on a number of function classes is the need to formally "place into service" which has a significant impact on commissioning and achieving HS2 start of service.

LEVEL

CAPABILITIES
(BUSINESS OUTPUTS)

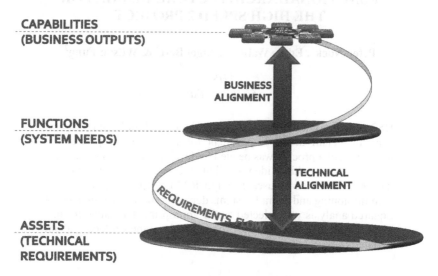

BUSINESS
ALIGNMENT

FUNCTIONS
(SYSTEM NEEDS)

TECHNICAL
ALIGNMENT

REQUIREMENTS FLOW

ASSETS
(TECHNICAL
REQUIREMENTS)

Figure 1.   HS2 Performance Hierarchy.

Table 1.   HS2 Function Classes.

Function Class	TSI Class	Formally Placed into Service (Y/N)
Civil Eng	Infrastructure	Y
Stations	Infrastructure	Y
Train Servicing and Storage	Infrastructure	Y
Rolling Stock	Rolling Stock	Y
Energy	Energy	Y
Track	Infrastructure	Y
Control & Communications	Trackside/On-Train Control command & Signalling	Y
Telematic Applications	Telematic Applications	N
Operational	Operations & Traffic Management	N
Maintenance	Maintenance	N
Passenger	None	N

Having identified a list of core functions required to deliver a HS2 operational train service, a formal approach (N Squared) has been applied to identify high-level function interfaces.

The functional interfaces for HS2 are currently being further developed and mapped to physical HS2 assets and locations along route. This will allow the minimum commissioning needs for sections of route to be identified.

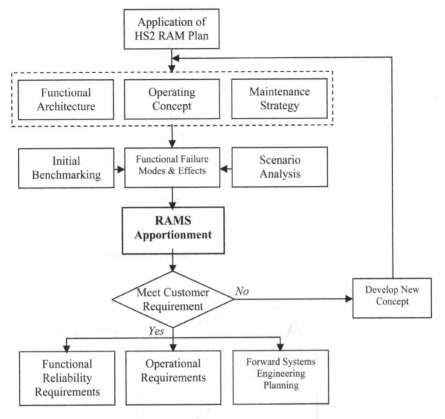

Figure 2.   HS2 Systems Approach 1.

Methodology

The definition of the functional architecture for HS2 is aligned to the broader systems engineering approach being progressed through the design by the HS2 Technical Directorate and aligned to achieving compliance against EN50126. This document resides within the functional architecture area of systems development as identified in Figure 2.

A high level HS2 functional architecture has been developed as a first step towards identifying and understanding key system requirements and interdependencies. This initial functional architecture has been developed by key discipline engineers using references to related architectures where appropriate.

Having deduced an initial set of functions for the railway, a structured N2 approach has been applied to identify the interfaces between functions. N2 (INCOSE, 2012) is a systematic approach to identify, tabulate, analyse and document functional interfaces. The approach can be highlighted via considering a simple functional flow block diagram (FFBD) as per Figure 3 and how each of the potential interface lines could be considered as per the N2 diagram in Figure 4.

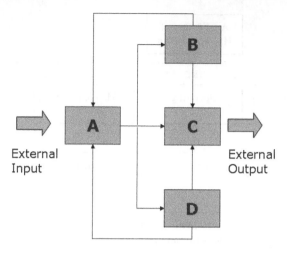

Figure 3.   Generic Functional Flow Block Diagram (FFBD).

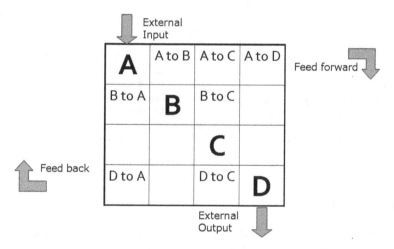

Figure 4.   N Squared Diagram.

Note the unusual positioning of the passenger in the function classes. This is a recognition that a high performing system needs to be designed taking account of the passenger in that system.

Ongoing functional architecture development

Several iterative steps are involved in the development of the architecture as outlined below:

1. Initial identification of functional architecture
2. Comparison of functions to other reference sources

3. N Squared diagram for each function
4. Cross comparison of function diagrams and resolution of conflicts
5. Vertical comparison of key functional flows
6. Formalise and test against operational scenarios
7. Control changes and manage via requirements

The definition of the functional architecture has been designed to align with the HS2 Programme Requirements Set and will be used as one mechanism for testing the completeness of the Requirement set as a whole.

It should be noted that as with other areas of the system, the functional architecture will continue to evolve as the design progresses, with the emphasis at this stage related to understanding the impacts on key corridor development decisions.

It is anticipated that as the design for HS2 progresses, the functional architecture will be further disaggregated and aligned to key procurement interfaces via the requirement set and used as the basis for gauging system completion and readiness for service.

The functional architecture is used throughout the design process, linking into the following key systems assurance areas;

- RAM planning and evaluation
- Safety planning and evaluation
- Requirements mapping and development
- Programme commissioning planning
- Overall systems demonstration & assurance

In terms of ergonomics, it will drive the requirements of the system and system safety approaches and inform task analysis, human reliability assessments and man/machine functional allocation.

In this way, the functional analysis becomes the fulcrum for whole system design.

Reference

INCOSE. 2012, *Systems Engineering Handbook: A Guide for System Life Cycle Processes and Activities*, version 3.2.2. (International Council on Systems Engineering (INCOSE), San Diego, CA, USA), INCOSE-TP-2003-002-03.2.2.

HUMAN FACTORS ASPECTS OF DECOMMISSIONING: THE TRANSITION FROM OPERATION TO DEMOLITION

Clare Pollard

AREVA RMC

Nuclear decommissioning is a complex activity requiring support from operators, regulators and specialist contractors. The systematic removal of engineered safety systems and the move to a reliance on administrative controls along with training, staffing, procedural and equipment modification presents licencees with significant safety challenges. It is vital that these challenges are identified, understood and managed carefully to ensure nuclear safety is maintained throughout the process. This paper discusses the transition from an operating facility to a demolished site in terms of the various human factors elements of decommissioning and suggests some areas of research to be undertaken in order to provide support to the site licensees throughout the process.

Introduction

This purpose of this paper is to discuss the transition from an operating facility to a demolished site from a HF perspective and to suggest some areas of research to be undertaken by the site licensees to ensure the decommissioning activities are appropriately managed. Most of the elements presented, unless referenced, are based on the author's own observations of decommissioning across a number of sites.

Decommissioning

Decommissioning is the process of processing redundant nuclear facilities that have reached the end of their operational life. The objective of decommissioning a nuclear installation is to return the site either to an unrestricted, de-licensed condition; or to a state where the land can be used for suitable alternative uses. Decommissioning is usually conducted in stages: post-operational clean out where the bulk of the radioactive material is removed from the facility; initial dismantling and removal of contaminated parts or care and maintenance to allow radioactive materials to decay– and finally dismantling of the facility; demolition of the structure; and remediation of land and water to meet an agreed end-state for future use.

There are a number of sites in the UK currently undergoing decommissioning, each with their unique challenges for their management teams. For some of these sites, decommissioning was given little thought at the time of their design and construction, leading to significant safety and technical challenges in current, more highly regulated, times.

Human Factors aspects of decommissioning

Decommissioning can be viewed as just another part of the plant lifecycle for which safety cases and assessments are needed to exert an appropriate level of control over the activities to be carried out. HF and the way the claims, arguments, and the evidence approach is developed and assessed for the safety case remains unchanged from the approach taken to supporting the earlier stages of a facility's lifecycle. The difference at this stage is the nature of the safety case itself and the additional demands and reliance that decommissioning places on humans due to vulnerabilities in the changing plant state and difficulties in implementing engineering controls (Internation Atomic Energy Agency, 2004a, b; Office for Nuclear Regulation, 2013; Energy Institute, 2010). Decommissioning can result in short-term increases in risk and hazard in order to achieve the longer-term overall reduction.

The nature of the decommissioning process requires the systematic withdrawal and removal of engineered safety systems that operators have worked alongside for many years during the operational life of the facility. Simultaneously, there is a shift to new administrative controls as the operator undertakes unfamiliar and often novel tasks. These shifts have to be appropriately defined and controlled by the decommissioning management team. The main differences between operation and decommissioning activities are illustrated in Table 1 taken from Internation Atomic Energy Agency (2004b).

The readiness of the site infrastructure to deal with the change is entirely dependent on the site and the type of facility being decommissioned, e.g. small facility on a large site versus the only facility on the site. The deluge of specialist contractors or new procedures may be overwhelming for a small organisation to manage, and provision of Work Supervisory Officers or the creation and approval of a revised safety management system requires significant effort and is predicated on the presence of available staff with a detailed understanding of the facility. However, as mentioned previously, the existing staff will not be familiar with the plant in its progressively decommissioned state and therefore staffing, training and competency management are key elements to be considered.

It is likely that new equipment may be required for aspects of decommissioning and the design of this equipment should be considered alongside existing equipment. When interfacing with existing equipment, consistency of interfaces will be a key design requirement. Some existing equipment may be decades old and so the most ergonomically designed equipment may not have the same control philosophy, thereby creating potential error traps. Therefore, meeting user expectations and having similar control philosophies may outweigh 'good' design to modern standards in ensuring human error is minimised. Also, we need to consider the

Table 1.   Comparison of the Operating and Decommissioning Regimes (Internation Atomic Energy Agency 2004b).

Operations	Decommissioning
Reliance on permanent structures for the operating life of the facility	Introduction of temporary structures to assist dismantling
Safety management systems based on an operating nuclear facility	Safety management systems based on decommissioning tasks
Production oriented management objectives (except perhaps in research facilities)	Project completion oriented management objectives
Routine training and refresher training	Retraining of staff for new activities and skills or use of specialised contractors
Permanent employment with routine objectives	Visible end of employment – refocus of the staff's work objectives
Established and developed operating regulations	Change of regulatory focus
Predominant nuclear and radiological risk	Reduction of nuclear risk, changed nature of radiological risk, significantly increased industrial risk
Focus on functioning of systems	Focus on management of material and radioactivity inventory
Repetitive activities	One-off activities
Working environment well known	Working environment unknowns possible
Routine lines of communication	New lines of communication
Low radiation/contamination levels relatively unimportant	Low radiation/contamination levels important for material clearance
Access to high radiation/contamination areas unlikely or for a short time	Access to high radiation/contamination areas for extended periods
Routine amounts of material shipped off-site	Larger amounts of material shipped off-site
Relatively stable isotopic composition	Isotopic composition changing with time

operators of the new equipment: will it be existing operators, or contractors who are new to the facility, or, more likely, a mix of the two?

The cultural aspects of decommissioning also need to be considered. Licencees may perceive HF aspects of work to be less important than during normal operations as the radiological hazards on site reduce. Workforce morale may also need to be considered as the decommissioning process places some reliance on a strong safety culture, which may be degraded as the operators face a future without a facility they may have worked in for the past 30–40 years. So, where decommissioning is essentially about putting yourself out of work, could there be a tendency to introduce delays, or will it be difficult to recruit and retain staff? A lack of continuity amongst the decommissioning team will negatively impact the retention of knowledge and the effectiveness of organisational learning that is possible.

Retaining workforce knowledge is a challenge for decommissioning projects. The workforces who designed, commissioned, constructed and operated the nuclear facilities are often different to those who will decommission them. This presents a threat to the effectiveness and safety of the decommissioning process as key information can be lost. Site experience presents a range of issues that have resulted from poor knowledge management practices. For example, plant drawings that show the facility as it was designed rather than as it was built, experts being sought from retirement for their unique plant and operational knowledge, and inadequate nuclear material storage records. Also, changes in nuclear safety culture since the facilities were originated may mean that decommissioning operations may face a degree of uncertainty that places increased demands on operators to interpret and respond to situations that differ from those found on operating facilities.

The consideration of all aspects of decommissioning management lies with the licensee. At times, the perception may be that the hazard is reducing and therefore attention to HF is not as warranted as for operations. The ONR research website defines a requirement for the licensees "to identify and develop ways to improve the management of decommissioning, including for example, managing the change from operations to decommissioning, putting in place processes which support work being undertaken, working methods to be appropriate for decommissioning (rather than operations) and improving the decommissioning culture".

Conclusion

Given the number of ageing facilities in the UK, nuclear decommissioning will become an industry, with the increasing need to capture expertise, define processes and best practice, state training requirements, etc. This paper illustrates that, despite the perception of reduced risk, the requirement for adequate consideration of HF within the decommissioning process remains a key safety requirement as it is for the operational phase of a facility. Decommissioning presents a different set of challenges to the HF practitioner, but the importance of HF to safely managing the process remains unchanged. So, what research could be undertaken to develop HF 'solutions' to these challenges?

- Research: development of a guide on decommissioning management good practice incorporating lessons learnt from the UK and beyond.
- Decommissioning practice: HF support to the development of novel approaches or technologies.
- Decommissioning practice/preparation: Effective knowledge capture, transfer and retention techniques.
- Research: culture/welfare – guidance on how to maintain safety when working towards unemployment.

In essence, relevant good HF practice equally applies to decommissioning tasks as it does for normal operations but we need to consider providing more support to key aspects of the decommissioning process and management.

References

Energy Institute 2010, *Guidance on Managing Human and Organisational Factors in Decommissioing*. March.

Internation Atomic Energy Agency 2004a, *Safety Reports Series No. 36. Safety Considerations in the Transition from Operation to Decommissioning of Nuclear Facilities*. Vienna.

Internation Atomic Energy Agency 2004b, *Technical Report Series No. 420. Transition from Operation to Decommissioning of Nuclear Installations*. Vienna. April.

Office for Nuclear Regulation 2013, *Nuclear Safety Technical Assessment Guide. Decommissioning. Revision 3*. May.

EAST: A METHOD FOR INVESTIGATING SOCIAL, INFORMATION AND TASK NETWORKS

Neville A. Stanton

Civil, Maritime, Environmental Engineering and Science Unit,
Faculty of Engineering and the Environment,
University of Southampton, Highfield, Southampton, SO17 1BJ, UK

This paper presents the Event Analysis of Systemic Team-work (EAST) method as a means for specifying, analysing and evaluation complex sociotechnical systems. The method is based on identification of task, social and information networks in systems as well as their interrelations. The origins of the method and some previous investigations are mentioned together with recent developments in the analysis approach. Implications for investigations into Distributed Situation Awareness and Resilience Engineering are presented. The paper concludes that EAST may be of most use in system design and development. It would support the rapid modelling and prototyping of organisations early in the system design life cycle.

EAST methodology

Stanton et al (2005) proposed Event Analysis of Systemic Team-work (EAST) as an integration of methods for analysing complex sociotechnical systems. Since its conception, EAST has been applied in many domains, including naval warfare (Stanton et al, 2006), aviation (Stewart et al, 2008), air traffic control (Walker et al, 2010a), emergency services (Houghton et al, 2006), energy distribution (Salmon et al, 2008), and railway maintenance (Walker et al, 2006). The approach is gaining momentum as well as showing its domain independence. The analysis has demonstrated how distributed cognition (Hutchins, 1995) for complex systems could be represented by networks, with the distinct advantage that networks enable both qualitative and quantitative investigations. It has been argued that the multifaceted nature of the different networks (i.e., social, task and information networks) has revealed the aggregated behaviours that emerge in complex sociotechnical systems (Stanton et al, 2008). This representation was proposed as an alternative to the reductionistic approaches often used to understand systems, which presented systems in their constituent parts but fail to capture the system as a whole. Walker et al (2010a) suggested that the insights gained by network modelling were superior to the traditional ethnographic narrative which has previously been used to describe distributed cognition because they present graphical models of systems. Griffin et al (2010) went further to show how the EAST method offers insight into system failure. Again, the cited advantage of the approach was the non-reductionist,

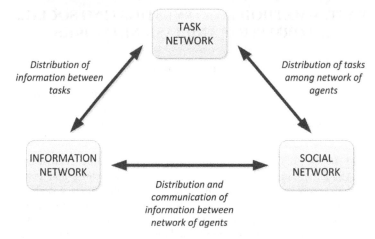

Figure 1.   Network of networks approach used by EAST.

non-taxonomic, method for analysing non-normative behaviour of systems. Whilst EAST does not employ taxonomies in the analysis, the resultant network structures may be classified into archetypes.

The systemic approach allows system interactions to be understood in their entirety (Plant and Stanton, 2012). EAST is underpinned by the notion that complex collaborative systems can be meaningfully understood through a network of networks approach (see Figure 1). Specifically, three networks are considered: task, social, and information networks. Task networks describe the relationships between tasks and their sequence and interdependences. Social networks analyse the organisation of the system (i.e., communications structure) and the communications taking place between the actors working in the team. Finally, information networks describe the information that the different actors use and communicate during task performance (i.e., distributed situation awareness). Each of these approaches have been presented independently in other papers. Farrington-Darby et al's (2006) presentation of task diagrams in a study of railway controllers was an example of a task network. Furniss and Blandford's (2006) presentation of communication channels in emergency medical dispatch teams was an example of a social network. Sanderson et al's (1989) analysis of verbal protocols for a process control task was an example of an information network. What EAST does is bring these three networks together into the same analysis framework.

The EAST framework lends itself to in-depth evaluations of complex system performance, examination of specific constructs within complex socio-technical systems (e.g. situation awareness, decision-making, teamwork), and also system, training, procedure, and technology design. Whilst not providing direct recommendations, the analyses produced are often highly useful in identifying specific issues limiting performance or highlighting areas where system redesign could be beneficial.

Social Network Analysis

Social Network Analysis (SNA) offers a means of analysing the network as a whole as well as the behaviour of individual nodes and their interactions. As such, SNA is potentially a very powerful tool for systems ergonomics. Whilst it has traditionally been applied to the analysis of social networks (as implied by the name of the method: Driskell and Mullen, 2005; Houghton et al, 2006), there is no reason why it cannot be applied to other networks, such as task and information networks. This is a new application for the method, but a potentially useful one. The method can also be applied to the design of anticipated networks, so that more effective task, social and information networks can be designed into new a system, which is another new avenue of research for systems ergonomics that would enable network resilience to be explored in a practical manner.

The first step in a SNA involves defining the network that is to be analysed. Once the overall network type is specified, the tasks, agents or information should be specified. Once the type of network under analysis has been defined, the scenario(s) within which they will be analysed should be defined. Once the network and scenario(s) under analysis are defined clearly, the data collection phase can begin. There are a number of metrics associated with the analysis of social networks, depending upon the type of evaluation that is being performed. The size of the network determines the number of possible relations, and the number of possible relations grows exponentially with the size of the network. This defines the network's complexity. The first step in the analysis is to calculate the statistics for each of the nodes in the network, of which a range can be produced to represent the metrics of distance (i.e., eccentricity), sociometrics (i.e., emission/reception and sociometric status) and centrality (i.e., centrality, closeness, farness and betweeness). The second step is to calculate statistics for the whole network (i.e., density, cohesion and diameter). The final step is to combine the networks for a qualitative assessment. In this analysis the networks can be considered in terms of their resemblance to basic archetypes such as the chain, circle, star, Y and all-connected as well as more advanced archetypes such as the mesh, bus, or a hybrid structure. Each of the archetypes has particular characteristics which may be more or less suited to any given scenario or circumstance.

Distributed situation awareness

Distributed situation awareness (DSA) is presented as an alternative way of thinking about SA in systems. As Hutchins (1995) advocated, the unit of analysis is not the individual person (as presented with the three-level model), but the entire system under investigation. This notion has since gained credence within human factors with Hollnagel (2003) even suggesting that, due to the complexity of modern day socio-technical systems, the study of information processing in the mind of individuals has lost relevance. In the original paper specifying the DSA theory and approach, Stanton et al (2006) indicate how the system can be viewed as a whole, by consideration of the information held by the artefacts and people and the way in which they interact. The dynamic nature of situation awareness (SA) phenomena means they change moment by moment, in light of changes in the task, environment and interactions (both social and technological). These changes need to be tracked in real time if the phenomena are to be understood (Patrick et al, 2006). DSA is considered to be activated knowledge for a specific task within a system at a specific time by specific agents. By agent, it is intended to mean either a human or non-human actor in a system. Thus, one could imagine a network of information elements, linked by salience, being activated by a task and belonging to an agent. To understand how this might work, one has to imagine a network where nodes are activated and deactivated as time passes in response to changes in the task, environment and interactions (both social and technological). Viewing the system as a whole, it does not matter if humans or technology own this information, just that the right information is activated and passed to the right agent at the right time. It does not matter if the individual human agents do not know everything, provided that the system has the information.

System resilience

Resilience engineering has emerging as a new paradigm in safety management, where "success" is based on the ability of organisations, groups and individuals to anticipate the changing shape of risk before failures and harm occur (Hollnagel et al, 2006). Resilient systems are able to withstand minor perturbations and are agile enough to adapt to major disturbances. Proactive resilient processes are the hallmark of such systems, which are in a state of continuous flux in anticipation of threat. Truly resilient systems are able to handle disruptions beyond those anticipated in the original design. These concepts have much in common with the socio-technical systems (STS) approach to systems, and viewing systems though network models offers one approach to testing resilience. Systems may be thought of multimodal interconnecting social (actor and artefact), task (goals and operations), action (procedures and rules), informational (data and knowledge) and cognitive (perceiving, remembering, recognising, deciding) networks. The structure and adaptability of these networks are likely to be indicators of the systems resilience, which can be subjected to non-destructive testing to identify weaknesses and strengths.

Past work has adapted the EAST method to analyse incidents of fratricide (Rafferty et al, 2012) although this has largely been conducted as retrospective and concurrent analyses. Similarly, 'broken links' approach (i.e., failure to communicate information from one agent to another in a system) has only been investigated by EAST analysts when looking retrospectively at accidents to identify underlying causes. Griffin et al (2010) demonstrated the 'broken link' between the engine vibration indicator and the pilots in the cockpit was a causal factor in their failure to shut down the correct engine in the Kegworth accident. If this information had been communicated more effectively it could have helped to prevent the crash. Stanton (2013) proposed that the networks could be used to assess the resilience of socio-technical systems to systemic failures. The twin challenges for this work are to develop a systems approach to risk and safety that can be prospective and embody the socio-technical systems perspective. It is considerable undertaking to incorporate all aspects of systems that have implications for risk and safety management (Stanton et al, 2010; Stanton et al, 2012a). More recent work has demonstrated that the broken link approach may be used to identify potential weaknesses in systems (Harvey and Stanton, 2013).

Discussion

EAST acknowledges that systems are inherently complex and multiple perspectives on the problem are required to more fully appreciate the relationships between the social and technical aspects of the system. EAST accepts that STS are intertwined and analyse the STS as a whole, rather than constituent parts. The research to date has shown that the network models are able to characterise the domain in different, but complementary, ways (Stanton et al, 2013). The seven outputs, namely the individual task networks, social and information networks (and associated metrics), and the combined networks (i.e., task and social network, information and social network, information and task network, and task, social and information network) offer a graphical representation of distributed cognition from different perspectives.

Stanton (2013) showed how SNA metrics can be applied to all three task, social and information networks to provide insights into the structural integrity and the relative contribution of each of the nodes. It has also been shown that it is possible to construct the networks directly from the observational data. Walker et al (2010a) demonstrated the benefits of the network representations over the traditional ethnographic narratives and pictures (Hutchins, 1995). It has been shown that the network models offer a useful way of considering distributed cognition in systems to reveal the interdependences between tasks, agents and information. There are some similarities between the approach taken by EAST and that of Actor Network Theory (ANT: Engestrom, 2000) but EAST represents the networks separately as well as together. Both EAST and ANT have, at their core, a conceptual triangulation between objects, actors and events. Both use networks of relationships to graphically display their analysis. Also, EAST goes further than ANT, to apply statistical analysis to the networks as well as identify network archetypes. Nevertheless, there are important similarities in the approach and representation. Arguably, ANT relies more heavily upon the skill of the analyst to identify themes in the networks, whereas the use of verbal protocol data and SNA metrics by EAST has reduced this potential bias. The use of social network statistics may be one way of examining the potential resilience of networks (Hollnagel et al, 2006) – which may be particularly useful as metrics of distributed cognition. This is a new concept and methodology, so further studies are required to test its efficacy.

The network representations reveal the clustering of tasks, social agents and information. These clusters show how the constituent parts of the networks have been bound together, either by chance or design. As STS develop, the social information and task networks reflect the current evolution, but that does not mean that they cannot be improved upon. The representations afford both quantitative and qualitative structural analysis. The quantitative analyses have been presented at some length and offer insights into the potential integrity and resilience of the system. As a method for resilience engineering, EAST can be used to assess the potential weaknesses and points of failure in socio-technical structures. The qualitative analyses enable the network structures to be classified into archetypes (e.g., chain, circle, tree, star, bus, mesh, small world and fully connected: Stanton et al, 2008; Stanton et al, 2012b). In these terms, the task network appears to be a hybrid of a chain and circle archetype, the social network appears to be a hybrid of a star and circle archetype, and the information network appears to be a small world archetype. This analysis is somewhat speculative at this point and further research is needed to understand the relationship between the metrics for network resilience and the archetypes of network structure. There are some early indications that the small world networks offer greatest resilience and efficiency (Stanton et al, 2012b).

EAST describes systems in terms of task, social and information networks as well as exploring the relationships between those networks. Individual networks describe the respective relationships between tasks (such as the task dependencies and sequences), social agents (such as sociometric status of agents based on communications), and information (such as the interdependences between the concepts discussed). The combined task and social networks show which roles perform tasks in series and parallel. The combined information and social networks showed which roles communicate information. The three integrated networks described how information is used and communicated by agents working together in the pursuit of tasks. Any new conceptualisation of a system will need to consider the likely changes on these socio-technical networked structures.

Conclusions

EAST can be applied to a complex socio-technical 'system of systems' to present 'networks of networks'. The networks show multiple perspectives on the activities in the system, which is a necessary requirement for socio-technical analysis. Further analysis should attempt to characterise future systems so that the multiple perspectives can be compared and 'so-what' questions can be asked as ideas for design of the social and technical aspects of the system co-evolve (Clegg, 2000; Walker et al, 2010b; Stanton et al, 2012b). EAST has the potential to map the task, social and information networks and their interdependencies. Adopting this socio-technical systems design approach would help to jointly optimise the whole system rather than the parts in isolation. This would require spending more time in the initial modelling and prototyping, working with end-users and SMEs with more focus on the social systems and ways of working (to redress the balance of focus on technical system development) than is currently the case.

References

Clegg, C. W. 2000, Sociotechnical principles for system design. *Applied Ergonomics* 31, 463–477.

Driskell, J. E and Mullen, B. 2005, Social network analysis. In: Stanton, N. A., Hedge, A., Salas, E., Hendrick, H. and Brookhaus, K. (Eds) *Handbook of Human Factors and Ergonomics Methods*. Taylor & Francis: London.

Engestrom, Y. 2000, Activity theory as a framework for analyzing and redesigning work, *Ergonomics* 43(7), 960–974.

Farrington-Darby, T., Wilson, J. R., Norris, B. J. and Clarke, T. 2006, A naturalistic study of railway controllers. *Ergonomics* 49(12–13), 1370–1394.

Furniss, D. and Blandford, A. 2006, Understanding emergency medical dispatch in terms of distributed cognition: a case study. *Ergonomics* 49(12–13), 1174–1203.

Griffin, T.G.C., Young, M.S. and Stanton, N.A. 2010, Investigating accident causation through information network modelling. *Ergonomics* 53(2), 198–210.

Harvey, C. and Stanton, N. A. 2013, Evaluation of STAMP and EAST for Systems of Systems Risk Analysis. *Paper in preparation*. University of Southampton, Southampton.

Hollnagel, E. 1993, *Human Reliability Analysis – Context and Control*. London: Academic Press.

Hollnagel, E., Woods, D. D. and Leveson, N. 2006, *Resilience Engineering: concepts and precepts*. Ashgate: Aldershot.

Houghton, R.J., Baber, C., McMaster, R., Stanton, N.A., Salmon, P.M., Stewart, R. and Walker, G.H. 2006, Command and control in emergency services operations: A social network analysis. *Ergonomics* 49(12&13), 1204–1225.

Hutchins, E. 1995, *Cognition in the Wild*. MIT Press: Cambridge.

Patrick, J., James, N. and Ahmed, A. 2006, Human processes of control: tracing the goals and strategies of control room teams. *Ergonomics* 49(12–13), 1395–1414.

Plant, K. L. and Stanton, N. A. 2012, Why did the pilots shut down the wrong engine? Explaining errors in context using schema theory and the perceptual cycle model. *Safety Science* 50(2), 300–315.

Rafferty, L. A., Stanton, N. A. and Walker, G. H. 2012, *Human Factors of Fratricide*. Ashgate: Aldershot.

Salmon, P.M., Stanton, N.A., Walker, G.H., Jenkins, D.P., Baber, C., and McMaster, R. 2008, Representing situation awareness in collaborative systems: A case study in the energy distribution domain. *Ergonomics* 51(3), 367–384.

Sanderson, P., Verhage, A. G. and Flud, R. B. 1989, State-space and verbal protocol methods for studying the human operator in process control. *Ergonomics* 32(11), 1343–1372.

Stanton, N. A. 2013, Representing distributed cognition in complex systems: how a submarine returns to periscope depth *Ergonomics* (in press).

Stanton, N.A., Baber, C. and Harris, D. 2008, *Modelling Command and Control: Event Analysis of Systemic Teamwork*. Ashgate: Aldershot, UK.

Stanton, N. A., Rafferty, L. A. and Blane, A. 2012, Human factors analysis of accidents in system of systems, *Journal of Battlefield Technology* 15(2), 23–30.

Stanton, N.A., Salmon, P.M., Jenkins, D.P. and Walker, G.H. 2010, *Human Factors in the Design and Evaluation of Central Control Room Operations*. CRC Press: Boca Raton, USA.

Stanton, N.A., Salmon, P.M., Walker, G.H., Baber C. and Jenkins, D. P. 2005, *Human Factors Methods: A Practical Guide for Engineering and Design* (first edition). Ashgate: Aldershot.

Stanton, N.A., Salmon, P.M., Rafferty, L. A., Walker, G.H., Baber C. and Jenkins, D. P. 2013, *Human Factors Methods: A Practical Guide for Engineering and Design* (second edition). Ashgate: Aldershot.

Stanton, N.A., Stewart, R., Harris, D., Houghton, R.J., Baber, C., McMaster, R., Salmon, P.M., Hoyle, G., Walker, G.H., Young, M.S., Linsell, M., Dymott, R. and Green, D. 2006, Distributed situation awareness in dynamic systems: theoretical development and application of an ergonomics methodology, *Ergonomics* 49(12–13), 1288–1311.

Stanton, N. A., Walker, G. H. and Sorensen, L. J. 2012, It's a small world after all: contrasting hierarchical and edge networks in a simulated intelligence analysis task. *Ergonomics* 55(3), 265–281.

Stewart, R., Stanton, N.A., Harris, D., Baber, C., Salmon, P.M., Mock, M., Tatlock, K. and Wells, L., 2008, Distributed situational awareness in an airborne warning and control aircraft: application of a novel ergonomics methodology. *Cognition, Technology and Work* 10(3), 221–229.

Walker, G.H., Gibson, H., Stanton, N.A., Baber, C., Salmon, P.M. and Green, D. 2006, Event Analysis of Systemic Teamwork (EAST): A novel integration of ergonomics methods to analyse C4i activity. *Ergonomics* 49(12&13), 1345–1369.

Walker, G.H., Stanton, N.A., Baber, C., Wells, L., Gibson, H., Salmon, P.M. and Jenkins, D.P. 2010a, From ethnography to the EAST method: A tractable approach for representing distributed cognition in Air Traffic Control. *Ergonomics* 53(2), 184–197.

Walker, G. H., Stanton, N. A., Salmon, P. M., Jenkins, D. P. and Rafferty, L. A. 2010b, Translating concepts of complexity to the field of ergonomics. *Ergonomics* 53(10), 1175–1186.

USER SYSTEMS ARCHITECTURES, FUNCTIONS AND JOB AND TASK SYNTHESIS

Mike Tainsh

Lockheed Martin, Reddings Wood, Ampthill, Bedford, UK

The paper addresses the inclusion of ergonomics within the systems engineering lifecycle based on Hierarchical Systems Description (HSD) (Tainsh, 2013). A three stage process is outlined for developing functional descriptions that support job and task design. The process stages are: requirements formulation, organisational modelling and implementation. Requirements are based on ISO 15288. For the organisational stage, a User Systems Architecture (USA) is described: the "V" model. This is exemplified in an implementation stage for job and task design. It is concluded that USAs are a useful means of representing some of the organisational aspects of functional HSDs to support job and task synthesis.

Introduction

The Hierarchical Systems Description (HSD) technique was first described by Tainsh (2013). It was linked to MODAF (DEFSTAN 00-250) and the relevant ergonomics viewpoints OV4 (Organisational Relationships Chart), OV5 (Operational Activity Model) and OV6 (Operational Rules, State Transitions, and Event-Trace Descriptions).

In line with MODAF, the HSD technique had three main levels: requirements, organization and implementation. This paper considers interdisciplinary HSDs with functional characteristics, and provides an example taken from current work on the design of Human Computer Interaction (HCI). The HCI is for equipment a vehicle application that is currently under development for the British Army.

The previous paper did not address the broader system aspects covering non-ergonomic aspects, nor did it deal with functional descriptions which are an important early stage in the process of task syntheses i.e. the creation of jobs and tasks functional elements i.e. activities and functions.

Stage 1: HSD statement of requirements/principle

The system engineering lifecycle is described in ISO 15288. It specifies the processes within which design occurs. It is composed from four streams of work which run throughout the lifecycle of the project, independent of duration. The streams of work are: agreement – acquisition and supply; organisational – project-enabling processes, including human resources and quality within the project; project – the process of coordinated activities which meet organisational requirements and goals; technical – the process to turn a requirement into a product or the supply of services This paper will concentrate on the technical stream only – where the design work occurs.

The design process includes the handling of requirements and the allocation of requirements to system elements (users, or equipment elements) across all disciplines which contribute. The concept of architecture is central to the technical work, and the design of the architecture is necessarily within the early stages. It includes the allocation of functions to elements. This will include the allocation of functions between users, and between users and the equipment.

The purpose of architectural design is to identify functions, allocate them to elements (or their parts) and hence influence design outcomes. The issue being addressed here is the development of a technique that includes the use of scenario and task descriptions, when moving from requirements, through architecture, to implementation and beyond. This means that the user task description needs to increase the validity of the engineering work, including architectural design and the allocation of functions, across the lifecycle and for all contributing disciplines.

During the formulation of requirements, ergonomists need to be assured that ergonomics requirements can be integrated into related disciplines and across the whole lifecycle. Hence it is essential to identify a set of possible scientific disciplines that might be contributing within the system lifecycle and may need integration with an ergonomics contribution. Then it is necessary to assess whether their contributions can be expressed in terms that enable a valid common representation.

Table 1 indicates a selection of the possible ergonomics contributions across a range of disciplines. It does appear we can come to an initial conclusion that requirements from a wide set of disciplines may be expressed in terms that are common with ergonomics. This will enable an expression of ergonomics requirements to be used across the lifecycle to ensure an integrated outcome. Experience, so far, suggests that this is a valid approach. This is compatible with other results from Human Factors Integration Programmes (Tainsh 2004).

Stage 2: HSD, the organisational model

The "V" model

While requirements may be multidisciplinary, it has been usual to maintain a separation between User task descriptions and other functional representations (Job Process Charts, which represent the interaction between User and equipment are an exception (Tainsh 1985)). The "V" model is proposed to integrate the contributions from disparate disciplines at the Organisational Stage.

Table 1. System lifecycle with contributing disciplines showing possible ergonomics contribution to task – based descriptions.

Discipline to which Task contribution is required	Requirements Definition and Analysis Architectural Design	Implementation Integration Verification Transition Validation	Operation Maintenance	Disposal
Control/communications/ Computing/IT	Information requirements	Information/communication presentation and flow	Information presentation and flow upgrades	Knowledge of communications systems
Anatomy/physiology	Work, manual handling constraints	Work system design, load carriage handling systems	Work system audits and upgrades	Disposal of equipment
Occupational sciences	Skills and knowledge requirements	Recruitment, selection and training policy and design	Manpower policy and operation	Retraining
Economics/statistical/ surveys	TAD, development of population characteristics	Understanding of populations characteristics and development	Understanding of populations characteristics	Training and education policy
Systems engineering, software development	Functions, performance requirements	Functional allocation and control system architecture and development	Performance audit	Disposal of equipment
Nuclear/chemical engineering	Hazards and risks, protection needs	Health hazard protection, PPE	Safety audits	Safety records
Architecture and buildings	Movement, space and form options	Buildings and town planning and construction	Architectural surveys	Local history
Mechanical engineering	Spatial arrangements, preferences and options	Equipment layouts and workstation architectures	Upgrades	Equipment disposal

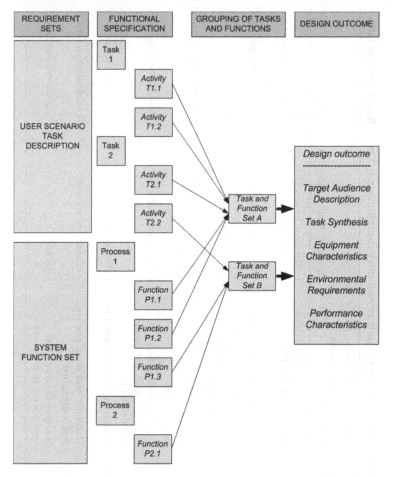

Figure 1.   This shows the "V" Model for Hierarchical System Description.

The "V" Model (Figure 1) includes: the User Scenario Task Description (USTD), and the System Function Set (SFS). The SFS is a functional description of the complete system (both User and equipment). The tasks and activities of the USTD are grouped together as understood by the Users. The system functions are allocated to sets and then to subsystems of equipment and users. The activities and functions are linked together "at the point of the V". This is the start of the matching process described in ISO 26800. These grouped sets of systems and task descriptions, are used to generate a design outcome.

The "V" diagram functional specification

The USTD and SFS are both functional descriptions which are parts of the high level specification of the organisational aspects of the system that is being produced.

Hence it is essential that it is understood how these descriptions should be handled within the systems lifecycle.

Functional descriptions and software code characteristics

Similarities are drawn between the generation of software code and functional descriptions in order to decide whether we have an effective set of characteristics or not. There are common characteristics between the development of software code and task descriptions. Both involve a set of descriptors for processes, and other properties, in terms of a language that acts as instructions to elements of the system.

The similarity in the development of software and task descriptions can be seen in the work of Rus, Lindvall and Sinha (2001), in a paper for software engineers on knowledge management. They discuss how User knowledge should be handled in order to develop software codes. It is a very similar process to the development of task descriptions by human factors specialists.

Further experience in the discipline of software engineering e.g. Rosenberg and Hyatt (1997), leads to a proposal for five general characteristics of software code: efficiency, complexity, understandability, reusability, and stability/maintainability. In another report (Kamer and Bond, 2004), there is evidence of the relatively advanced state of the discussion in that discipline (when compared to human factors). They consider the relationship of the dimensions used to characterise software products to various metrics. These include: programmer productivity, errors relative to size, requirements stability and time spent fixing faults.

It is believed that human factors can build on the experience gained in software engineering. We can use the results above to create an initial list of characteristics to assess functional descriptions and help to establish their benefits, while being aware that the validity of these characteristics still needs development.

Hence, it is inferred that an HSD which covers functional descriptions needs to have the following characteristics to enable it to be used alongside functional descriptions from other disciplines:

(a) The use of common concepts to enable common handling techniques. This will enable an efficient handling of information on complex projects.
(b) The use of a common language for representing items within the "V" model, to support a common understanding.
(c) Use of description techniques that enable description through a number of levels to gain both high level statement and detail according to requirements.
(d) Ease of documentation/recording/storage. This includes being open to configuration control using appropriate tools, to help ensure (b) and (c).
(e) The use of description techniques that enable the reuse and maintenance of information both within and across system developments.

Outcome

In Figure 1, a selection of from the USTD task and activities, and SFS Processes and functions could be allocated to one User while a second selection could be allocated to another User e.g. a Maintainer, for example. Hence it appears that The "V" diagram enables the bringing together of sets of functional descriptions from all project disciplines integration with the ergonomics contribution to provides a representation supporting an organisational description. As such it enables a development of the architectural process described in ISO 15288 and supports the allocation of functions to equipment and users alike

Stage 3: HSD, allocation functions to users and equipment sets. An implementation example for HCI

The high-level organisational analysis and description in Stage 2 leads to a further consideration of personnel with their jobs and tasks within a scenario.

The sets of functions need be grouped together to enable a consideration of:

(a) Functions associated with individual items of equipment, as a subsystem with user involvement
(b) Functions grouped within items of equipment by, for example, location, or page.

These groupings are essential so that assessment can be made to:

(a) Ensure all tasks within the scenario are fully supported (or otherwise if required).
(b) All functions are allocated so that they may be executed according to the system requirements.

The design outcome for human computer interaction may be as represented in Figure 2 where the USTD and SFS are listed on the left and relevant function and activities are linked to support a design activity. In turn the design can be assessed against the USTD, personnel and equipment requirements.

Conclusion

The final question then: Is this a User System Architecture that is shown in Figure 1? Certainly, it appears to have some of the architectural characteristics required of ISO 15288:

(a) Its composition includes representations of people and equipment including the activities and functions.
(b) It appears applicable across the lifecycle including all the potential contributing disciplines.

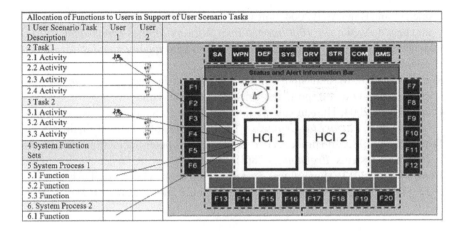

Figure 2. HSD "V" model in implementation for Human Computer Interaction.

(c) It enables the representation of the results of allocation of function between users and equipment elements.

(d) The allocation of tasks is not simply to user or equipment elements, but combinations and parts

(e) It supports work on integration. Particularly as it seems potentially capable of providing a complete system description including users, equipment and tasks.

Work on the validity of the approach supporting the development of HSDs appears as being in early days. However there appears to be the opportunity to learn from work in cognate disciplines such as software engineering. It can be inferred that this HSD approach results in a User Systems Architecture (USA) – a complete functional description from a User perspective. This will support the overall Systems Architecture. Experience at Lockheed Martin on major projects supports its effectiveness as a human factors tool that can be used by human factors and other professionals.

References

DEFSTAN 00-250 2008, *Human Factors for the Designers of Systems* (Parts 0, 1, 2, 3, 4), MoD, UK.

DEFSTAN 23-09 2011, *Generic Vehicle Architecture,* Version 2.

de Winter, J. C. F. and Dodou, D. 2011, Why the Fitt's List has persisted throughout the history of function allocation. *Cognition, Technology and Work.*

ISO 15288 2001, *Information Technology – Life Cycle Management – System Life Cycle Processes.*

ISO 26800 2011, *Ergonomics – General Approach, Principles and Concepts.*

Kamer, C. and Bond, W. P. 2004, Software engineering metric: what do they measure and how do we know? *10th International Software Metrics Symposium*.

Rosenberg, L. H. and Hyatt, L. E. 1997, Software quality metrics for object-orientated environments. *Cross Talk Journal*.

Rus, I., Lindvall, M. and Sinha, S. S. 2001, *Knowledge management in software engineering*. DACS Report. The University of Maryland.

Tainsh, M. A. 2004, Human Factors Integration. In C. Sandom and R. S. Harvey. (ed.), *Human Factors for Engineers*. The Institution of Electrical Engineers.

Tainsh, M. A. 1985, Job process charts and man-computer interaction within naval command systems. *Ergonomics* 28(3): 555–565.

Tainsh, M. A. 2013, Hierarchical System Description (HSD), MODAF and ISO 15288. *Contemporary Ergonomics*, (Taylor and Francis, London), 219–227.

VEHICLE DESIGN

BETTER VEHICLE DESIGN FOR ALL

Sukru Karali, Diane Gyi & Neil Mansfield

Loughborough University
Loughborough Design School

With increases in life expectancy, there is a growing population of older drivers. Many cars have not been designed to meet the needs of people with age-related disabilities. A questionnaire survey of older and younger drivers (paper/online version $n = 903$) and supplementary interviews with drivers aged ≥ 65 years ($n = 15$) were conducted. Questions concerned musculoskeletal symptoms, driving exposure, the vehicle seat, access to in-vehicle controls/features, ingress/egress, driving performance and driving behaviour. Compared to younger drivers (<65 years), older drivers (≥ 65 years) reported more difficulties (e.g. parallel parking and driving in fog), and significantly more discomfort in the hips, thighs, buttocks and knees.

Introduction

Vehicle design and performance are constantly being developed with vehicles becoming smarter and more sophisticated. Contemporary vehicles are now equipped with many characteristics formerly constrained to the luxury market: for example, technologies to assist the user with specific driving tasks such as intelligent automated parking systems (Bradley et al., 2008). The population of older people is increasing worldwide, particularly in developed countries (Meyer, 2009). There are now more than 15 million people aged over 60 with a driving license; more than 1 million of these are over 80 (IAM, 2012). However, many cars do not meet the needs of people with age-related disabilities (Herriotts, 2005). Vehicle manufacturers claim to provide a positive driving experience, but do they really meet the requirements of users of all ages?

Methodology

A questionnaire survey was conducted within the UK with a large sample of car drivers. The aim was to understand the experiences of car drivers of different ages and identify some of the key challenges for car design. The questionnaire was available as both paper and online versions and included: musculoskeletal symptoms, driving exposure, the vehicle seat, access to in-vehicle controls/features, ingress/egress, driving performance and driving behaviour.

Table 1. Age and gender distribution (whole sample).

Age category of participants	Age	Male n	%	Female n	%	Total n	%
Younger (<65 years)	20–34	70	7.8	66	7.3	136	15.1
53.5%	35–49	62	6.9	53	5.9	115	12.7
	50–64	144	15.9	88	9.7	232	25.7
Older (≥65 years)	65–79	216	23.9	140	15.5	356	39.4
46.5%	≥80	43	4.8	21	2.3	64	7.1
Total		535	59.2	368	40.8	903	100

Many factors were considered during the design and development of the survey, for example, questions had to be to be specific, short and easy to read/understand for older people. Questions from a survey conducted by Sang et al. (2009) were included in this survey with slight modifications. Where appropriate, Likert scales were incorporated with specific statements and tick boxes.

A stratified sampling technique was used, whereby the population is divided into various subgroups/strata. Once the strata were determined, a simple random sample was taken from each stratum individually. For the questionnaire survey, the sample was arranged in a number of sub-groups focusing on age and gender.

Major organisations within the UK were consulted for the distribution of the questionnaire to the target audience. These are well-known institutes, voluntary action groups, and charity and motoring organisations. Agreement was obtained for the distribution of questionnaire (data collection from e-mails, interviews and personal contacts). For the on-line survey several techniques were used to increase the number of responses such as; snowballing technique. Initially, the target sample size for the survey was 600; this was thought to be a reasonable number in order gain a robust data set for statistical analysis. Supplementary interviews ($n = 15$) were conducted with a further sample of drivers aged 65 years and over by using the questionnaire in a structured interview format.

Results

In total 903 people took part; 53.5% were younger drivers ($n = 483$, <65) and 46.5% were older drivers ($n = 420$, ≥65). Drivers over 80 years represented 7.1% ($n = 64$) of the whole sample. 59% of participants were male and 41% were female (Table 1).

Musculoskeletal symptoms: high levels of musculoskeletal symptoms were reported in the lower back, knees, neck, shoulders and elbows. In general, younger participants reported higher levels of musculoskeletal symptoms in the neck ($p < 0.01$), shoulders ($p < 0.05$) and middle back ($p < 0.001$). These symptoms are likely to be related to their level of activity (e.g. work) compared to older participants. However, significantly more discomfort was reported by older individuals in the hips/thighs/buttocks and knees ($p < 0.05$).

Table 2. Driving behaviour (whole sample).

Answer options	Agree (%)	Neither agree or Disagree (%)	Disagree (%)
Other drivers' lights restrict my vision when driving at night	46.7	19	34.2
I have difficulty turning my hand and body around when reversing	24.5	14.2	61.4
Operating entertainment systems distract me from driving (e.g. playing radio)	23.7	38	38.3
Operating navigation systems distract me from driving (e.g. looking at sat-navigation)	22.9	29.5	47.6
My reactions are slower than they used to be (e.g. braking in an emergency situation)	22.8	20.3	56.9

Driving behaviour: half of all respondents (46.7%) reported that other drivers' lights restricted their vision when driving at night; more females (53.3%) than males (42.5%) reported this ($p < 0.001$; Table 2). Older drivers (31.7%) reported more difficulties than younger drivers (18.4%) with turning their head and body around during reversing ($p < 0.001$). Similarly, older drivers reported their reactions were slower than they used to be (e.g. braking in an emergency situations) compared to younger drivers.

Driving performance: Differences were found in age and gender. Older drivers (25.3%) reported more difficulty driving on a foggy day ($p < 0.01$) than younger drivers (16.8%). Similarly, with parallel parking 16.9% of older drivers compared to 12.3% of younger drivers reported difficulty ($p < 0.01$). With regard to gender, females reported more difficulty than males with parallel parking ($p < 0.001$). A higher proportion of younger drivers (25.5%) than older drivers (19.5%) reported being distracted by navigation systems but this was not significant. Reasons for this may be that older drivers are more experienced, travel shorter distances, use familiar routes and are therefore less likely to use these technologies.

Adjusting the seat features: 10.5% of respondents reported that they were dissatisfied with adjusting specific seat features, namely the head rest height, head rest distance from the head and setting the seat belt height. Females reported more difficulty than males with adjusting the head rest height ($p < 0.001$). Reasons given for this difficulty may include reaching, accessing and operating the controls while seated. No age differences were found with adjusting the seat features.

Supplementary interviews: the main findings are summarised as follows:

- Females (specifically older females) reported more difficulty than males with adjusting seat features, namely the head rest height/distance and seat belt height. Reasons given for this include reaching, accessing and operating the controls while seated.

- When reaching and pulling the boot door down to close, older females reported having less mobility and reduced reach. A reason may be that older females are shorter in stature.
- Older drivers are less likely to drive at night.
- Older drivers generally avoid using navigation systems due to their short travelling distances.
- The most commonly used entertainment system among older drivers is the radio.

Conclusion

This study has provided data to understand the key issues experienced by drivers of all ages. Some are common for all ages, and some are age related. The future direction of this research will focus on understanding how design of the vehicle cab impacts posture, comfort, health and wellbeing in older drivers.

References

Bradley, M., Keith, S., Kolar, I., Wicks, C., Goodwin, R., 2008. *What do older drivers want from new technologies?* [online]. SPARC, Reading. [viewed 23/11/2013]. Available from: http://www.sparc.ac.uk/media/downloads/executivesummaries/exec_summary_bradley.pdf

Herriotts, P., 2005. Identification of vehicle design requirements for older drivers. *Applied Ergonomics* 36(3), 255–262.

IAM, 2012. *More than a million drivers now aged over 80.* [online]. Institute of Advanced Motorists, London. [viewed 29/11/2013]. Available from: http://www.iam.org.uk/component/content/article?id=983

Meyer, J., 2009. *Designing in-vehicle technologies for older drivers.* [online]. National Academy of Engineering, Washington. [viewed 29/11/2013]. Available from: http://www.nae.edu/Publications/Bridge/TechnologiesforanAgingPopulation/DesigningIn-VehicleTechnologiesforOlderDrivers.aspx

Sang, K., Gyi, D. and Haslam, C., 2009. Musculoskeletal symptoms in pharmaceutical sales representatives. *Occupational Medicine* 60(2), 108–114.

THE MOTORCYCLE: A HUMAN OPERATOR INFLUENCED WORKSTATION

M.I.N. Ma'arof[1], H. Rashid[1], A.R. Omar[1], S.C. Abdullah[1], I.N. Ahmad[1] & S.A. Karim[2]

[1]*Faculty of Mechanical Engineering, Universiti Teknologi MARA, 40450 Shah Alam, Selangor, Malaysia*
[2]*MASMED, Intekma Resort, Universiti Teknologi MARA, 40450 Shah Alam, Selangor, Malaysia*

The objective of this study was to investigate how the motorcycle, as a workstation, is affected by the integration of the human operator. The study showed significant variations of load distributions on the motorcycle once the human operator was integrated into the system with varying riding postures. The load distribution biases result in variations in term of the motorcycle's manoeuvrability and riding characteristics. It is concluded that the motorcycle is a unique case in which the workstation is being influenced by the human operator. It is recommended that the influences given by the human operator on various other workstations are investigated in future research.

Introduction

Ergonomics and workstation design

Various studies have suggested that workstation designs have a critical influence to human operators, from both physical and physiological viewpoints (Robertson et al., 2013; Cimino et al., 2009; Hipper and Tavangarian, 1998; Das' and Segupta, 1996). Hence, workstations are designed to 'fit' the human operators. However, in certain unique cases the situation is reversed – the workstation is influenced by the human operator. If this is so, then what is the impact made by the human operator on the characteristics of the workstation?

The objective of this study was to discuss the situations and conditions where the workstation is affected by integration of the human operator. For the scope of this study, the system comprised the human operator and the workstation. The third element, environmental elements, was not considered for this study.

Motorcycle as a workstation

Based on Robertson et al. (2013), Cimino et al. (2009) and Das' and Segupta (1996), this study defined workstation as 'a place, positioning, location or platform where the human operator would perform their designated tasks, commonly accompanied with equipment or tools which are necessary or needed to perform the designated tasks efficiently and safely'. The motorcycle was chosen as the workstation of interest for this study – a concept documented by Robertson and Minter (1996). The motorcycle is interesting since it could be considered as both the workstation and the tool which made motorcycling possible. This study has followed the same definition of motorcycle as given by Ma'arof and Ahmad (2012a). The motorcycle was chosen for the following reasons:

1. Motorcycling requires a very high level of human operator involvement to safely and efficiently perform the complex real world motorcycling 'work task' (Ma'arof and Ahmad, 2012a) as opposed to other 2- or 4-wheeled road-legal vehicles. All body parts have a designated job and have to work together in unison – collectively known as "Uno Body Motion" (UBM) (Ma'arof et al., 2012).
2. The motorcycle is a unique workstation since it could be categorized as both a static and dynamic workstation. It is a static workstation when the motorcyclist first gets seated on the motorcycle (assuming the riding posture) and becomes dynamic when the motorcycle starts to move due to the input by the motorcyclist. These two stages were noted by Ma'arof et al. (2012) as the stages of human-motorcycle interface, which are the "Human-Motorcycle-Interface" (HMI) and "Human-Machine-Environment-Interface" (HMEI). The findings are analysed in both of these stages.
3. The motorcycle was also selected due to its globally high numbers of human users. Various motorcycle manufacturers as well as mega oil and gas organizations have anticipated growth in the motorcycle market in the coming years (Shell, 2013; Honda Annual Report, 2012).

Motorcycle riding postures

Although motorcycle riding posture is commonly determined by the 'design' of the motorcycle; motorcycle riding posture is still significantly influenced by the anthropometric characteristics of the motorcyclist. Two motorcyclists with different anthropometric characteristics, for instance, the length of the arms, would result in two different riding postures for the same motorcycle. Thus, the best way to assess the motorcycle or motorcycling in term of ergonomics is via the riding postures assumed by the motorcyclist. Assessment via motorcycle designs would be lengthy and complex due to their diversity (Teoh and Campbell, 2000). Also, the need to understand and know numerous motorcycle terminologies, such as 'superbike', 'streetfighter' or 'supersport', could be confusing for most laymen. In using the riding posture as reference, the aforementioned terms could be replaced by just one referred riding posture: 'Type 1' (Ma'arof and Ahmad, 2012b).

For the application of the motorcycle as a workstation concept, the 'Riding Posture Classification' or 'RIPOC' system by Ma'arof and Ahmad (2012b) was used to provide analytical platform for the investigation. The system a method to segregate motorcycles from the perspective of ergonomics via the riding postures assumed by the human operator.

Types of riding postures according to the "RIPOC" System

| Type 1: Forward lean riding posture | Type 2: Upright riding posture | Type 3: Seatback-leg-forward riding posture | Type 4: Double forward riding posture |

Figure 1. Riding postures as designated by the "RIPOC" System (Ma'arof et al., 2012b) (modified from source (Type 1-3): cycleergo.com, Type 4: Breselec, 2009).

Ma'arof and Ahmad (2012b) have designated four types of riding posture according to the RIPOC system (see Figure 1). For this study, the assessments were made on all riding postures.

Research methodology

The following research methodology was formulated for this study. Each research method has its own pros and cons, hence, cumulatively would aid in achieving the aims of this study.

Field research

The field research serves as the core for this study by providing vital information on real world motorcycling. It consists of hands-on real world motorcycling assessments by the authors (observer type observations). The information gathered by the authors was recorded by way of data collection check sheet and travel-log. The authors have riding experience from 2 to 38 years of active motorcycling. The results from this method were presented and interpreted from the perspective of researcher motorcyclist.

Survey research

Semi-structured interviews with experts from the field of motorcycling and ergonomics were conducted. This was to acquire supporting evidence from the perspective of non-motorcycle-researcher public motorcyclist and experts. This method provided critical neutral insights on the topic, complementing the result from the field research.

Mathematical calculation and software development

The authors developed software to calculate the forces directed throughout the motorcyclist body based on physics and with reference to selected literature:

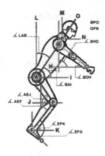

Figure 2. Free-body diagram for riding posture load calculation (source: Ma'arof et al., 2012).

Resnick (2013), Ma'arof et al (2012) and Swat and Krzychowicz (1996). Assumptions and limitations were as follows:

1. The spine and both arms are straight and make 90 deg angle with the torso.
2. The load distributions ratio between the motorcycle's frontal and rear section (the engine is the centre – regardless of motorcycle) without the human operator is 50:50.
3. The motorcyclist is seated very close to the centre of the motorcycle (at the seat-pan) if not at the centre of the motorcycle.
4. The calculation is limited to a two-dimensional (2D) rendition of the motorcyclist-motorcycle integration during the 'HMI' stage (Ma'arof et al., 2012) with the motorcyclist assuming a static posture.
5. The riding postures are assumed according to the "riding posture establishment conditions" with the first three conditions fulfilled and the seat-pan is flat (Ma'arof and Ahmad, 2012b).

Results and discussion

The aim of this study was to show the influences of the working postures assumed by the motorcyclist (human operator) on the motorcycle (workstation). The human operator would determine the weight distribution of the man-machine integration. The term 'weight' in this context is the force loading exerted by the motorcyclist to the motorcycle as a result of body mass and gravitational pull. Weight distribution is among the main factors that affect motorcycle control and manoeuvrability (Cossalter, 2006; Glimmerveen, n.d.). Hence, the motorcycle is designed with variable weight distributions between the front and rear sections in accordance to its purposes (Cossalter, 2006; Glimmerveen, n.d.; Munoz and Rougler, 2011). Nevertheless, this issue is rarely presented in the specification sheet of motorcycles (V-Twinforum.com, 2007). Thus, for the purpose of this study only, it was assumed that the weight distributions to be equal for both front and rear section (50:50) without the human operator.

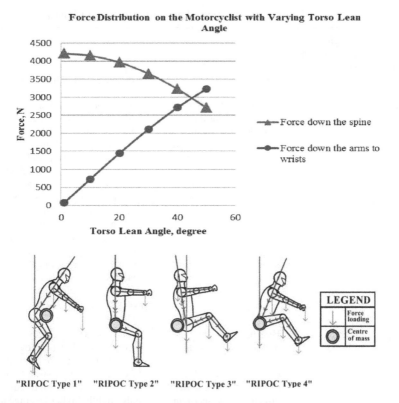

Figure 3. Force distribution with varying torso lean angle and with increasing positive angle back inclination on a 70 kg motorcyclist (left) and forces distribution of the four (4) "riding postures" (Ma'arof and Ahmad, 2012b).

The mathematical calculation result showed (see Figure 3) that the static loading from sitting (force going down the spine) was inversely proportional to positive flexion and extension of the spinal column. Upon assuming the riding posture, positive flexion of the spinal column (forward lean of torso) results in more forces being transferred down the arms instead of the spine. These forces would be channelled directly to the frontal section of the motorcycle (e.g. handlebar, forks). Thus, adding more loading and transfers the resultant centre of mass (COM) of both motorcyclist and motorcycle to the frontal section of the motorcycle. The case was almost similar for positive extension of the spine (torso reclines), though the forces were now transferred to the rear section of the motorcycle. This would result in the resultant COM (both motorcyclist and motorcycle) being transferred to the rear section of the motorcycle. At zero lean angle (torso perpendicular to the seat-pan) the static load would be directed directly down the spine to the buttocks and the seat-pan. Thus, concentrating all the forces to the seat-pan and maintaining the resultant COM (both motorcyclist and motorcycle) at the centre of the motorcycle. These results show that variable riding postures result in different form of force

Table 1.   Weight Distributions According to Riding Posture.

Riding Posture	Weight Distribution Bias on the Motorcycle	Description	Advantages and Disadvantages
Type 1	Forward biased	By leaning the torso forward, the motorcyclist provides significant load on the frontal section of the motorcycle as opposed to the rear section. The regions affected are the handlebar and forks.	The motorcycle is responsive and manoeuvrable especially when changing directions due to physics in which high force is utilized to move a body (in this case the steering head which connects the handlebar and forks). Less leverage is needed to manoeuver the motorcycle. Only averagely size and width tubular handlebars or clip-ons are needed. The motorcyclist could easily perform the "UBM" (Ma'arof et al., 2012) and the motorcycle could effectively receive and exert the necessary output.
Type 2	Centre biased	By sitting upright, the motorcyclist provides significant load on the centre section of the motorcycle. The region affected is the seat-pan.	A centre-biased motorcycle shows a more balanced overall characteristic due to the even force (or weight) distributions. The motorcyclist would require only slightly higher leverage as opposed to the forward-biased to manoeuvre the motorcycle. The motorcyclist could perform the task with ease, though not as effectively as the forward-biased motorcycle.
Type 3	Rearward biased	By inclining the torso rearward, the motorcyclist would provide load on the rear section of the motorcycle.	The motorcycle is less responsive due to minimal force channelled to the handlebar and forks. The motorcycle cannot receive and exert the necessary output to manoeuvre. The motorcyclist has to rely solely on the leverage provided by the handlebar of the motorcycle. Hence, the handlebar design for this specific type of motorcycle is crucial to provide the necessary leverage. Usually, wide handlebars are used.
Type 4	Centre biased	By leaning the torso forward and extending the legs forward, the motorcyclist balances force distribution into the centre of the motorcycle	Similar to Type 2.

exertion, overall weight distributions and resultant COM (both motorcyclist and motorcycle) allocations of the system. The findings are summarized in Table 1.

In summary, the riding posture assumed by the motorcyclist determined the characteristics of the motorcycle via the overall weight distribution of the motorcycle ergonomics system. Ultimately, this determined the functional characteristics of the motorcycle especially in term of manoeuvrability and control, hence, influencing the overall safety level in real-world motorcycling.

Conclusion

The motorcycle is a good example of a workstation that is influenced by the human operator. The riding posture assumed by the human operator influences the characteristics of the motorcycle via the overall weight distribution of the system. This determines the effectiveness of the input exerted (work tasks performed by the motorcyclist) and how the motorcycle (workstation) conforms to generate the output. Ultimately, these affect the overall safety of motorcycling. For these reasons, motorcycles have to be designed in such a way that the motorcyclist can ride at the within an acceptable range of efficiency and effectiveness. In addition, the calculation made in this study was limited to the data gathered from a static riding posture and calculation on 2D analysis of the motorcyclist-motorcycle integration. For future research, 3D analysis or assessment using force plates could show a more diverse distribution of dynamic path loading on the motorcycle due to the physical activities performed by the motorcyclist. Also, this study was limited by the motorcycles and the assumptions made. In addition, it should also be noted that other settings on the motorcycles such as the suspension (or shocks or absorber) or tyre pressure could also affect the weight distribution and manoeuvrability of the motorcycle. Henceforth, further studies are required to support the findings for other two-wheeled vehicles and workstations.

Acknowledgement

The authors express their gratitude to the staffs of the Faculty of Mechanical Engineering, Universiti Teknologi MARA, Malaysia (UiTM) and the members of the Human Factors and Ergonomics Malaysia (HFEM) Society who had directly or indirectly contributed to this study.

References

Cimino, A., Longo, F. and Mirabelli, G. 2009, A multimeasure-based methodology for the ergonomic effective design of manufacturing system workstations. *International Journal of Industrial Ergonomics* 39, 447–455.
Cossalter, V. 2006, *Motorcycle Dynamics* (2nd edition.), Lulu.

Das, B. and Sengupta, A. K. 1996, Industrial workstation design: a systematic approach. *Applied Ergonomics* 27(3), 157–163.

Glimmerveen J., n.d., Retrieved February 3rd, 2013, from About.com: http://classic motorcycles.about.com/od/technicaltips/ss/MotorcycleWeightDistribution.htm

Kroemer, K., Kroemer, H. and Kroemer-Elbert, K. 2000, Ergonomics: how to design for ease and efficiency (second edition)", *Prentice Hall International Series in Industrial & System Engineering*, 1.

Ma'arof, M. I. N. and Ahmad, I. N. 2012b, Proposed standardized method for motorcycle nomenclature system. *Southeast Asian Network of Ergonomics Societies Conference (SEANES), Road User Ergonomics*, Paper 60.

Ma'arof, M. I. N. and Ahmad, I. N. 2012a, A review of ergonomics and other studies on motorcycles. *Southeast Asian Network of Ergonomics Societies Conference (SEANES), Road User Ergonomics*, Paper 59.

Ma'arof, M. I. N., Ahmad, I. N., Abdullah, N. R. and Karim, S. A. 2012, Motorcycling riding issues: understanding the phenomenon and development of ergonomics intervention in improving perceived comfort for prolonged riding, *iDECON*.

Munoz, F. and Rougler, P. R. 2011, Estimation of center of gravity movements in sitting posture: application to trunk backward tilt. *Journal of Biomechanics* 44, 1771–1775.

Resnick, E. Motorcycle Ergonomics Simulator, Retrieved September 26th, 2013, from cycle-ergo official website: http://cycle-ergo.com/

Robertson, M. M., Ciriello, V. M. Garabet, A.M. 2013, Office ergonomics training and a sit-stand workstation: Effects on musculoskeletal and visual symptoms and performance of office workers. *Applied Ergonomics* 44, 73–85.

Robertson, S. A. and Minter, A. 1996, A study of some anthropometric characteristics of motorcycle riders. *Applied Ergonomics*, 27(4), 223–229.

Swat, K. and Krzychowicz, G. 1996, ERGONOM: Computer-aided working posture analysis system for workplace designers. *International Journal of Industrial Ergonomics* 18, 15–26.

Teoh, E. R. and Campbell, M. 2010, Role of motorcycle type in fatal motorcycle crashes. *Journal of Safety Research* 41, 507–512.

V-Twinforum.com, 2007, rear/front weight distribution – StreetBob? Retrieved February 1st, 2013 from V-TwinForum.com official website: http://www.v-twinforum.com/forums/harley-davidson-dyna/110177-rear-front-weight-distribution-streetbob.html

MOTORCYCLIST MUSCLE FATIGUE INDEX: AN EFFORT TO HELP REDUCE MOTORCYCLE ACCIDENTS

H. Rashid[1], M.I.N. Ma'arof[1], A.R. Omar[1], S.C. Abdullah[1], I.N. Ahmad[1] & S.A. Karim[2]

[1] *Faculty of Mechanical Engineering, Universiti Teknologi MARA, 40450 Shah Alam, Selangor Darul Ehsan, Malaysia*
[2] *MASMED, INTEKMA Resort, Universiti Teknologi MARA, 40450 Shah Alam, Selangor Darul Ehsan, Malaysia*

Globally, 23% of all traffic deaths are attributed to motorcycle riders that lost their lives as a result of severe injuries from motorcycle accidents (World Health Organization, 2013). Amongst the factors causing such accidents is muscle fatigue experienced by motorcyclists, especially during prolonged riding. This paper elaborates on an effort to help reduce motorcycle accidents through the development of a muscle fatigue indexing system. The method used to capture muscle fatigue and develop the muscle fatigue index is discussed. It is considered that motorcycle accidents could be reduced by managing motorcyclist muscle fatigue.

Introduction

Motorcycle accidents have risen from year to year in most countries in the world, and motorcyclists involved in road accidents often die due to severe injuries. A report by the New Zealand Ministry of Transport stated that from the year 2007–2011, in 96% of fatal crashes involving motorcyclists, the motorcyclist or the pillion rider was among those who died (New Zealand Ministry of Transport, 2011). In the 2009 Royal Malaysia Police Annual Report, 60.3% of total deaths due to road accidents in Malaysia are from motorcycle casualties (Malaysia, 2009; World Health Organization, 2013). Malaysia holds the highest road fatality risk (deaths per 100,000 population) among ASEAN countries, with more than 50% involving motorcyclists (Abdul Manan & Várhelyi, 2012; Jacobs, Aeron-Thomas, & Astrop, 2000). Asian countries rank the top highest average number of motorcycle ownership ratio in the population, with 7 times the average of the rest of the world (Abdul Manan & Várhelyi, 2012; Senbil, Zhang, & Fujiwara, 2007). This calls for many researchers to investigate and carry out more studies regarding this alarming issue towards reducing the number of motorcycle fatalities on the roads globally and in Malaysia specifically.

Research reviews

Current research in motorcycle ergonomics

From numerous studies involving motorcycle accidents either involving a single motorcycle crash or a two-vehicle motorcycle crash, many factors are related to the cause of those fatal accidents (Keall & Newstead, 2012; Schneider, Savolainen, Van Boxel, & Beverley, 2012). Besides the common causes, classical causes identified were violating the motorcyclist right-of-way (ROW) (Clarke, Ward, Bartle, & Truman, 2007; H.H. Hurt, Ouellet, & Thom, 1981; Pai, 2011) and also loss of control on bends or curve (Sexton, Fletcher, & Hamilton, 2004), with twice the risk of death of the motorcyclist or pillion in an accident (Clarke et al., 2007).

However, looking at the trend of existing studies involving motorcycle safety and accidents, the majority of researchers focus more on studying the external elements that interacts with a motorcyclist, for example the use of helmets, consumption of alcohol and drug abuse, not having legal riding license and many more (Abdul Manan & Várhelyi, 2012; Jou, Yeh, & Chen, 2012; Keall & Newstead, 2012; Schneider et al., 2012). Very little research explores the interactions of the motorcyclist with the motorcycle either physically or physiologically, for example motorcyclist riding posture, thermal comfort and vision ability. This leaves a gap in this area of research, although other transport ergonomics research has been established since 25 years (Ma'arof & Ahmad, 2012b; Robertson, Stedmon, Stedmon, & Bust, 2009).

Looking into how ergonomics aligns with the effort in reducing motorcycle accidents, there is relatively little previous research that incorporates ergonomic elements. However elements of ergonomics must also be considered when studying a field that involves human-machine-environment interaction (HMEI), including motorcycle research (Ma'arof, Ahmad, Abdullah, & Karim, 2012; Ma'arof & Ahmad, 2012b; Robertson et al., 2009; A. W. Stedmon, Brickell, Hancox, Noble, & Rice, 2012). A specific Special Interest Group (SIG) in Motorcycle Ergonomics was established in the UK in 1999 under the Institute of Ergonomics and Human Factors (IEHF) to concentrate on research concerning motorcycle ergonomics and rider performance. Hence, effort in considering ergonomics elements into motorcycle research has already been initiated, which provides room for more study regarding risks, causes and interventions in motorcycle accidents. Amongst research and knowledge gaps being identified are lack of research specifically on rider fatigue, importance of rider fatigue – prevalence, subjective states, and effects on performance, contribution of rider fatigue to crashes, operational definitions and assessment methods for fatigue, causes and effects of rider fatigue, rider recognition of fatigue, and rider fatigue countermeasures (Horberry, Hutchins, & Tong, 2008).

Motorcyclist fatigue

Based on the aforementioned research gaps identified, most of them necessitate research in motorcyclist fatigue including muscular fatigue. Although there is little prior research studying motorcyclist fatigue, most of the results and conclusions were drawn from self-reported data and does not have a well-designed and controlled methodology (Halim & Omar, 2012a; Haworth & Rowden, 2006; Horberry et al., 2008; Ma, Williamson, & Friswell, 2003). This is where this paper will discuss on how to better capture and define fatigue with an indexing system to identify the risk that a motorcyclist may encounter with a certain level of fatigue being experienced.

In narrowing the fatigue type being experienced by the motorcyclist, muscle fatigue is extensively explored in this research. The build-up of body fatigue is mostly contributed by a number of muscles in the upper abdomen of the motorcyclist body (Karuppiah, Salit, Ismail, Ismail, & Tamrin, 2012; Velagapudi, Balasubramanian, Adalarasu, Babu, & Mangaraju, 2010). This is why this research mainly investigates muscular fatigue experienced by the motorcyclist besides literatures stating that muscle fatigue causes of some major motorcycle accidents (Haworth & Rowden, 2006; Horberry et al., 2008; Williamson et al., 2011).

The research mainly focuses on which muscles are involved, how long a motorcyclist can ride a motorcycle before experiencing the first sign of muscle fatigue and how indexing muscle fatigue may help the motorcyclist to increase their awareness of risk factors in prolonged riding that may lead them to fatal road accidents. Similar to previous referred literatures on prolonged sitting and standing (Carvalho, 2008; Halim, Omar, Saman, Othman, & Ali, 2012; Sugiyama et al., 2012), when muscle fatigue starts to build up in the body of a human worker or operator, it will tend to cause other effects on the human sense, body reaction and metabolism. Therefore, how the motorcyclist's muscle fatigue is measured and how the results are conveyed in a sense that the motorcyclist can decide when to take a rest in a prolonged riding needs to be clarified and documented.

Methodology and research framework

The methodology used in this research was mainly referred and replicated from an existing established conceptual framework involving muscle fatigue in prolonged standing by Isa Halim et al. (2012), as shown in Figure 1. This is because previous literature on prolonged riding of motorcycle riders tempt to rely on driver's literatures reported by Horberry et al., 2008 including the duration of sitting time by assuming prolonged sitting to be very much similar to prolonged riding in terms of sitting condition. Some experiments on motorcyclist prolonged riding design the duration of the experiments to last from 1 to 2 hours of riding (Karuppiah et al., 2012; Velagapudi et al., 2010). This is supported by literature on prolonged sitting that defines prolonged sitting to be a minimal duration of 2 hours (Aota et al., 2007; Karuppiah et al., 2012; Schinkel-Ivy, Nairn, & Drake, 2013; Sugiyama et al., 2012). The conceptual framework for this research is shown in Figure 2.

Motorcycle test rig

In capturing the muscle fatigue experienced by motorcyclists in this research, a motorcycle test rig was designed using state-of-the-art CAD software, CATIA

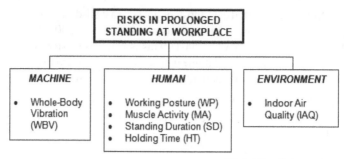

Figure 1.   Conceptual framework of investigating muscle fatigue in prolonged standing (Halim & Omar, 2012a, 2012b).

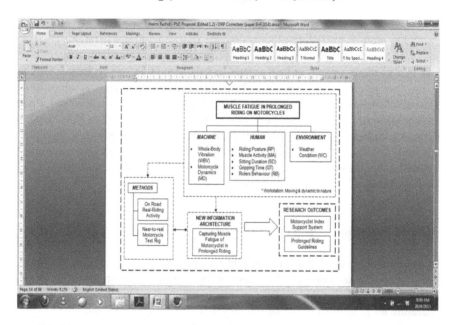

Figure 2.   Conceptual framework of investigating muscle fatigue in motorcycle prolonged riding.

V5R20 and filed for patent. This motorcycle test rig is only treated as a low cost test rig without assuming it as a high-end simulator. CAD models are generated referring to an established Riding Posture Classification (RIPOC) System (Ma'arof & Ahmad, 2012a) that replicates motorcyclist riding postures on a motorcycle. This motorcycle test rig provides adjustable attributes to allow various riding postures to mimic the type of riding postures established by the RIPOC system. Besides that, in narrowing validity and fidelity issues arising from research involving simulators (Naweed, Hockey, & Clarke, 2013; A. W. Stedmon et al., 2012; A.W. Stedmon, M.S.Young, & Hasseldine, 2009; Weir, 2010), this motorcycle test rig is designed by integrating the elements of human-machine-environment interactions (HMEI) that offers a more near-to-real riding experience. Being operated in a controlled laboratory setup, the motorcycle test rig also gives a higher safety precaution for both researchers and motorcyclist subjects during experimentation.

Muscle fatigue data extraction

In having to capture and look into the correlations between the muscles and the muscle fatigue build-up of motorcyclist in prolonged riding, a surface electromyography (sEMG) device is extensively used. The procedures to measure muscle activities on the motorcyclist subjects have been approved by the Research Ethics Committee (REC), Universiti Teknologi MARA, Malaysia. The muscle types involved are mainly from the trunk of the human body consisting of the erector spinae muscle group which is an important muscle for human to maintain an upright posture of the trunk (Sung, Lammers, & Danial, 2009). However, when riding motorcycle with RIPOC Type 1 and 4 (Ma'arof & Ahmad, 2012a), the erector spinae muscle that are attached to the lumbar vertebrae (Sung et al., 2009) will be affected and may cause low back pain (LBP) and eventually leads to muscle fatigue to motorcyclist. Besides the erector spinae muscle, other muscles investigated in this research are the Flexor Carpi Ulnaris (forearm muscle), Trapezius type II fibers (shoulder/neck) and Latissimus Dorsi (middle back) (Velagapudi et al., 2010).

Results and discussion

Muscle fatigue experienced by motorcyclist especially in prolonged riding needs to be thoroughly studied in the effort to reduce the number of motorcycle accidents and fatalities, which keeps increasing from year to year globally. By having a muscle fatigue indexing system, the motorcyclist shall be equipped with appropriate information on their level of fatigue either before, during or after riding. This gives them an option and guideline on when they should rest after a few hours of riding or how long they can ride with a certain level of muscle fatigue experienced. Moreover, this decision support approach provides motorcyclist with better awareness of risk that may occur when they tend to neglect information provided by the indexing system. However, further studies in testing the effectiveness of this indexing system are required and needs to be performed for the betterment and improvement of the whole outcomes of this research specifically and for the motorcyclist target group generally.

Conclusion

Through this research, an established experimentation research facility that includes a patentable motorcycle test rig, wireless/wired sEMG device, modified RULA, REBA and WERA assessment and survey sheet and also validated procedures and protocols in measuring muscles activities for motorcyclist in prolonged riding are being setup at the Ergonomics Laboratory, Faculty of Mechanical Engineering, Universiti Teknologi MARA, Malaysia. These facilities will facilitate the motorcycle ergonomics research group in developing a muscle fatigue indexing system for motorcyclists, especially in prolonged riding. It can also facilitate future research in this area of motorcycle ergonomics and help to form a Special Interest Group (SIG) of motorcycle ergonomics under the Human Factors and Ergonomics Malaysia (HFEM) Society in the near future. By having this indexing system, it is hoped that motorcyclists will be more aware of their readiness and health condition before they start to ride their motorcycles, to help reducing the increasing numbers of motorcycle accidents that caused many deaths. This is not only for the sake of their safety, but also for the benefit of other road users that share the same infrastructure and risk together every day.

Acknowledgement

The research team would like to highly acknowledge the staffs of the Ergonomics Laboratory, Faculty of Mechanical Engineering, Universiti Teknologi MARA, Malaysia and also members of the Human Factors and Ergonomics Malaysia (HFEM) Society for their fervent support and also to those involved either directly or indirectly towards this research.

References

Abdul Manan, Muhammad Marizwan and Várhelyi, András. 2012. Motorcycle fatalities in Malaysia. *IATSS Research* 36(1), 30–39.

Aota, Yoichi, Iizuka, Haruhiko, Ishige, Yusuke, Mochida, Takashi, Yoshihisa, Takeshi, Uesugi, Masaaki and Saito, Tomoyuki. 2007. Effectiveness of a Lumbar Support Continuous Passive Motion Device in the Prevention of Low Back Pain During Prolonged Sitting. *SPINE* 32(23), 674–677.

Carvalho, Diana Elisa De. 2008. Time Varying Gender and Passive Tissue Responses to Prolonged Driving. (Master of Science), University of Waterloo, Waterloo, Ontario, Canada.

Clarke, D. D., Ward, P., Bartle, C. and Truman, W. 2007. The role of motorcyclist and other driver behaviour in two types of serious accident in the UK. *Accident Analysis and Prevention* 39(5), 974–981.

H.H. Hurt, Jr., Ouellet, J.V. and Thom, D.R. (1981. Motorcycle Accident Cause Factors and Identification of Countermeasures (T. S. Center, Trans.) (Vol. 1): University of Southern California.

Halim, Isa and Omar, Abdul Rahman. 2012a. Development of Prolonged Standing Strain Index to Quantify Risk Levels of Standing Jobs. International *Journal of Occupational Safety and Ergonomics* 18(1), 85–96.

Halim, Isa and Omar, Abdul Rahman. 2012b. Prolonged Standing Strain Index (PSSI): A Proposed Method to Quantify Risk Levels of Standing Jobs in Industrial Workplaces. *Advanced Materials Research* 433–440, 497–506.

Halim, Isa, Omar, Abdul Rahman, Saman, Alias Mohd, Othman, Ibrahimand Ali, Mas A. 2012. Analysis of Time-to-Fatigue for Standing Jobs in Metal Stamping Industry. *Advanced Materials Research* 433–440, 2155–2161.

Haworth, Narelle and Rowden, Peter. 2006. Fatigue In Motorcycle Crashes: Is There An Issue? Paper presented at the *Australasian Road Safety Research, Policing and Education Conference*, Gold Coast, Queensland.

Horberry, T., Hutchins, R. and Tong, R. 2008. *Motorcycle Rider Fatigue: A Review.* London: Department for Transport.

Jacobs, G., Aeron-Thomas, A. and Astrop, A. 2000. *Estimating global road fatalities.* (Crowthorne, Berkshire, England: Transport Research Laboratory).

Jou, R. C., Yeh, T. H. and Chen, R. S. 2012. Risk factors in motorcyclist fatalities in Taiwan. *Traffic Injury Prevention* 13(2), 155–162.

Karuppiah, K., Salit, M. S., Ismail, M. Y., Ismail, N. and Tamrin, S. B. M. 2012. Evaluation of motorcyclist's discomfort during prolonged riding process with and without lumbar support. *Anais da Academia Brasileira de Ciencias* 84(4), 1169–1188.

Keall, M. D. and Newstead, S. 2012. Analysis of factors that increase motorcycle rider risk compared to car driver risk. *Accident Analysis and Prevention* 49, 23–29.

Ma'arof, M. Izzat Nor, Ahmad, Ismail N., Abdullah, Nik R. and Karim, Shamsury A. 2012. Motorcycling Riding Issues: Understanding the Phenomenon and Development of Ergonomics Intervention in Improving Perceived Comfort for Prolonged Riding. Paper presented at the *International Conference on Design and Concurrent Engineering* (iDECON 2012), Melaka.

Ma, Theree, Williamson, Ann and Friswell, Rena. 2003. A Pilot Study of Fatigue on Motorcycle Day Trips: NSW Injury Management Research Centre.

Ma'arof, M. Izzat Nor and Ahmad, Ismail N. 2012a. Proposed Standard Method for Motorcycle Nomenclature System. Paper presented at the *2012 Southeast Asian Network of Ergonomics Societies Conference (SEANES)*, Langkawi.

Ma'arof, M. Izzat Nor and Ahmad, Ismail N. 2012b. A Review of Ergonomics and Other Studies on Motorcycles. Paper presented at the *2012 Southeast Asian Network of Ergonomics Societies Conference (SEANES)*, Langkawi.

Malaysia, Polis Diraja. 2009. *Laporan Tahunan Polis Diraja Malaysia 2009.* Polis Diraja Malaysia.

Naweed, A., Hockey, G. R. and Clarke, S. D. 2013. Designing simulator tools for rail research: the case study of a train driving microworld. *Applied Ergonomics* 44(3), 445–454.

New Zealand Ministry of Transport 2011. *Motorcyclist Crash Statistics for the Year Ended 31 December 2011.* New Zealand Ministry of Transport.

Pai, C. W. 2011. Motorcycle right-of-way accidents – a literature review. *Accident Analysis and Prevention* 43(3), 971–982.

Robertson, S., Stedmon, A.W., Stedmon, D.M. and Bust, P.D. 2009. Motorcycle Ergonomics: Some Key Themes in Research. *Contemporary Ergonomics* (Taylor & Francis, London), 432–441.

Schinkel-Ivy, Alison, Nairn, Brian C. and Drake, Janessa D. M. 2013. Investigation of trunk muscle co-contraction and its association with low back pain development during prolonged sitting. *Journal of Electromyography and Kinesiology* 23(4), 778–786.

Schneider, W. H. th, Savolainen, P. T., Van Boxel, D. and Beverley, R. 2012. Examination of factors determining fault in two-vehicle motorcycle crashes. *Accident Analysis and Prevention* 45, 669–676.

Senbil, Metin, Zhang, Junyi and Fujiwara, Akimasa. 2007. Motorization in Asia – 14 Countries and Three Metropolitan Areas. *IATSS Research* 31(1), 46–58.

Sexton, Barry, Fletcher, John and Hamilton, Kevin. 2004. *Motorcycle Accidents and Casualties in Scotland 1992–2002 Transport Research Series*. Scottish Executive Social Research.

Stedmon, A. W., Brickell, E., Hancox, M., Noble, J. and Rice, D. 2012. MotorcycleSim: A user-centred approach in developing a simulator for motorcycle ergonomics and rider human factors research. *Advances in Transportation Studies* (27), 31–48.

Stedmon, A.W., M.S. Young and Hasseldine, B. 2009. Keeping It Real or Faking It: The Trials and Tribulations of Real Road Studies and Simulators in Transport Research *Contemporary Ergonomics 2009*, (Taylor & Francis, London).

Sugiyama, T., Merom, D., van der Ploeg, H. P., Corpuz, G., Bauman, A. and Owen, N. 2012. Prolonged sitting in cars: prevalence, socio-demographic variations, and trends. *Preventive Medicine* 55(4), 315–318.

Sung, P. S., Lammers, A. R. and Danial, P. 2009. Different parts of erector spinae muscle fatigability in subjects with and without low back pain. *Spine* 9(2), 115–120.

Velagapudi, S. P., Balasubramanian, V., Adalarasu, K., Babu, R. and Mangaraju, V. 2010. Muscle fatigue due to motorcycle riding. SAE Technical Papers.

Weir, David H. 2010. Application of a driving simulator to the development of in-vehicle human–machine-interfaces. *IATSS Research* 34(1), 16–21.

Williamson, A., Lombardi, D. A., Folkard, S., Stutts, J., Courtney, T. K. and Connor, J. L. 2011. The link between fatigue and safety. *Accident Analysis and Prevention* 43(2), 498–515.

World Health Organization. 2013. *Global Status Report on Road Safety 2013 – Supporting A Decade of Action*. World Health Organization.

WORKLOAD & VIGILANCE

SPEAKING UNDER WORKLOAD IN FIRST AND SECOND LANGUAGES: IMPLICATIONS FOR TRAINING AND DESIGN

Chris Baber & Hong Liu

Electronic, Electrical and Computer Engineering, University of Birmingham

An emphasis on multinational cooperation (e.g. Air Traffic Control, Disaster Response) can require people to communicate in a second language. There is little work on the interaction between workload and speech production in a second language. This paper explores the potential impact of workload on the production of speech in a Stroop task. A group of Chinese students completed the task in Mandarin (L1) or English (L2) and their performance was compared with a group of students speaking English. The results suggest that changes in workload affect speech production, and that these effects are increased when responding in a second language. Responding in a second language is cognitively demanding and can be easily disrupted by increases in workload.

Introduction

This paper considers the impact of workload on the production of speech in one's first (L1) or second (L2) language. There are two reasons why such a study is of interest to the Human Factors community. First, there are many applications in which multi-national cooperation requires people to communicate in languages which are not their first. Often (but not always) this second language is English. It is plausible to assume that communication errors might arise, particularly when the speaker is placed under conditions of stress or workload; they might know a particular word in their first language but might struggle to recall the correct translation. Second, while there are continued developments of speech recognisers for a variety of languages, many systems are designed to work with English and people speaking English as a second language might have problems in terms of pronunciation. Such problems could be exacerbated when the speakers are placed under stress.

Previous work has shown that the performance of automatic speech recognition systems can be significantly impaired by speakers adjusting their speech to compensate for background noise (the Lombard Effect (Hansen, 1996)) and that human speech production is also affected by other forms of stressors (Baber and Noyes, 1996). The effects could range from acoustic effects, such as changes in fundamental frequency or shifts in formant frequency, or could arise from changes in the timing of speech, e.g. hesitant or rushed, or from errors in the spoken response,

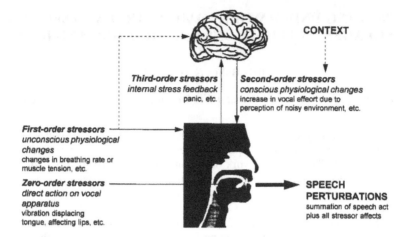

Figure 1.   Working model of stress and speech [Murray et al., 2006].

e.g. wrong word being spoken. These ideas are summarized in figure 1. This distinguishes between effects which are produced by direct action of speech production (zero-order stressors) from those which are mediated, either in response to unconscious changes in speech production (first-order and third-order stressors) or as a result of conscious adaptation to the environment (second-order stressors). It is proposed that while this distinction from zero-order to third-order stressors provides a convenient means of classifying results from the literature, it is likely that such results are likely to arise from a combination of stressors. Thus, for example, Lombard effects might combine first-order and third-order stressors and result in increases in pitch (as a response to environmental demands) and increases in vocal effect and timing (as a result of deliberate changes to communication strategy). The question of how people respond to changes in workload, which is the focus of this paper, is likely to be predominantly second-order.

Lively et al. (1993) analysed speech produced by participants performing a compensatory visual tracking tasks, and found that *some* talkers exhibited changes in speaking rate and acoustic parameters. They suggested that people compensate for task demands by altering their speech production in order to maintain intelligibility. Baber et al. (2006) showed that increasing time pressure or introducing a secondary task can impair the performance of Automated Speech Recognisers. Workload effects have been identified in other domains, e.g. simulated cockpits (Chen and Masi, 1999; Brenner et al. (1994) and Armoured Fighting Vehicles (Noyes et al., 2000). These studies imply that the cognitive component of producing speech can suffer when paired with other tasks (even when these are not directly related to verbal activity). The question to be considered in this study is whether this impairment is worse when participants are using their second language. Chen (2001) compared performance for participants speaking Swedish or English and found workload effects. In this study, we consider the effects of speaking English or Mandarin on Chinese participants performing a simple speech production task.

The Stroop task and speech production

The Stroop task (Stroop, 1953) is a common experimental task in cognitive psychology. It requires participants to say the colour of the ink used for the letters spelling a colour name, i.e., the word **white** should produce the response 'black' (because it is written in black ink). The Stroop task provides a simple and easy to replicate paradigm for the study of speech production (Chen et al., 2012). Not only does it require a small vocabulary (assisting comparison between conditions because one can assume that the same words were spoken in all of the conditions) but it also has minimal language requirements, in that participants can be expected to know the names of the primary colours (in this experiment, we used red, blue, yellow, green, and black). While there are several theories to explain how this effect might work, the most common assumes that interference occurs because people are more familiar with reading aloud the word than naming colours. This leads to the need to exercise attentional control to inhibit the word naming and excite the colour naming (Cohen et al., 1990).

In terms of producing responses in L2 there could be an additional challenge of managing the activation of both L1 and L2 words. For example, when an object is presented, the bilingual speaker *could* have both L1 and L2 words activated and would need to select for the desired response (Green, 1986; Green and Abutalebi, 2013). This would imply that interference between L1 and L2 could be a source of additional workload for the speaker (which could delay response time or lead to mispronunciation). From this perspective, one might anticipate problems to arise if one is performing the Stroop task in a language other than one's first language. If one assumes that object naming (and by implication, the Stroop task) relies on a common underlying semantics which is accessed by both language, then problems could arise with the need to suppress the irrelevant word name and activate the relevant colour name. One could consider this in terms of the notion of asymmetrical switching proposed by Allport et al. (1994), in that there is a possible cost (in processing overheads) between selecting and speaking the colour word and suppressing the word name, in one's first or second language. Roselli et al. (2002) found some evidence that colour-naming was marginally slower in bilinguals. In a study using numbers rather than colours, Campbell (2005) asked participants to respond in English or Chinese when cued using Mandarin or Arabic numerals. He found that performance varied with presentation format, i.e., asymmetrical perseveration of English when using Arabic numerals, presumably due to the familiarity of naming the stimulus in that language. This effect persisted for speaking the answers to simple arithmetic problems.

Experiment

Participants

18 people took part in the study. 9 were English (6 male, 3 female), 9 were Chinese (4 male, 5 female). All Chinese students have lived in the UK for at least 6 months, and had a proficient level of spoken English (defined as IELTS level 6.5 or higher).

Equipment and set-up

The experiment took place in a sound-proofed booth (of dimensions approximately 3 m × 3 m × 3 m). The Stroop application (written in VisualBasic) involved 5 colours (red, blue, black, yellow, green) paired randomly with the colour words. When each colour-word pair was presented, an auditory 'beep' was passed (via line one) to a DAT recorder. Participants spoke the word into a Shure SM150 microphone, connected on line two to the DAT recorder. The timing between words was controlled by the application, with a variation of 15% around the mean time for each trial. This was designed to prevent anticipation of presentation.

Procedure

Participants were given an explanation of the aims of the experiment and asked to complete a consent form. They were seated in the booth and asked to put on the headset and adjust it until it was comfortable. Recorder levels were set by asking the participant to repeat strings of words. The experiment began with a familiarisation trial, in which the first block of words were presented in English in the correct colour. For the Chinese participants, this was repeated with the colours were presented in Standard Mandarin and the trial repeated. This allowed recording levels to be checked. A practice trial, involving 3 blocks of 10 words with random colour-word matching followed. Participants were given feedback following this practice trial. The experiment consisted of 5 blocks of 10 words presented under six levels of workload: normal (one word per second); medium (one word per 0.6 second); fast (one word per 0.3 second), with an without an additional tapping task. The tapping task required participants to tap their index finger on the table top in a 'waltz' pattern. This was intended to have an additional demand on participants with the intention that tapping could load the spatial component of working memory without loading verbal performance (Farmer et al., 1986).

Metrics

The study uses three metrics for evaluating performance: Correct Response; Reaction Time; and Change in Fundamental Frequency relative to practice. A Correct Response was defined as the person speaking the correct colour of the ink for a given presentation. If the person said the wrong word or failed to say a word, then this was counted as an incorrect response. If the person stumbled on the word, e.g. saying/r/.../blue/, then this was accepted as a correct response. Reaction Time is defined as the time, in seconds, from the appearance of the stimulus to the start of speaking. This was determined through manual inspection of spectrograms (such as those shown in figure 1) and measuring the end of the 'beep' to the start of the speech. While such an approach could be prone to experimenter error (for example, mistaking the start of speech for some other noise), the analysis was repeated three times and, if a discrepancy arose, this was analysed by another person. Fundamental Frequency varies across individuals. Consequently, the analysis was performed by first, subtracting the 'practice' F0 from each set of data. This ensured that all

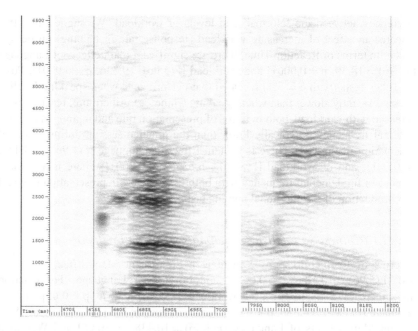

Figure 2.   Spectrograms of the word "BLUE" being spoken by English male (left) and Chinese male (right) speakers during the Practice trial.

analysis would be performed on the relative change in F0 as a result of the different workload levels (rather than as a straightforward comparison of individual F0s). For the purposes of this analysis, we have averaged these data across a condition rather than reporting them for individual performance. The reason for doing this is that previous work, mentioned in the Introduction, identified fundamental frequency as an approximate index of workload under different conditions.

All speech recordings were imported into the Speech Filing System (SFS) analysis toolkit. From this each word spoken could be analysed in terms of acoustic characteristics and reaction time, i.e., the time between the 'beep' (indicating the stimulus appearing) and the beginning of the spoken response. Figure 2 shows traces of the spectrographs produced, for the English word /blue/ by male English and Chinese speakers.

Results

L1 vs L2 for Chinese participants

A two-way Analysis of Variance Correct responses, over the levels of workload, for Chinese students speaking Mandarin (CM) or English (CE) show little effect of language, but a significant main effect of workload [$F(5,96) = 9.968$, $p < 0.0001$]. Post-hoc analysis (using Tukey tests) indicates that this effect arises from significant differences between the 'Normal' and levels of workload. We suggest that this indicates an effect of increasing workload (response rate or tapping) on performance. In terms of Reaction Time, there are significant main effects of Language $(F(1,96) = 18.39, p = 0.0001]$ and Workload $[F(5,96) = 9.336, p = 0.0001]$. Post-hoc Tukey tests (with $\alpha = 5\%$) revealed that performance when speaking English was significantly slower than when speaking Chinese. Furthermore, reaction time increased with workload, both in terms of presentation rate and tapping task. Fundamental frequency is typically lower than the practice session during 'normal' presentation of the task. There is a Main Effect of Language $[F(1,96) = 10.438, p < 0.001]$, and of Workload $[F(5,96) = 6.833, p < 0.0001]$. There is a marked increase in fundamental frequency with both the increase in presentation rate and the introduction of the additional tapping task.

L2 vs L1 (Chinese vs English participants speaking English)

In terms of Correct Response, there is a Significant Main Effect of Workload $[F(5,96) = 10.189, p < 0.0001]$ and but not of Language. For reaction time, there is a main effect of workload $[F(5,96) = 3.37, p < 0.05]$ and of Language $[F(1,96) = 4.886, p < 0.05]$. For Fundamental Frequency, there are Significant Main Effects of Language $[F(1,96) = 10.438, p = 0.05]$ and Workload $[F(5,96) = 6.884, p < 0.0001]$. Across all workload conditions, L2 shows a higher increase in F0 than L1.

L1 vs L1 (Chinese vs English)

For correct response, there was a main effect on Workload $[F(5,96) = 9.42, p < 0.0001]$ but not of language. For RT, there was a main effect of Workload $[F(5,96) = 4.77, p < 0.001]$ but not of language. For F0, there was a main effect of Workload $[F(5,96) = 10.354, p < 0.0001]$ and also of language $[F(1,96) = 6.54, p < 0.05]$.

Discussion

There are significant effects of workload (Chinese participants speaking Mandarin or English participants speaking English [L1] or Chinese participants speaking English [L2]). The pacing of the response (from slow to fast) and the requirement to tap while responding lead to reductions in the number of correct responses, increases in reaction time and changes in fundamental frequency. Thus, even for so simple a task as colour naming, the requirement to perform under time pressure or with an additional task impaired performance. This replicates the findings of previous studies reviewed in the introduction.

As previously studies showed, reaction time for L2 was significantly impaired in comparison to use of L1. Thus, Chinese participants produce slower responses when speaking English than when speaking Mandarin or when English participants spoke English. This suggests that there are cognitive costs when people use a second language to respond – whether these costs are sufficient to cause errors or other performance impairment remains to be investigated.

In terms of technology adapting to changes in speech, workload leads to a measurable shift in F0 (in both L1 and L2). While it might be possible to create adaptive speech recognisers which respond to shifts in F0, the fact that there is a greater shift for L2 than L1 production of English responses suggests that simply adapting to frequency shifts might not be sufficient and that differential adaptation might be required for people speaking second language to control speech systems (which could, for instance, be used in cockpit voice systems).

References

Allport, A., Styles, E.A. and Hsieh, S., 1994, Shifting intentional set: exploring the dynamic control of tasks, In C. Umilta and M. Moscovitch (eds.) *Attention and Performance XV: Conscious and Unconscious Information Processing*, Hillsdale, NJ: Erlbaum, 421–452

Baber, C., Mellor, B., Graham, R., Noyes, J.M. and Tunley, C., 2006, Workload and the use of automatic speech recognition: the effects of time and resource demands, *Speech Communication, 20*, 37–53

Brenner, M., Doherty, E.T. and Shipp, T., 1994, Speech measures indicating workload demand, *Aviation, Space and Environmental Medicine, 65*, 21–26

Campbell, J.I.D., 2005, Asymmetrical language switching costs in Chinese-English bilinguals' number naming and simple arithmetic, *Bilingualism: Language and Cognition, 8*, 85–91

Chen, F., 2001, The effect of time stress on automatic speech recognition accuracy when using second language, *InterSpeech '01*

Chen, F. and Masi, C., 1999, Effect of noise and workload on an automatic speech recognition system, In J. Abeysekara, E.M. Lonnroth, D.P. Piamonte and H. Shahnavaz, *Proceedings of the 10th Anniversary of the Ergonomics International Conference*, Lulea, Sweden: Lulea University 286–292

Chen, F., Ruiz, N., Choi, E., Epps, J., Khawaja, M.A., Taib, R., Yin, B. and Wang, Y., 2012, Multimodal behavior and interaction as indicators of cognitive load, *ACM Transactions on Interactive Intelligent Systems, 2*, 22–36

Cohen, J.D., et al., 1990, On the control of automatic processes: a parallel distributed account of the Stroop effect, *Psychology Review, 97*, 332–361

Farmer, E., Berman, J.V.F. and Fletcher, Y.L., 1986, Evidence for visuo-spatial scratch-pad in working memory, *The Quarterly Journal of Experimental Psychology Section A: Human Experimental Psychology, 38*, 675–688

Francis, W., 1999, Cognitive integration of language and memory in bilinguals: semantic representation, *Psychological Bulletin, 125*, 193–222

Green, D.W., 1986, Control, activation and resource: a framework and a model for the control of speech in bilinguals, *Brain and Language, 27*, 210–223

Green, D.W. and Abutalebi, J., 2013, Language control in bilinguals: the adaptive control hypothesis, *Journal of Cognitive Psychology*, http://dx.doi.org/10.1080/20445911.2013.796377

Hansen, J.H.L., 1996, Analysis and compensation of speech under stress and noise for environmental robustness in speech recognition, *Speech Communication, 20*, 151–173

Le Heij, W., De Bruyn, E., Elens, E., Hartsuijker, R., Helaha, DS. And Van Schleven, L., 1990, Orthographic facilitation and categorical interference in a work-translation variant of the Stroop tasks, *Canadian Journal of Psychology, 44*, 76–83

Lively, S.E., Pisoni, D.B., van Summers, W. and Bernacki, R.H., 1993, Effects of cognitive workload on speech production: acoustic analysis and perceptual consequences, *J Acoust Soc Amer 93*, 2962–2973

Meuter, R.F.I., and Allport, A., 1999, Bilingual language switching in naming: asymmetrical costs of language selection, *Journal of Memory and Language, 40*, 25–40

Mohamed Zied, K., Phillipe, A., Pinon, K., Havet-Thomassin, V., Aubin, G., Roy, A. and Le Gall, D., 2004, Bilingualism and adult differences in inhibitory mechanisms: evidence from a bilingual stroop task, *Brain Cognition, 54* 254–256

Murray, I.R., Baber, C. and South, A., 2006, Towards a definition and working model of stress and its effects on speech, *Speech Communication 20*, 3–12

Noyes, J., Baber, C., and Leggatt, A.P., 2000, Automatic speech recognition, noise and workload, *Ergonomics for the New Millennium. Proceedings of the XIVth Triennial Congress of the International Ergonomics Association*, 762–765

Roselli, M., Ardilla, A., Santisi, M.N., Arecco, M.R., Salvatierra, J., Conde, A. And Lenis, B., 2002, Stroop effect in Spanish-English bilinguals, *Journal of the International Neuropsychological Society, 8*, 819–827

Stroop, J.R., 1935, Studies of interference in serial verbal reactions, *Journal of Experimental Psychology, 18*, 643–662

ALARM VIGILANCE IN THE PRESENCE OF 80 dBA PINK NOISE WITH NEGATIVE SIGNAL-TO-NOISE RATIOS

Buddhika Karunarathne[1], Richard H.Y. So[1] & Anna C.S. Kam[2]

[1]*Department of IELM, Hong Kong University of Science and Technology*
[2]*Department of Otorhinolaryngology, Head & Neck Surgery,
Chinese University of HK*

Workers often have to be vigilant for critical auditory signals in the presence of loud noise. However this phenomenon appears to have received relatively less attention especially when the signal-to-noise ratios (SNRs) are less than unity (or −ve dB). In this study we focus on alarm vigilance in the presence of loud pink noise (80 dBA) and with SNR of −18, −21, −24 and −∞ dB. The results show that people with no known hearing impairments, were able to detect a 56 dBA alarm in the presence of a noise level of 80 dBA (i.e., a SNR of −24 dB). The findings can help to establish threshold boundaries for audible alarm signal in the presence of loud noise.

Introduction

Auditory alarms have a wide variety of applications and the "better safe than sorry" principle has discouraged the use of alarms with loudness lower than the background noise (i.e., with negative signal-to-noise ratios, SNRs). Consequently, past studies on alarm perception with negative SNRs are few. On the other hand, after an alarm has been installed in the industry, levels of background noise could have increased over the years and gone beyond the original estimated levels resulting in negative SNRs. Authors of this paper were involved in an industrial consulting project where workers claimed to be able to detect alarms in which A-weighted sound pressure levels (in dBA) were 20 dB lower than that of the background noise. The testing was done in an actual industry setting. With the lack of literature studying the perception of alarm, with negative SNRs, in the presence of loud (80 dBA) noise, the authors decided to conduct their own studies, hence this paper. In this study, both the alarm and the pink noise had similar spectral characteristics with those observed in the industrial study.

Alarm perception has been the subject of many studies. Guillaume et al. (2002) reported that a loud alarm is not necessarily a good alarm. Edworthy, Hellier and their colleagues investigated the effects of acoustic properties, such as pitch and harmonics on levels of perceived urgency of alarm in the absence of loud noise (Edworthy et al., 1991, Hellier et al., 1993). Carter and Beh (1987) reported that high level of background noise (92 dBA) has adverse effects on performance of vigilance tasks.

Patterson studied detection of tones in the presence of noise and found out that repeated signals are more detectable in uncorrelated noise than in repeated noise when the signal is 1.6 kHz (Patterson et al., 1982). Studying detection of auditory signals in reproducible noise, Pfaflin and Mathews identified that the main determinant of detection behaviour was the energy increment produced by the signal (Pfaflin and Mathews 1965). However in these cases, the effect of SNR level on detectability has not been specifically looked into.

Pure tone detection (Elliot and Katz 1980) and warning signal detection (Haas and Casali 1995) with positive SNRs have been studied. However, studies focusing on the vigilance of alarms with negative SNRs in the presence of loud (80 dBA or above) background noise could not be found. Therefore it is important to conduct more experiments on perception of alarms with negative SNRs. Data collected would be vastly useful in establishing the threshold boundaries within which an alarm is detectable in the presence of loud (about 80 dBA) noise.

A recent industrial consulting study conducted by the authors on workers' vigilance provided motivation for this study. According to data collected from the field studies, even when an alarm was with a SNR of −18 dB; listeners were able to successfully detect the alarms. Therefore the idea of determining the negative boundary of SNR level in which listeners start failing to detect the signals, formed the motivation of this study.

Methodology

Objective

The objective of this study was to determine the ability of the listeners to detect an alarm signal in the presence of loud noise (80 dBA) with less than unity SNR.

Design, apparatus, stimuli and procedures

The study had four stimuli conditions: (i) 80 dBA pink noise plus an alarm at 62 dBA (SNR = −18 dB); (ii) 80 dBA noise plus an alarm at 59 dBA (SNR = −21 dB); (iii) 80 dBA noise plus an alarm at 56 dBA (SNR = −24 dB); and (iv) 80 dBA noise only (the control condition with SNR = −∞ dB). In order to compare the spectra of the alarm and the noise directly, dB levels after filtered with a 1/3 octave band centred at 2 kHz were measured for the noise (77 dBA) and the alarms (59, 56 and 53 dBA for SNR levels of −18, −21 and −24 dB). Each stimulus lasted for five seconds. The three conditions with alarm were repeated 16 times and the control condition was repeated 12 times. This gave 60 stimuli and they were presented consecutively (in random order) without breaks. The total duration of the presentation was 300 seconds (5 seconds × [(3 alarm conditions × 16 repeats) + (control condition × 12 repeats)]). The order of presenting each stimulus was randomized. In this study, data from the repeats were averaged to get better mean estimations. We acknowledge if the control condition was also repeated 16 times would make the design full factorial.

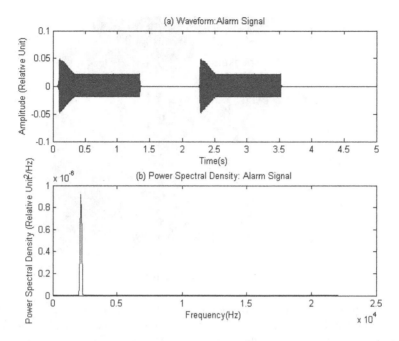

Figure 1.   Temporal and spectral characteristics of the alarm signal (Power spectrum used FFT length of 512 and Hanning window).

The alarm was a recording of a real alarm used in the industry and the noise level of 80 dBA was comparable to the background noise in the real situation. The choice of pink noise was made so that the spectrum of the noise was similar to that measured in the industrial consulting project. Figure 1(a) and 1(b) shows the waveform and the frequency spectrum of the alarm signal respectively.

Subjects were seated in front of a computer screen and listened to the audio clip presented. While listening, subjects used a hardware slider bar to rate the perceived loudness of the alarm in a scale from 0 to 100 (Figure 2). The position of the slide bar was sampled at 10 Hz using the Phidget® system. As the alarm signals changed once every 5 s (i.e., 0.2 Hz), 10 Hz sampling should be fast enough to measure the perceived changes of loudness as represented by positions on the sliding bar. Before the experiment, subjects were trained to anchor the 0 position to $-\infty$ dB condition in the presence of 80 dBA noise and the 100 position to the -18 dB condition in the presence of noise. The screen was only used to present the instructions before the experiment and we acknowledge that the screen can be replaced by a paper instruction.

Three speakers were used to present the audio stimuli. Two speakers placed on the left and right sides broadcasted the 80 dBA pink noise (the 80 dBA level was measured at the two ears of the subjects) and the third speaker in the front broadcasted the alarm signal (Figure 2). The positioning of the speakers was reconstructed from a real industry setting taken from a consulting study.

Figure 2.   Experiment setup.

Twelve subjects (seven male and five female) with no known hearing impairments participated in the experiment. The average age of the subjects was 25 years. They were tested to have hearing thresholds at 20 dB or below at 500 Hz, 1 kHz and 2 kHz for left and right ears, respectively. The deviation of hearing threshold of both ears was 5 dB or less for all subjects.

Results

Data obtained from the first eight participants are analysed and reported in this paper. The full data set will be presented at the conference.

Figure 3 shows the average positions of the slide bar during the five second presentation of each of the four conditions. The starting positions for each condition were different because there was no break between the 60 stimuli and the starting position of the slide for each stimulus would be the ending position of the immediate preceding stimulus. Inspections of Figure 3 show that the positions for each condition asymptotically approached a final level.

To test the effects of convergence, average slider positions measured within each 0.5 second segments were extracted and compared using Friedman two-way ANOVAs and Wilcoxon signed ranked tests. Friedman and Wilcoxon tests were used so that data collected from the same subjects but in different time segments were directly compared with each other. Results of Friedman tests indicated that, at all levels of SNRs, time of measurement (within the 5 second measurement period) had significant main effects on the slider positions. This is consistent with Figure 3.

Results of Wilcoxon tests indicated that the slider positions collected in the last 0.5 second period were not significantly different from those collected between 4 and

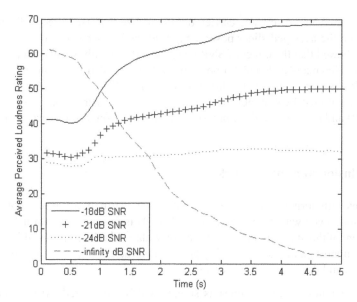

Figure 3. **Average changes of slider positions (step responses) during the 5 second alarm stimulus at SNR levels of −18 dB, −21 dB, −24 dB and −∞ dB.**

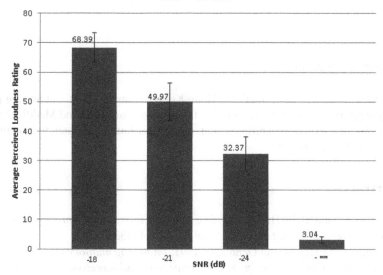

Figure 4. **Average perceived loudness ratings as measured by the averaged slider positions in the last second of the 5 second measurement period as a function of different SNR levels (−18, −21, −24 and −∞ dB).**

4.5 seconds and between 3.5 and 4 seconds. This suggests that the slider positions had reached their asymptotical steady levels in the last one second. This is also consistent with Figure 3. Consequently, the slider positions of the last one second were extracted to determine the effects of SNR levels (Figure 4).

A one-way ANOVA test was carried out to analyze the main effects of four SNR levels on the averaged slider positions during the last second of the stimuli. The result showed that the effect of SNR levels was statistically significant ($p < 0.05$). Student-Newman-Keuls (SNK) post hoc testing revealed that data from all four SNR levels were significantly different from each other. This suggests that, in the presence of 80 dBA pink noise, subjects were able to hear an alarm with SNR of −24 dB and discriminated the presence of an alarm from the absence of an alarm.

Conclusion and future work

Persons with normal hearing are able to detect an alarm in the presence of 80 dBA pink noise even with a SNR level of −24 dB. This result helps towards the establishment of thresholds boundaries for audible alarms in the presence of loud noise. This is important in industrial safety procedures where the background noise levels are high and detecting alarms is critical.

Further studies with pure tones as signals will be conducted. Also an effort to simulate the reported phenomenon using an auditory model developed by Professor Raymond Meddis is continuing (Meddis, 2006a, b). The objective of the simulation is to examine the role of efferent feedbacks in alarm detection in the presence of loud noise.

Acknowledgement

The authors would like to thank the Hong Kong Research Grants Council for partially supporting this study. We are also thankful to Prof. Raymond Meddis for his kind comments on the manuscript and support for latter computational analyses and modelling of the data.

References

Carter, N.L. and Beh, H.C. 1987, *The effect of intermittent noise on vigilance performance*. J. Acoust. Soc. Am., Vol. 82, No. 4, 1334–1341.

Edworthy, J., Loxley, S. and Dennis, L. 1991, *Improving auditory warning design: relationship between warning sound parameters and perceived urgency*. Human Factors, Vol. 33, 205–231.

Elliot, L.L. and Katz, D.R. 1980, *Children's pure tone detection*. J. Acoust. Soc. Am. Vol. 67, No. 1.

Guillaume, A., Drake, C., Rivenez, M., Pellieux, L. and Chastres, V. 2002, *Perception of urgency and alarm design*. Proceedings of the 8th International Conference on Auditory Display (pp. 357–361), Kyoto, Japan.

Haas, E.C. and Casali J.G. 1995, *Perceived urgency of and response time to multi-tone and frequency-modulated warning signals in broadband noise.* Ergonomics, 1995, Vol. 38, No. 11, 2313–2326.

Hellier, E. J., Edworthy J. and Dennis I. 1993, *Improving auditory warning design: quantifying and predicting the effects of different warning parameters on perceived urgency.* Human Factors, Vol. 35, No. 4, 693–706.

Meddis, R. 2006a, *Auditory-nerve first-spike latency and auditory absolute threshold: A computer model.* J. Acoust. Soc. Am. 119, 406–417.

Meddis, R. 2006b, *Reply to comments on 'Auditory-nerve first-spike latency and auditory absolute threshold: A computer model'.* J. Acoust. Soc. Am. 120(3), 1192–1193.

Patterson, R. D., Milroy, R. and Lutfi, R. A. 1982, *Detecting a repeated tone burst in repeated noise.* J. Acoust. Soc. Am., Vol. 73, No. 3, 951–954.

Pfafflin, S.M. and Mathews, M.V. 1965, *Detection of auditory signals in reproducible noise.* J. Acoust. Soc. Am. 39(2), 340–345.

CONTINUOUS DETECTION OF WORKLOAD OVERLOAD: AN fNIRS APPROACH

Horia A. Maior[1], Matthew Pike[2], Max L. Wilson[2] & Sarah Sharples[1,3]

[1] *Horizon Doctoral Training Centre, University of Nottingham*
[2] *Mixed Reality Lab, University of Nottingham*
[3] *Human Factors Research Group, Faculty of Engineering,*
University of Nottingham

Functional Near-Infrared Spectroscopy (fNIRS) is a brain imaging technique that offers the potential to provide continuous, detailed insight into human mental workload, enabling an objective means of detecting overload conditions during complex tasks. When compared to other brain imaging techniques, fNIRS provides a non-invasive, portable and reliable measure that lends itself towards more ecologically valid settings. Our findings confirm a correlation between fNIRS and NASA-TLX subjective workload questionnaire. Our results provide insights into fNIRS and its relation to mental workload, and we propose the use of fNIRS as a continuous objective tool for detecting task overload situations.

Introduction

Understanding and identifying users' limitations has always been a challenge within work contexts, our aim is to advance the current understanding in one particular respect – workload overload.

Mental workload (MWL) is described by Hart and Staveland (1988) as the relationship between the mental processing capabilities and the demands imposed by a task. Non-optimal MWL level will result in human performance issues such as slower task performance and errors such as slips, lapses or mistakes. Measurements such as primary task performance, secondary task performance, and subjective ratings are commonly used methods of assessing MWL. Subjective ratings are usually obtained after the task has been completed, potentially missing essential information about a user's experiences during the task.

To address this issue, we propose the use of a non-invasive, real time brain monitoring technique called functional Near Infrared Spectroscopy (fNIRS) to objectively measure participants' physiological changes (indicative of brain activity and MWL) during tasks. fNIRS is a neuroimaging technique used for monitoring brain activation (Villringer, 1993). It has the properties of being non-invasive, portable, inexpensive and suitable for periods of extended monitoring relative to other neuroimaging techniques. fNIRS measures the hemodynamic response – the delivery of blood to active neuronal tissues and it is designed to be placed directly upon a participants scalp, typically targeting the prefrontal cortex (PFC). While some other brain sensing techniques like functional Magnetic Resonance Imaging (fMRI) require minimal or no movement from users, fNIRS can be successfully used while seated naturally at a computer (Solovey, 2009) as well as in car settings (Solovey, 2012). Further, because fNIRS is an optical based technology rather than electrical (such as Electroencephalography (EEG)), it permits more natural movements such as those associated with using a computer without introducing significant artefacts to the data.

Measuring mental workload

To use fNIRS as a tool for measuring MWL, we first need to understand what cognitive processes fNIRS is detecting. Peck (2013) identified three reasons that support fNIRS being an appropriate tool for measuring MWL:

1. fMRI studies (D'Esposito, 1999) have confirmed that a decrease in deoxygenated hemoglobin indicates an increase in brain activity. When a brain region becomes active, it requires more oxygen. To meet these demands, there is an increase in oxygenated hemoglobin, resulting in a decrease of deoxygenated hemoglobin.
2. Peck (2013) found a correlation between the NASA-TLX (Hart and Staveland, 1988) subjective questionnaire and deoxygenated hemoglobin levels in fNIRS data during a visual task.
3. Peck (2013) successfully managed to distinguish between various levels of n-back visual tasks with fNIRS, suggesting that the different levels of n-back tasks induce different levels of MWL. Confirming with the fMRI findings, fNIRS deoxygenated hemoglobin levels had lower values during 3-back task compared with 1-back task.

We believe that the inclusion of this novel measurement complements existing task evaluation measures such as NASA-TLX. We must note the potential negatives associated with this type of technology. fNIRS is an emerging technology therefore does not have the associated supporting research proving its correctness. Studies have correlated the measurements to those observed with fMRI (Toronov, 2001), specifically the blood-oxygen-level dependent signal. In addition, in the current generation of measurement technology, fNIRS can only be used to detect a level of workload (high or low), leaving a distinct lack of mapping between the readings recorded with fNIRS and the actual cognitive or emotional states.

Detecting overload with fNIRS

MWL can be described as the amount of resources an operator uses when performing a specific task. These resources are limited; therefore, a problem arises when a task requires the operator to use more resources than are maximally available. This state is known as a human operator overload, and normally results in a significant

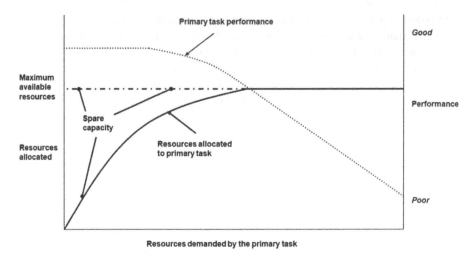

Figure 1.   Limited Resource Model.

drop in performance. Therefore, it is important to consider the operator's optimum level of workload (not overloaded) throughout the task.

One theoretical model of workload presented by Megaw (2005) is the Limited Resource Model (adapted from Kahneman (1973) and Wickens (Holland and Wickens, 1999)). One of the model's assumptions is that the "limited processing capacity" (which Kahneman refers to as "attentional resources") has to be shared between a number of physiological processes such as perception, cognition and response. The Limited Resource Model graph (see Figure 1) describes the relationship between the task demands (Resources demanded by the primary task, the x-axis on the graph), the resources allocated to the task (the left y-axis on the graph), and the impact on task performance (the second y-axis on the graph). When task demands increase, more resources need to be allocated (therefore the spare capacity decreases). When allocated resources reach a point near the maximum available resources, a drop in performance is expected as the operator cannot cope with the task demands.

We propose using an fNIRS device as a continuous measure for detecting Resources allocated in accordance with the Limited Resource Model. Specifically we are interested in using fNIRS for detecting overloaded situations (the states where resources allocated are near the maximum available resources and the spare capacity is minimal).

Methodology

The aim of this study is to identify how fNIRS is suitable in the application of detecting MWL (specifically MWL overload). We aim to validate the measures obtained with fNIRS by correlating it against measures recorded via the subjective MWL questionnaire NASA-TLX. With these aims in mind, we formulate the following research questions:

1. How can we identify MWL using fNIRS?
2. Can we identify a correlation between fNIRS and NASA-TLX (as identified by Peck (2013))?

To answer these research questions we devised a task that tested participants at solving mathematical problems of varying difficulty under two separate conditions, baseline (C1) and verbal (C2). The baseline required participants to simply solve the mathematical problem, whereas the verbal condition introduced a continuous nonsense verbal utterance ("Blah") that participants were required to verbalise whilst solving the mathematical problem.

From the research questions above, we propose the following hypotheses:

A. There will be a significant difference in performance between conditions C1 and C2.
B. There will be a relationship between the NASA-TLX ratings and fNIRS brain data.
C. There will be a difference in MWL between conditions C1 and C2.

Task

A mathematical problem was chosen as it allowed us to vary difficulty (and therefore resource requirements). This problem is a variation on what is commonly known as the countdown numbers game. Participants have sixty seconds to get as close to a target number as possible, using six given numbers. Each number may only be used once and participants can use addition, subtraction, multiplication, and division to reach the target number. There was no requirement to use all six numbers.

Sixteen versions of the task were generated at varying difficulties across the two conditions. Difficulty was classified by one researcher and two independent judges. Difficulty was judged in four categories: easy, quite easy, quite hard, and hard. Inter-rater agreeability was confirmed with a Cohen's Kappa test, where the researcher achieved scores of 0.6419 (substantial agreement) with the first independent judge, and 0.8571 (almost perfect agreement) with the second.

Participants

Twenty participants (14 Males, 6 Females) with an average age of 28.55 years were recruited to take part in the study. All participants had normal or corrected vision and reported no history of head trauma or brain damage. Participants provided informed consent, and were compensated with £15 gift vouchers.

Procedure

Participants were fitted with the fNIRS brain imaging device, which was placed upon their forehead targeting the PFC. Participants performed two practice runs of the task (under baseline condition) to familiarise themselves and reduce the impact of learning under the first condition.

Conditions were counterbalanced and each condition included eight of the experimental tasks described previously. For each of the eight tasks in each condition, participants were given sixty seconds to attempt the problem. All calculations were performed mentally and no pen and paper was provided. After the sixty seconds had elapsed, participants were prompted to enter the number they had achieved during the calculation period.

To avoid participants simply entering the target number, they were prompted to recall their solution. The solutions provided by participants were recorded by the researcher on pen and paper. After each condition, participants completed a standard NASA TLX form to subjectively rate their MWL during the task.

We collected a variety of data during the study:

1. Performance – We measured primary task performance according to distance from the target answer as a percentage, where 100% was the perfect answer and 0% was an absence of an answer or 0.
2. NASA-TLX – We used a weighted NASA-TLX questionnaire, based on the weighted average ratings of six subscales including, in order: Mental Demand, Physical Demand, Temporal Demand, Performance, Effort and Frustration.
3. fNIRS Data – fNIRS data was recorded using a fNIRS300 setup and the associated COBI Studio recording software provided by Biopac Systems inc. 16 channels of data were recorded, providing oxygenated (HbO) and deoxygenated (Hb) hemoglobin readings in each of the channels. Oxygenation values were calculated using the Modified Beer-Lambert Law (MBLL) (Villringer and Chance, 1997) and the data was pre-processed using the NIRS-SPM software package (Ye, 2009).

Results

The aim of this study is to identify whether fNIRS is suitable in the detection of MWL, with the eventual aim of using the imaging technique to detect workload overload. Our hypotheses state that MWL detection in fNIRS should correlate with the measurements observed in the NASA-TLX scale, and thus show that fNIRS is indeed capable of identifying MWL. We report below results that agree with these hypotheses.

The aim of hypothesis A was to identify if there was a significant difference in performance between conditions C1 and C2. In other words, we investigate whether or not participants felt overloaded during condition C2 (as it places additional demands on users through the continuous verbalisation of nonsense words). We did not identify any significant differences in performance between the two conditions (Pair 8 in Table 1). We can attribute this to the findings of Geddie (2001), who in Chapter 3, states: "two systems with the same level of overall performance may impose quite different levels of workload on operators". In relation to MWL, and specifically overload situations, it is not necessarily the case that our task was not demanding enough to elicit an overload state. Rather, since the performance measure used here is an average across all problems in each condition, some of the overload situations may be hidden through averaging.

Hypothesis B states that there is a statistical relationship between NASA-TLX and the measures obtained from the fNIRS device. Correlations were found to support this hypothesis:

- A Pearson correlation ($r = -0.340$, $p = 0.03$) exists between overall deoxygenated hemoglobin and the mental effort subscale measure of NASA-TLX. This finding agrees with (Peck, 2013) who found that decreases in deoxygenated hemoglobin correlated with increased mental effort in NASA-TLX.
- A Spearman test ($r = -0.352$, $p = 0.02$) identified a negative correlation between the Total oxygenation (HbO + Hb) and Mental Demand subscale from NASA-TLX questionnaire.

Hypothesis C stated that a difference in MWL would be observed between conditions C1 and C2 in the study. We found a significant amount of evidence to support this hypothesis. A t-test on the mental demands subscale from NASA-TLX reveals a significant difference between C1 and C2 ($p = 0.025$, Pair 2 in Table 1). Similarly, for the physical demands and mental effort scales from NASA-TLX, C2 was significantly more demanding ($p = 0.012$) and required more effort ($p = 0.04$) than C1 (Pair 3 & 6 in Table 1). This supports Geddies' (2001) findings, showing that despite there being no impact in performance, there was a difference in participants' MWL.

Despite not finding any significant differences with fNIRS data alone, we found fNIRS to be complementary to existing measures such as NASA-TLX. We believe that increasing the number of participants would increase power, reduce type II error and positively impact our findings. Also we note that overload situations might be hidden through averaging the fNIRS data over conditions. Data that cannot be identified with the NASA-TLX questionnaire might be detected using fNIRS.

Conclusion

The aims of this research were to investigate whether fNIRS is a suitable technique for detecting MWL. To do this we devised a mathematical task with varying difficulties to elicit different workload requirements from participants. Additionally, we introduced another condition that included a nonsense utterance, requiring additional, non-complementary resources.

Table 1.   Study Findings.

Paired Samples Test				
		t	df	Sig. (2-tailed)
Pair 1	Weighted Nasa Score C1 – WeightedMWL2 C2	−1.042	19	.310
Pair 2	Nasa-TLX Mental Demands C1 – Nasa-TLX Mental Demands C2	−2.437	19	.025
Pair 3	Nasa-TLX Physical Demands C1 – Nasa-TLX Physical Demands C2	−2.785	19	.012
Pair 4	Nasa-TLX Temporal Demands C1 – Nasa-TLX Temporal Demands C2	.980	19	.339
Pair 5	Nasa-TLX Performance C1 – Nasa-TLX Performance C2	−1.045	19	.309
Pair 6	Nasa-TLX Mental Effort C1 – Nasa-TLX Mental Effort C2	−2.204	19	.040
Pair 7	Nasa-TLX Frustration C1 – Nasa-TLX Frustration C2	−.071	19	.944
Pair 8	Average Distance From Target C1 – Average Distance From Target C2	.005	19	.996
Pair 9	Average Time Spent C1 – OverallOxy C2	1.547	19	.138
Pair 10	OverallOxy C1 – OverallOxy C2	−1.643	19	.117
Pair 11	OverallDeOxy C1 – OverallDeOxy C2	−.940	19	.359
Pair 12	OverallTotal C1 – OverallTotal C2	−1.468	19	.158
Pair 13	OxyL C1 – OxyL C2	−1.163	19	.259
Pair 14	DeOxyL C1 – DeOxyL C2	−1.021	19	.320
Pair 15	TotalL C1 – TotalL C2	−1.730	19	.100
Pair 16	OxyR C1 – OxyR C2	−.586	19	.565
Pair 17	DeOxyR C1 – DeOxyR C2	.114	19	.910
Pair 18	TotalR C1 – TotalR C2	−.348	19	.731

The correlation of fNIRS and NASA-TLX provides an insight into fNIRS ability to detect MWL. Coinciding with the findings from (Peck, 2013) and (D'Esposito, 1999), we believe that fNIRS is in fact capable of detecting MWL. This demonstrates the need to continue researching how to detect specific workload states such as MWL overload. We believe that the combination of complementary measures (NASA-TLX, fNIRS and other measures of MWL) will provide greater insight into MWL.

Our future work will look at expanding on these findings with the aim of being able to detect varying degrees of MWL in accordance with the Limited Resource Model. We have proposed to look at problems on an individual basis. fNIRS property of being a continuous measure enables the detection of MWL states that are not observable in NASA-TLX data alone.

References

D'Esposito, M., Zarahn, E., & Aguirre, G. K. (1999). Event-related functional MRI: implications for cognitive psychology. *Psychological Bulletin*, **125**(1), 155.

Geddie, J. C., Boer, L. C., Edwards, R. J., Enderwick, T. P., & Graff, N. (2001). *NATO Guidelines on Human Engineering Testing and Evaluation* (No. RTO-TR-021). Nato Research and Technology Organization Neuilly-Sur-Seine (France).

Hart, S. G., & Staveland, L. E. (1988). Development of NASA-TLX (Task Load Index): Results of empirical and theoretical research. In P. Hancock & N. Meshkati (Eds.) *Human Mental Workload*, 139–183.

Hollands, J. G., & Wickens, C. D. (1999) *Engineering Psychology and Human Performance*. New Jersey: Prentice Hall.

Kahneman, D. (1973) *Attention and effort*. Prenctice Hall, Englewood Cliffs, NJ.

Megaw, T. (2005) The definition and measurement of mental workload. In E. N. Corlett & J. R. Wilson (Eds.) *Evaluation of Human Work,* 525–551. Taylor & Francis, London.

Peck, E. M. M., Yuksel, B. F., Ottley, A., Jacob, R. J., & Chang, R. (2013) Using fNIRS brain sensing to evaluate information visualization interfaces. In *Proceedings of the SIGCHI Conference on Human Factors in Computing* Systems (pp. 473–482). ACM.

Solovey, E. T., Girouard, A., Chauncey, K., Hirshfield, L. M., Sassaroli, A., Zheng, F., & Jacob, R. J. (2009) Using fNIRS brain sensing in realistic HCI settings: experiments and guidelines. In *Proceedings of the 22nd annual ACM symposium on User Interface Software and Technology* (pp. 157–166). ACM.

Solovey, E.T., Bruce M. & Reimer, B. (2012) Brain Sensing with fNIRS in the Car. In *Proceedings of the 4th Annual Automotive User Interfaces and Interactive Vehicular Applications* (pp. 21–22). ACM.

Toronov, V., Webb, A., Choi, J. H., Wolf, M., Michalos, A., Gratton, E., & Hueber, D. (2001) Investigation of human brain hemodynamics by simultaneous near-infrared spectroscopy and functional magnetic resonance imaging. *Medical Physics*, **28**, 521.

Villringer, A., Planck, J., Hock, C., Schleinkofer, L., & Dirnagl, U. (1993) Near infrared spectroscopy (NIRS): a new tool to study hemodynamic changes during activation of brain function in human adults. *Neuroscience Letters*, **154**(1), 101–104.

Villringer, A., & Chance, B. (1997). *Non-invasive optical spectroscopy and imaging of human brain function*. Trends in neurosciences, 20(10), 435–442.

Ye, J. C., Tak, S., Jang, K. E., Jung, J., & Jang, J. (2009). NIRS-SPM: statistical parametric mapping for near-infrared spectroscopy. *Neuroimage*, **44**(2), 428–447.

POSTER

PROPOSED DECISION MODEL FOR CONTEMPORARY MEMORIALS

Philippa Davies[1], Claire Williams[2] & Patrick W. Jordan[3]

[1] De Montfort University
[2] University of Derby
[3] City University London

The last few decades have seen changing death practices in the UK, with a wider movement towards post-modern approaches to death. Considering memorials as products, the paper proposes a decision model for contemporary, domestic cremation memorials, from the perspective of the user. The proposed model is based on thematic analysis of 12 semi-structured interviews. The model is a start-point in conveying the complex and interrelated factors of user needs and the corresponding decision-processes. It highlights the central role played by the 'Decision-Maker(s)'.

Introduction

Memorials pose a unique product set, whose primary functions relate to the psychological meaning 'endowed' by their users. The traditional view of a memorial is closely tied to the concept of memory, acting as a memory aid and 'relieving us of the burden of remembering' (Socolovsky, 2004). Memorialisation can also serve a number of different purposes: to mark the location of the deceased; to continue connection with the dead; to provide a tangible focus; to 'honour' the deceased; or to act as a tool through which people can communicate with others, both dead and alive (Woodthorpe, 2010: p. 122).

Cremation is now the most common form of disposal in the UK (The Cremation Society of Great Britain, 2011) and the trend for removing ashes from the crematorium has increased in the UK over the last 30 years. In 2009, 66% of people chose to take ashes home from the crematorium, in comparison to 12% in 1970 (The Cremation Society of Great Britain, 2010). New cremation practices should be viewed in the context of a wider movement towards post-modern, post-death rituals, hallmarked by 'choice and personal expression' (Walter 1996: 194). While studies have been conducted into these emerging post-death rituals (e.g. Roberts, 2010; Kawano, 2004 and Prendergast et al, 2006), few have considered their impact on the formation of physical memorials.

Most of the previous research surrounding contemporary memorials has focused on post-creation analysis as a reflection of the varying cultural, political and historical narratives associated with death and memory. Interpreting memorial forms, locations and make-up, these studies provide commentary based mainly on the wider cultural context, rather than as a reflection of the individual's decisions. This is particularly evident in the literature associated with memorials which have a public dimension; either official or spontaneous. It seems it is the public quality that provides their fascination from a cultural context; but what about more 'domestic' memorials? A memorial that represents a death that does not necessarily 'share the characteristics of being lived collectively as traumatic' (Margry & Sanchez-Carretero, 2007) or hold 'national or international significance' (Gibson, 2011). These 'everyday' commemorations have received comparatively less attention than studies devoted to public forms of memorials (Hallam & Hockey, 2001: 1).

The proposed decision model is based on a study that investigates the decision-making processes involved with choosing permanent, domestic memorials associated with cremations, from the perspective of the user. The model is a start-point in representing the factors behind users' decisions and consequently how the design of future memorial options and services can be improved to facilitate the needs of a modern UK society.

For the purpose of this paper it is important to differentiate between the uses of the term 'memorial'. The current model is intended to take a design discipline perspective by narrowing the focus on the noun definition of 'memorial'. The noun memorial is defined as '*a statue or structure established to remind people of a person or event*' (Oxford Dictionaries, 2012). However, mapping anecdotal examples from current literature against this definition suggested it may be too narrow, reflecting only traditional concepts of memorials, such as gravestones and mausoleums. Popular options such as placing a bench or planting a tree (Holloway, 2007: p. 161) would not, it is argued, necessarily be considered a memorial under this definition. Therefore, while the study allowed the participants to define what constitutes a memorial, a working definition was proposed; 'Object or space (Virtual or Real) created or altered to 'embody' memory'.

Methodology

Data collection

Due to the sensitive nature of the subject matter in this study and resource limitations, a self-selection sampling approach was chosen. An invitation email asking for volunteers was sent to two university mailing lists. This approach was supplemented, distributing the invitation through the author's professional and social networks, resulting in a convenience sample.

Semi-structured interviews were conducted with participants, guided by a pre-determined list of topic areas and questions. This included open topics relating to the decision-making process behind the memorial choices, and questions associated with the characteristics of the stakeholders and the context of the choices. Influence factors identified from previous ash disposal studies (Roberts, 2010; Kawano, 2004;

ESRC, 2012) were used as a basis for some of the topic points. All interviews were audio recorded with the consent of the participant.

Participants

All participants, with the exception of two, were selected using the following criteria; they had been involved in deciding upon a memorial for somebody who had died in the last 10 years in the UK, the ashes of whom had been removed from the crematorium. In regard to the two exceptions, one interview related to two burials as well as one cremation, and the second related to events which had occurred in the United States. While neither of the contexts completely fulfilled the inclusion criteria, it was felt the unusual circumstances in both cases, and therefore potential learning opportunity warranted their inclusion within the study.

Data analysis

The interviews were transcribed verbatim from the audio data collected. Thematic Analysis was conducted on the qualitative data following guidelines laid out by Braun & Clarke (2006). Since minimal research has been conducted for the subject area, there is no existing theoretical framework in which to generate coding, therefore the coding and subsequent themes in this study were primarily data-driven.

Methodological considerations

While the selected sampling approach was considered the most feasible it is also acknowledged that this approach would be unlikely to produce a representative sample of the wider UK population. Characteristics of the sample population will be skewed due to relationship with the university settings and to the author's networks. The self-selecting nature of the sampling approach also resulted in some established memorial manifestations, such as websites, not being represented within the data set. While a sample size of 12 is not suitable for predictive empirical generalisations or intended to indicate the prevalence of factors, it was felt that it would provide adequate insight to begin to identify a decision framework.

Proposed model

Primary decision

Prior to the memorial decision process (and separate to the proposed model) the interviews suggest the choice between burial and cremation is the primary decision. In nearly all of the interviews, this choice was taken by the deceased and clearly communicated to others prior to death. Since the study relates principally to cremations, this primary decision is considered separately to the model. However, assuming a similar framework would apply for either disposal approach, the repercussions of this choice are likely to influence the options stage in the model. For example, regulations restricting the location of a burial may limit the memorial options.

User(s) vs Decision-Maker(s)

In all interviews a Decision-Maker or Makers was indicated in each memorial process. The choice of person generally reflected the hierarchy of next-of-kin, for example partner then child etc. of the deceased. However, other factors also influenced who took the role/s of Decision-Maker(s);

- Emotional relationship with deceased (i.e. perceived closeness)
- Deceased's character, likes/dislikes
- Capability
- Knowledge
- Locational proximity
- Willingness

While other people may 'use' the memorial, the interviews suggest an acceptance to defer to the main Decision-Maker when choices need to be made.

User needs

It is the Decision-Maker(s) who appears to decide on the specific set of user needs to be met by the memorial, if a memorial is perceived to be needed at all. Several different needs were noted. The majority related to the creation of a 'connection';

- Connect with different types of audience; either family, wider community or future generations
- Facilitate an opportunity for continued connection with deceased
- Connection to memories
- Connection between the deceased and people who had died previously

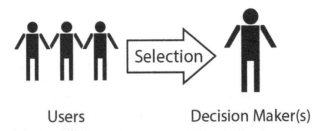

Figure 1.   Selection of Decision-Maker(s).

Influence factors

The Decision-Maker(s) appeared to balance different 'influence factors' to decide upon the specific user needs, either consciously or subconsciously. Within the interviews the predominant influence factors identified included;

- Precedents, e.g. previous memorial choices for other members of the family
- Deceased's character, likes/dislikes.
- Circumstances of death
- Relationships between stakeholders and deceased

While similar factors appeared in multiple interviews, the apparent weighting given to each influence varied between scenarios.

Memorial options

While theoretically the memorial options available to the Decision Maker are endless, the study suggests the final memorial in most instances is selected from a limited pool of options. This pool of options represents those the Decision-Maker is aware of and/or those presented by a third party, for example a funeral parlour. Only in certain cases, particularly where there is a very strong set of user needs, do Decision-Maker(s) seem to look outside the available pool of options.

Memorial options appear to be comprised of two main elements – location and form. This study suggests location is generally considered the most important element in the options, with the form (or physical manifestation) considered secondary. The location and form are tightly related, with the choice of location having a level of dictation over the form (and to a lesser degree vice-a-versa); thereby limiting the available option combinations.

Both the location and the form are subject to external restrictions which have a bearing on the available options. For example, restrictions placed on the process by external organisations. The majority of these restrictions related to rules put in place by the organisation responsible for the memorial's location; for example, memorial forests forbidding the placement of objects next to a memorial tree. External factors also include the wider context of the location/form, such as the available access. The study suggests access to the memorial is considered an important factor when choosing a location, both in terms of convenience and feasibility.

Choice

In making a final choice regarding the memorial realisation, the Decision-Maker(s) appear to weigh the memorial options against the determined user needs. The success of this process depends on the associations attributed to the options by the Decision-Maker(s) and whether these are in line with the perceived user needs.

These associations are drawn from the users' connection to the options, and the characteristics of the options. Within the study, the strongest associations related to

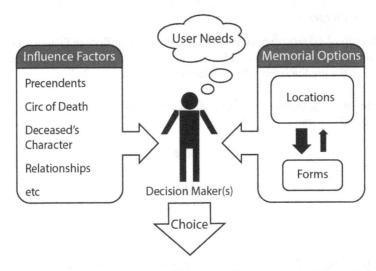

Figure 2.   Proposed Decision Model for Contemporary Memorials.

the memorial's location rather than its form. Association with the location included the proximity to family (living and deceased), happy memories and location of death.

Discussion

Removal of ashes from crematoria theoretically provides users with endless options for the memorials to the deceased, including the choice not to have one at all. However, the reality appears to be more limited, particularly by users' own awareness of potential options. While some of the interviews showed Decision-Makers actively seeking out bespoke elements for the memorial, a lot of decisions came down to a limited list of presented options that the Decision-Maker(s) weighed against the user needs.

The model highlights the central role the Decision-Maker plays in the subsequent decision process. Though others may use the memorial, key Decision-Makers were identified at the start of all the processes. While not present in the study, this stage has the potential for conflict, particularly if multiple users identify themselves as Decision-Makers. It is through these Decision-Makers that the available influences seem to be synthesised into specific user needs for that situation. Therefore, multiple Decision-Makers could formulate different and potentially competing user needs. It is also through these Decision-Makers that the user needs are weighed against the available options, via perceived associations, to decide upon a specific memorial choice.

Previous research has suggested that, historically, religious affiliation widely influenced the choice of memorial design within cemeteries (Buckham, 2003). While some religions are opposed to cremation, for example Islam Muslims and Orthodox Judaism (Davies & Mates, 2005), and therefore would not necessarily be represented in the study, the sampling approach taken did result in bias towards participants from a Christian background. Further research needs to be undertaken to establish whether the different religious and cultural affiliations which represent the UK's diverse population give rise to different user needs and influences memorial choices.

The model begins to convey the complex and interrelated influences behind user needs and the corresponding decision-process for contemporary memorials. It also emphasises the importance of not viewing a memorials' physical form in isolation, as its success as a product relies on a complex web of factors, specific to each user. Connection to location appears paramount in achieving user needs for most Decision-Makers, with the physical manifestation of the memorial chosen to fit within the location-specific restrictions. Though providing endless options may seem the easiest way to accommodate varying user needs, this approach may overwhelm users at a time of emotional stress. As Walter (1999; p. 208) notes, the bereaved are 'likely to cling to the first lifebelt thrown them' therefore users need to be presented with appropriate design options, for their specific needs. Though this paper makes no claim about the model's current generalisation, it is hoped it will begin to provide reference points for further research into the design of contemporary memorials which support the needs of 21st Century users. These designs need to walk a delicate line between the flexibility to address a diverse range of user needs, and overwhelming with choice. Indeed, the most effective immediate action may be the creation of a service to support Decision-Makers synthesise all the information, thereby maximising the success of a memorial choice. Experts are readily available to facilitate the bespoke design of a user's kitchen, so why not a memorial?

References

Braun, V. & Clarke, V. 2006, Using thematic analysis in psychology. *Qualitative Research in Psychology*, 3(2), 77–101.

Buckham, S. (2003). Commemoration as an expression of personal relationships and group identities: a case study of York cemetery. *Mortality*. 8(2).

Davies, D. J. & Mates, L. H. 2005, *Encyclopedia of Cremation* (Aldershot: Ashgate).

Economic and Social Research Council (ESRC). 2012, *Environments of Memory: Changing Rituals of Mourning*. Retrieved August 15, 2012 from http://www.esrc.ac.uk/my-esrc/grants/R000239959-A/read.

Gibson, M. 2011, Death and Grief in the Landscape: Private Memorials in Public Space. *Cultural Studies Review*. 17(1), 146–161.

Hallam, E. & Hockey, J. 2001, *Death, Memory & Material Culture* (Oxford: Berg).

Holloway, M. (2007). *Negotiating death in contemporary health and social care*, Bristol: Policy Press.

Kawano, S. 2004, Scattering Ashes of the Family Dead: Memorial Activity among the Bereaved in Contemporary Japan. *Ethnology.* 43(3), 233–248.

Margry, P.J. & Sánchez-Carretero, C. 2007, Memorializing traumatic death. *Anthropology Today. 23*(3), 1–2.

Oxford Dictionaries. 2012, *Memorial.* Retrieved May 08, 2012 from http://oxforddictionaries.com/definition/memorial?q=memorial.

Prendergast, D., Hockey, J. & Kellaher, L. 2006, Blowing in the wind? Identity, materiality, and the destinations of human ashes. *Journal of the Royal Anthropological Institute.* 12, 881–898.

Roberts, P. 2010, What Now? Cremation without Tradition. *OMEGA.* 62(1), 1–30.

Socolovsky, M. 2004, Cyber-Spaces of Grief: Online Memorials and Columbine High School Shooting. *JAC. 24*(2). 467–489.

The Cremation Society of Great Britain. 2010, Disposition of Cremated Remains in Great Britain. *Pharos International.* 76(4). 21.

The Cremation Society of Great Britain. 2011, *National Cremation Statistics 1960-2010.* Retrieved May 06, 2012 from http://www.srgw.demon.co.uk/CremSoc4/Stats/National/2010/StatsNat.html.

Walter, T. 1996, Facing Death without Tradition. In G. Howarth & P. C. Jupp (Eds.), *Contemporary Issues in the Sociology of Death, Dying and Disposal.* (pp. 193–204) (Basingstoke: Macmillan Press).

Walter, T. 1999, *On Bereavement: The Culture of Grief* (Buckingham: Open University Press).

Woodthorpe, K. 2010, Private Grief in Public Spaces: Interpreting Memorialisation in the Contemporary Cemetery. In J. Hockey, C. Komaromy & K. Woodthorpe (Eds.), *The Matter of Death: Space, Place and Materiality.* (pp. 117–132) (Basingstoke: Palgrave Macmillan).

RELATIONSHIP BETWEEN VISUAL AESTHETICS EVALUATION AND SELLING PRICE TO KIMONO OF YUZEN-ZOME

Takashi Furukawa[1], Yuka Takai[2], Akihiko Goto[2], Noriaki Kuwahara[1] & Noriyuki Kida[1]

[1] *Graduate School of Kyoto Institute of Technology*
[2] *Osaka Sangyo University*

"Yuzen-zome" is a traditional but still popular method of dyeing fabrics in Japan. The products using the Yuzen-zome method and manufactured in Kyoto city are called "Kyo-Yuzen-zome." The dyeing method of Yuzen-zome can be dividing into 10 procedures. A specialized craftsman is in charge of each procedure. During the paste application (Nori-oki) procedure, the expert applies a starch paste or a rubber paste to a fabric. The two pastes create different effect on the dyed fabric. At market, the fabric with a starch paste application is perceived to have a higher value than that with a rubber paste. We developed a questionnaire to determine why two materials have different evaluations and different price determinations.

Introduction

There are several methods to dye a kimono in Japan, such as Yuzen-zome (Yuzen dyeing), Shibori-zome (tie dyeing), Ai-zome (indigo dyeing), and Rou-zome (batik dying). Among them, the representative example is the hand-drawn Yuzen, for which whole production procedure is performed by hand. The dyeing process of hand-drawn Yuzen consists of ten procedures. Figure 1 shows the dying process and Figure 2 shows picture of the paste application procedure.

Materials used for a paste application procedure are a starch paste, a rubber paste, and flour. The typical materials are a starch paste and a rubber paste, which consists of over 99% of the material used in the process. There is research on paste application and dyeing technology. In 1984, Kyoto City Senshoku Shikenjou (presently, the Kyoto Municipal Institute of Industrial Technology and Culture) published the

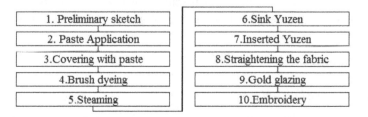

1. Preliminary sketch	6.Sink Yuzen
2. Paste Application	7.Inserted Yuzen
3.Covering with paste	8.Straightening the fabric
4.Brush dyeing	9.Gold glazing
5.Steaming	10.Embroidery

Figure 1. Flow chart of dying process.

Figure 2.   Picture of the paste application procedure.

"Technology and Techniques of Hand-Drawn Yuzen-Zome". The editor was the Committee of the Technology and Techniques of the Hand-Drawn Yuzen-Zome. The viscosity of starch paste is 30 times higher than that of rubber paste (Furukawa, 2012). Consequently, the starch paste is harder to work with than the rubber paste. While rubber paste is used by all paste application craftsmen, the starch paste is used only by a certain craftsmen. On the other hand, in the Japanese dress industry, it is said that dyeing method using a starch paste during the paste application procedure gives the end products a greater impression of softness and depth compared to rubber paste. Consequently, in many cases, the product using the starch paste tends to be sold at a higher price than the rubber paste.

The price setting of the kimono using the starch paste is not based on a quantitative analysis of the impression differences for the kimono using two different materials but on the prevailing notion in the Japanese dress industry. Thus, it is unclear if the products dyed using different materials, such as the starch paste and the rubber paste, are evaluated appropriately to set the price, or if the product using the starch is, in fact, better than the one using the rubber pastes. The latter matches the prevailing notion, the higher value for the end product using the starch paste, to set the price. The purpose of this study is to examine the impression differences of the kimono dyed with a starch paste and with a rubber paste. In addition, we examined the relationship between the impression differences and the price differences. A survey was conducted to evaluate the kimono product and to quantify the differences in visual aesthetics by checking the kimono using two different materials in the paste application procedure. This is the first research to analyze the evaluation of kimono quantitatively as the Japanese dress industry has been using an unclear standard.

Methodology

Sample and participants

Two hand-drawn Yuzen-zome *some obi* (i.e., dyed obi) were used as a sample in this study. The material of *some obi* is a silk fabric. Those who use *some obi* are

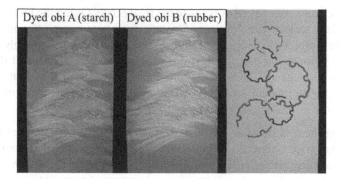

Figure 3.   Sample pictures and Dye at a standard price.

typically females who frequently wear kimono. Figure 3 shows the samples used in this questionnaire research and dye at a standard price. Two samples of dyed obi were prepared; one used a starch paste and the other used a rubber paste during the paste application procedure. Both samples used the same pattern, called "Ryusui itome age". This pattern expresses flowing water by the traces of "itome" method of paste application with either a starch paste or a rubber paste. This pattern was chosen so that only the paste application is the factor for participants' impressions and judgments; other factors such as colouring do not need to be considered. The participants of this study were 71 people (64 males, 7 female) who work in the Japanese dress industry, and the age ranges from the late 20s to 70s. They understood the purpose of the questionnaire and agreed to our investigations. The participants' forms of engagement in the industry are as follows: 39 people work for production stage, 25 people for the wholesale and the distribution stages, and seven people for the retailing. We categorized the participants into the two groups to get the "maker's" and "seller's" perspectives in the Japanese dress industry; 39 people in the production stage were categorized into the "maker"; and for 32 people in the wholesale, distribution, and retail stages into the "seller."

Survey content and method of implementation

A visual aesthetics rating scale and a selling price evaluation were given to the participants. A pilot study was conducted to examine the items for the visual aesthetics rating scale; the participants were university teaching staff and doctoral students who are specialized in visual aesthetics evaluation, and people in the Japanese dress industry. Based on the results, four out of ten items were chosen and added two more by employing the expressions used in the evaluation of visual aesthetics for the materials of traditional craft. Therefore, six items were used to develop a scale. The words used are 'sharpness', 'elegance', 'brightness', 'warmth', 'richness', and 'depth'. 'Sharpness', 'brightness', and 'depth' are the words used by the manufacturers when they evaluate the end product. 'Elegance', 'warmth', and 'richness' are the words mainly used by distribution and retail industry to evaluate final product.

The participants were asked to evaluate the final product using the following five-point Likert scale: strongly agree (5-point), somewhat agree (4-point), neither agree nor disagree (3-point), somewhat disagree (2-point), and strongly disagree (1-point). Prior to the evaluation, it was explained to participants that the words 'sharpness', 'elegance', 'warmth', and 'richness', were used to evaluate the state of dyeing for the product while 'brightness' was used to evaluate if the product was dyed clearly.

Regarding the evaluation of the selling price, a standard dyed obi was prepared which retail price is about 80,000 yen in the Japanese dress industry (Figure 3). The dyed obi at a standard price has the same manufacturing process with the two dyed obi for evaluation. Using the selling price of the standard dyed obi (i.e., 80,000 yen) as a benchmark, the participants were asked to evaluate the selling price of the dyed obi for the evaluation in increments of 1 yen. The price investigated here is not a price that the consumer wants to buy the product at, but the seller's retail price.

Next, research was conducted in the building of a distributor of the Japanese dress manufacturer in Kyoto. In order to suppress the effect of time, climate, and variability of the light source due to variation of the participants' locations, the examination was conducted under a white fluorescent lamp, to reduce influence of outside light. The luminance level was measured by the digital illuminometer (51001, Yokogawa Meters & Instruments Corporation) and adjusted to be in the range of 1,490~1,500lx in order to control the environment in each case. The starch paste and rubber paste obis were randomly presented to the participants without disclosing the paste materials and the differences between the samples. In addition, when presenting the two samples, a 10-second interval was given for the participants to conduct an absolute evaluation for each sample. After evaluating two samples, the participants were told the differences between the materials, and which sample used the starch paste and the rubber paste. Then, the participants were asked to answer the selling price again in increments of 1 yen.

Results

Evaluation for visual aesthetics and of the selling price

The results of the factor analyses (maximum-likelihood method, promax rotation) of the items showed that two factors emerged based on the decrease in the Eigen value in the initial factor solution and the interpretability. Table 1 shows the factor loading matrix. Factor loading of all items were above .50, and no cross-loaded items were found to be over .20. Therefore, after examining the items that constitute each factor, the first factor has a relevance to the quality such as 'richness' and 'elegance'; therefore, we named the factor 'refinement'. The second factor has a relevance to the clarity of the dyed obi such as 'sharpness' and 'brightness'; this factor was named 'vividness'. Cronbach's α coefficients of the subscales ranged from $\alpha = .830$ to $\alpha = .602$, indicating that subscales show a certain degree of internal consistency. The correlation between the factors was low at .091; therefore, the average of the items of the first factor was the 'refinement' score and that of the second factor

Table 1.   Factor loading matrix.

	factor 1	factor 2		factor 1	factor 2		factor 1	factor 2
Refinement (Cronbach α=.830)			Vividness (Cronbach α=.602)			correlation between factors		
· Warmth	.869	-.127	· Sharpness	-.094	.851		1	2
· Richness	.859	.105	· Brightness	.153	.589			
· Depth	.769	.027				1	-	.091
· Elegance	.585	.039				2		-

Table 2.   Scores of the visual aesthetics rating scale by a form of engagement.

	Maker		Seller		main effect		interaction
	Starch paste	Rubber paste	Starch paste	Rubber paste	pes of indust	material	effect
	M(SD)	M(SD)	M(SD)	M(SD)	F-value	F-value	F-value
Refinement score	3.58(0.73)	2.94(0.56)	3.46(0.82)	2.88(0.66)	0.61	25.34 **	0.08
Vividness score	3.12(0.62)	3.56(0.79)	2.78(0.63)	3.55(0.78)	2.34	23.88 **	1.63

*, $p<.05$; **, $p<.01$

was the 'vividness' score. In the following analysis, the visual aesthetics rating scale consisting of these two scores are used. Table 2 shows the average scores of the subscales of the visual aesthetics rating scale based on different materials used during the paste application procedure. Two-way factorial analysis of variance was conducted with the material (starch paste, rubber paste) as the within-subjects factor, and the forms of engagement as the between-subjects factor. There was no significant interaction. There were significant main effects of the materials in the refinement score and the vividness score. For the refinement score, the starch paste was significantly higher ($F(1, 70) = 25.34$, $p < .01$), and for the vividness score, the rubber paste was significantly higher ($F(1, 70) = 23.88$, $p < .01$).

Regarding the changes of the price between the maker and the seller, comparisons were made between before and after the presentation of the names of the materials. The seller did not show much difference before and after presenting the names of the materials for both starch paste and rubber paste. On the other hand, the maker did not show much difference in the prices for the rubber paste; however, an increase in the price was found for the starch paste. For both samples, the maker set the prices higher than the seller.

The relationship between visual aesthetics evaluation and the selling price evaluation

Next, the impact of visual aesthetics evaluation on the selling price was examined. The multiple regression analysis was conducted to examine how two variables

Table 3.   Multiple Regression Analysis.

	Before material presentation	After material presentation
	β	β
Refinement score	.219 **	.266 **
Vividness score	-.080	-.085
Proportion of variance explained (R2)	.037 *	.061 **

*, $p<.05$; **, $p<.01$

(i.e., 'refinement', 'vividness'), consisting of "visual aesthetics rating scale" predict the evaluation of the selling price, with the selling price as a dependent variable and scores of two subscales as independent variables. Multiple regression analysis was conducted both for before and after the presentation of the material names (forced entry method). Table 3 shows the obtained standard partial regression coefficient. The results show that both before and after presenting the materials, the proportion of variance explained was significant. Regarding the standard partial regression coefficient, the refinement score was significant.

Discussion and conclusion

Sensitivity evaluation

In this study, items of the visual aesthetics rating were uniquely developed and used to analyze the impact of using different materials during the paste application procedure on the impression of the final product and the evaluation of the selling price for the hand-drawn Yuzen of Kyo-Yuzen-zome. Six items were uniquely developed based on the words used to evaluate kimono in the Japanese dress industry or the expressions used to evaluate visual aesthetics of traditional craftwork material (Ryouke, 2008). The factor analysis showed that two factors emerged, 'refinement' and 'vividness'. Research targeting the traditional craftwork other than Yuzen-zome confirmed the visual aesthetics rating scale for 'contrast', 'colour tone' and 'stately', and 'earnest' (Ryouke, 2008), while the evaluation of the impression for the kimono includes terms for visual aesthetics such as 'flashy', 'somber', 'soft', 'hard', 'refined', and 'baseness' (Ishihara, 2008). Therefore, because the visual aesthetics rating scale consists of the following items; 'elegance', 'warmth', 'richness', and 'depth' (from the 'refinement' subscale), and 'sharpness' and 'brightness' (from the vividness), this scale can be considered to have content validity.

When the starch paste was used in the paste application procedure, the score of 'refinement' was high; the following four items of the refinement subscale showed the high scores, 'elegance', 'warmth', 'richness', 'depth'. On the other hand, when the rubber paste was used in the paste application procedure, significantly high scores were found for the 'vividness' subscale; the items of 'sharpness' and 'brightness' showed significantly high scores. A study examining the viscosity of the material (Furukawa, 2012) found that the starch paste is thirty times as thick as the rubber paste. As a result of observing the struck surface of section, it was evident that the rubber paste penetrates more deeply on the stick than the starch paste does, which leads to the conclusion that the rubber paste has a higher resting effect. Consequently, when using the rubber paste, the boundary between the dyed and the undyed sections were clearer than when using the starch paste. This may explain the significantly higher evaluation of 'vividness' in the impression evaluation for the final product. On the other hand, the characteristics of the starch paste is that higher viscosity and lower resistance to the dyes than that of rubber paste. Therefore, the boundary between the dyed and the undyed sections do not become clear when the fabric is dyed (Furukawa, 2012). Therefore, when starch paste was used, the final product is likely to attain the high level of refinement.

Evaluation of the selling price

Examining the impact of the material differences used in the paste application procedure on the selling price evaluation (regardless of the form of the engagement [maker, seller], and the timing of presenting the material [before and after]), the dyed obi using the starch paste is acknowledged to have a higher product value; therefore, a high price can be set. The multiple regression analysis showed that no significant correlation between the selling price and the 'vividness' score, a significant positive correlation was found between the selling price and the 'refinement' score. Consequently, the impression that influences the selling price is 'refinement'.

In the Japanese dress industry, a mildly soft and refined final product is preferred to a vivid final product (Kyoto municipal institute of industrial technology and culture, 1984). The study also showed that refinement, not vividness, is reflected in the price evaluation. In this study, as a result of informing the participants of the material used in the paste application procedure, both maker and seller did not change their price evaluations for the dyed obi using rubber paste. Some makers increased their price evaluation up to 7,000 yen when they understood the dyed obi was using the starch paste. However, no price change was found among the sellers. Regarding the differences of price evaluation between the two dyed obi, the maker indicated 7,000 yen and the seller indicated 12,000 yen before the presentation of the material names. After presenting the materials, the maker indicated 13,000 yen while seller indicated about 14,000 yen. An increase in maker's price occurred due to an expectation of the craftsman's payment for the fabrication at the paste application procedure as the number of the craftsman using the starch paste is limited in comparison to the ones using a rubber paste. The cost of the starch paste itself is higher than that of the rubber paste. For these reasons, the participants felt the need to set a higher selling price, knowing that the product has the starch paste application. The reason why a seller set different prices with a wide discrepancy even before presenting the materials can be considered that they can judge the product if the starch paste was applied based on their impression, and they pass on the scarcity to the price higher than the level of its impression. Furthermore, the maker sets the price higher than the seller for both the starch paste and the rubber paste, both before and after the material presentation. This is because the seller has more contacts with customers than the maker, and they may evaluate the price at not which they want to sell, but at which they can sell by considering society's deflationary trend.

This study used a pattern of 'ryusui itome age' for the dyed obi. Therefore, future research is needed to examine if the result would be replicated when we increase the number of patterns, and use the different ground colour of the silk fabric. Furthermore, the reason for this can be considered that people in the Japanese dress industry understand the number of products applying a starch paste is limited and is sold at a high price. It is necessary to gather information on how many participants realized if the sample product was applied either a starch paste or a rubber paste before presenting the materials. Currently, it is difficult to develop a new rubber paste that has a similar impact of the starch paste on the final product. Consequently, the material that can express the refinement is a starch paste only. We will investigate the material characteristics that a starch has and its visual aesthetics value expressing refinement hereafter.

References

Furukawa, T. (2012). Relationship between structure and kansei evaluation of Kyo-Yuzen-zome with different pastes. The 56th Proceedings of Japan Congress on Materials Research, 295–296

Ishihara, H. (2008). The color factor to be involved in the image of the person wearing a kimono and hakama. *Journal of Nagoya Women's University 54*, Home economics, natural science.

Kyoto municipal institute of industrial technology and culture. (1984). The technique and the skill of Tegaki-Yuzen-zome (dying), 67–100

Ryouke, M. (2008). Kansei evaluation and analysis of traditional craft. *International Journal of Affective Engineering 7*, 633–640

LEVEL OF COMFORT OF CUSTOMERS AFFECTED BY EXPERT MANICURIST AND NON-EXPERT MANICURIST

Sojun Isobe, Atsushi Endo, Tomoko Ota & Noriaki Kuwahara

Nailsalon Today
Kyoto Institute of Technology
Chuo Business Group
Kyoto Institute of Technology

This study performed electroencephalogram (EEG) measurement of models before and after nail art to examine psychological changes with manicures by expert and non-expert manicurist. There was no big change on the EEG measurements; it only showed that some distribution of the alpha wave share became small. However, expert manicurist makes redaction of timewise burden and might make improvement of customer satisfaction because expert manicurists are able to paint manicure in shorter time.

Introduction

The beginning of nail art is considered to be the ancient Egypt era. There was a custom to dye nails by using the juice of the henna flower for noble people. In Japan, it is considered that nail art began in Asuka and Nara era. Nail art was only for people of noble social position, like the ancient Egypt era and according to the record, nails were painted red with plant dye.

The current nail art service of Japan developed as a technique based on a manicure technology transmitted from France in the Meiji era, which began to spread in the Showa era.

In recent years, with the social advancement of women, nail care and nail art is considered "the second face". The nail salon has an important needs to guarantee a high quality service and satisfy customer needs as nail care and nail art needs increase. As the price is expensive, some have nail art by unskillful manicurists.

The difference between manicurists' techniques affects customer satisfaction and so quantitative evaluation is important to ensure high-quality service. While studies by electroencephalogram (EEG) measurements for cosmetic therapy and elderly care by make-up therapy have been reported, there are no reports of such evaluation of nail care. Accordingly, this study attempted to perform EEG measurement of models before and after nail art by Expert and non-expert manicurist to examine any psychological change.

Table 1.   Information of subjects.

Subject		A	B	C	D	E	F
Sex		Female	Female	Female	Female	Female	Female
Age		32	33	26	25	27	24
Experience of nail art	Painting by theirselves	10	15	10	10	10	10
	Painting by manicurist	5	13	0	0	0	0

Measurement

Subject

There were six models as subjects, showed in Table 1. All were female, and the average age was 28 ± 4 years old. They had painted their own nails since an average of 11 ± 2 years ago. Subject A had manicure at nail salons from five years ago and subject B had manicure at nail salons from 13 years ago. This experiment was the first time that the others had manicure at nail salon.

EEG measurements

The expert manicurist and non-expert manicurist manicured the right and left hand of each subject. After having finished painting with one pick, the manicurist were changed and painted with the other pick. Subjects had brain waves measured before and after that had manicure by a portable EEG (Digital Medic Company Co., Ltd.). Subjects closed their eyes and rested for the 30-seconds while brain waves were measured after doing two-figure addition mentally for 30-seconds. After that, subjects were manicured naturally, and when finished, they closed their eyes and rested. Then, they had brain waves measured again. The share of alpha wave was computed about the analysis of brain waves.

Results

The time of having manicure

Figure 1 shows the time that each manicurist painted five nails. The expert manicurist took 238 seconds on average; the non-expert manicurist took 311 seconds on average.

Brain wave before and after having manicure

Figure 2 shows the share of alpha wave before and after each subject had a manicure by the expert manicurist. Figure 3 showed the alpha wave occupancy before and after each subject had manicure by non-expert manicurist. In both cases, the share of alpha wave is about 70%, and there was no big change before and after.

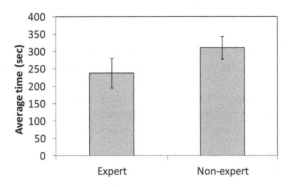

Figure 1.   Average time of painting.

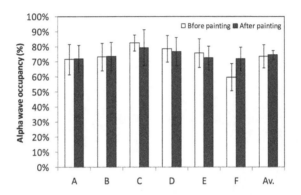

Figure 2.   Alpha wave occupancy in expert's painting.

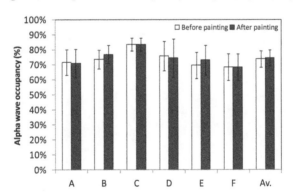

Figure 3.   Alpha wave occupancy in non-expert's painting.

Discussion and conclusion

There was no big change in EEG measurements; it only showed that some distribution of the alpha wave share became small. However, expert manicurists are able to paint manicure in shorter time, which might improve customer satisfaction.

References

Machida, A., Shirato, M., Takata, S. and Yagi, T. 2012, Evaluation of the effect of cosmetic therapy on elderly people on electroencephalogram. *Japanese Journal of Geriatric Psychiatry Vol. 23, No. 8*, pp. 978–987

Yamamoto, A., Machida, A., Shirato, M., Takata, S. and Yagi, T. 2010, Quantitative Evaluation of Cosmetic Therapy for Alzheimer Disease with the Use of EEG. *Proceedings of the Symposium on Biological and Physiological Engineering Vol. 25*, 37–40

ERGONOMICS OF PARKING BRAKE APPLICATION – FACTORS TO FAIL?

Valerie G. Noble, Richard J. Frampton & John H. Richardson

Design School, Loughborough University

The vehicle parking brake may be foot, lever or electronically operated but its function remains the same: i.e. to hold the vehicle stationary whether on an up or down gradient even in the absence of the driver. Failure to do so can have serious consequences. Around 8% of drivers surveyed reported experience of "vehicle roll away" with around 55% citing the causative factor to be mechanical or system related (Noble, Frampton, Richardson, 2013). The ergonomic factors associated with the effective application of the mechanical lever operated parking brake are explored and the potential contribution to failure considered.

Introduction

The function of the vehicle parking brake system is to "hold the vehicle stationary on an up or down gradient even in the absence of the driver" (UNECE, 2008). Failure of the unattended parked vehicle to remain stationary can result in minor property damage, injury or even fatality. Although not considered to be a significant problem, evidence suggests that "vehicle roll away" is not a rare phenomenon. However, determining the extent of these incidents is difficult as they are not consistently reported and those that are may not be reliably recorded.

A task analysis of applying the parking brake to hold the vehicle stationary indicates there is a complexity of potential factors that can contribute to a successful or unsuccessful outcome. These may include driver behaviour and demographics, training and instruction, mechanical or system and environmental factors. This paper illustrates an initial exploration of physical and mechanical factors to be considered when the driver interacts with the lever operated parking brake system.

The manually operated lever parking brake (handbrake) employs a simple ratchet and pawl arrangement. This type of mechanism is considered to be of relatively low cost, and has been regarded as reliable with the ability to carry a large force in relation to its size. It must be capable of holding a laden vehicle stationary on a 20% up or down gradient and the force applied to operate it should not exceed 40 daN (400 N) (UNECE, 2008). However, this system holds the potential for problems with wear, control and stability due to its impacting mechanism and requires the driver to effect considerable bio-mechanical effort. Parking brake related faults resulted in 29 vehicle recalls by manufacturers in a 5 year period 2006–2010 (VOSA, 2011).

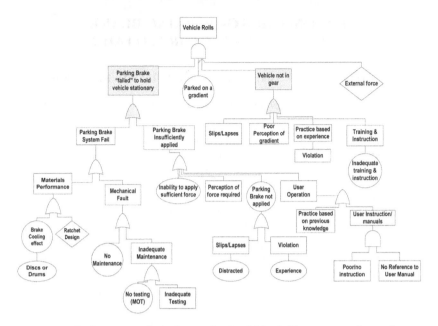

Figure 1.   Fault tree analysis for parked vehicle failing to remain stationary.

The parking brake system is a secondary system applied independently of the service brakes and may utilise a drum or disc design. The drum design is considered to be ideal as a parking brake due to a higher brake factor in relation to the friction coefficient (Limpert, 1999). With an increase in temperature, the disc brake assembly expands. As the system cools and returns to ambient temperature the discs and pads contract with a potential loss in braking force. In drums, the drum diameter increases as the temperature increases but cooling has little or no reduction in the friction coefficient (McKinlay, 2007).

The complexity of a control action that is regarded as relatively simple is illustrated in Figure 1. The driver must apply the parking brake before leaving the vehicle and if parked on an incline he/she should select the appropriate gear (if manual) or park (automatic) and turn the wheels of the car in the appropriate direction (DFT & DSA, 2007). If this is incomplete and the parking brake's holding capability is compromised the risk of the vehicle failing to remain stationary is increased. The fault tree analysis depicts an incident where a "vehicle rolls" and identifies areas for further investigation and data analysis.

Materials and methods

Online survey

An on-line survey to explore the question: *What is the driver's perception and experience of the parking brake system?* (effort, usability, vehicle roll away) was

Figure 2.   Handbrake lever and load cell in casing.

distributed through a local health care organisation, professional networks and social media sites.

Participants

One hundred and thirty eight drivers aged 20–80 years (mean 52.7, SD 13.6) with a 3:1 ratio of male to female and around 91% reporting over 10 years experience responded. All but 6 drivers passed their driving test in the UK and 14 (10.69%) of the respondents reported regularly driving a left hand drive vehicle. The driving environment represented was reasonably evenly spread across motorway, rural and urban categories with least responses to city driving (14%).

Observation of driver interaction in static assessment rig

An established static was used to observe and evaluate the interaction of individual drivers with a lever operated parking brake in a controlled environment and served as a pilot for a study of drivers in their own vehicles. The following questions were explored: *What force do individuals apply to the handbrake lever? How does the interaction with the geometric layout affect the force applied?*

Participants

Twenty eight licensed drivers aged 21–59 years with varying levels of driving experience were recruited from the staff of the Cornwall Disability Assessment Centre and medical students on clinical placement who were familiarised with the operation of the rig.

Equipment and materials

A Novatech F268 handbrake load cell was fixed to the parking brake lever with plastic tie wraps. The force applied was recorded using custom made data acquisition hardware powered by 2 cell batteries connected to Focus Oscilloscope software (Figure 2). Calibration was performed using a known weight and testing of the system indicated the combination was appropriate for data collection.

Procedure

Pre test measurements. Static anthropometric measurements recorded for the seated participant included: Body height, shoulder height, upper arm length, lower arm length, elbow height, grip length, hand length, hand thickness, thumb length. Observations were conducted with participants wearing indoor clothing and comfortable driving shoes.

Test Scenario. Participants were instructed to adjust the seat to their preferred driving position.The body landmark locations and seat position in relation to the parking brake were recorded. Six static road scenes were displayed in front of the driver as a visual cue to either park or drive. The driver was instructed that for the parking cue they were to stop and park as if they were leaving the vehicle. When a picture of a driving scene was displayed, the driver was instructed to release the parking brake and drive on. Each cue was displayed for 20 seconds and in the following order: drive scene, car park symbol, drive scene, parking on a hill, drive scene, parking in a supermarket car park.

Observation of Parking on an Incline. To observe driver interaction in a "real life" setting, a 2 part study was developed where drivers parked their own vehicle on a 20% incline to address the following questions: *What force does the driver apply to the parking brake to hold the vehicle stationary? What are the mechanical considerations-is there any difference in application and performance between brake types?*

Participants

Twenty three female and 14 male (2 left hand dominant) drivers aged between 21 and 70 (mean 48.4, S.D. 13.4) with an average driving experience of 25 years participated in part one of the incline study. In part 2, 16 vehicles registered 1999–2013 were tested before and after driving a set route. Nine of the vehicles were fitted with disc brakes and 7 with drums type.

Procedures

In part one, drivers were asked to reverse onto a 20% incline and park their vehicle. As part of the risk assessment, chocks were positioned in front of the rear wheels. The Novatech F268 load cell was applied to the parking brake lever and connected through the custom made data acquisition hardware to Focus Oscilloscope software on a Toshiba Portege laptop. The drivers were then requested to release and re-apply the parking brake. Anthropometric measurements were recorded to explore the geometric layout and hand grip force tested using a dynamometer. Drivers were questioned about their driving experience and vehicle controls and asked to rate their confidence in the parking brake holding capability on a scale of 1 to 5 (not all confident to extremely confident).

In part two, the minimal force required to hold the vehicle stationary, the travel distance of the parking brake and the temperature of the rear brakes was recorded before and after driving a pre-determined route. Force application was recorded as before and the temperature of the brakes was recorded at 5 minute intervals up to 15 minutes using a handheld infrared pyrometer. Any creaking and settling of the brakes was noted and a "roll" recorded when the vehicle rolled forwards and made contact with the chocks.

Results

Survey results

A hand lever operated parking brake was employed in the vehicles of 105 (88.24%) respondents with 8 foot activated and 22 electronic parking brakes in the remainder. Contrary to most manufacturer's owner manuals, 84.7% reported pulling the lever up while pushing the button in. Twenty seven (25.7%) of the respondents operating a lever hand brake reported the effort required to be somewhat hard or heavy on a perceived exertion scale (Borg, 1998).

Around 5% of respondents reported returning to their car to find the parking brake was not applied. Eleven (8%) respondents reported an experience where their vehicle had rolled. and in 4 of these the parking brake had not been applied. Five (45%) respondents documented distraction or error as a reason and 6 (55%) indicated the causative factor to be mechanical or system related

Application of force

The mean forces recorded in response to visual cues for parking in the car park, parking on a hill and parking at the supermarket were 100.8 N, 145.8 N and 94.5 N respectively. The mean figures recorded in relation to shoulder height are illustrated in Figure 3. It would appear that drivers apply an increased force to the parking brake when they perceive there to be an incline. In the *incline study*, the recorded force that the driver applied to the parking brake ranged from min 152 N to max 432 N (mean 255 N, S.D. 93.05). Twenty five (67.7%) of the drivers parked in gear. The male driver recording the maximum force reported that he never parks in gear and has not experienced a roll away. However, his wife cannot release the handbrake when he has applied it. The female driver recording the smallest force reported she always parks in gear as "sometimes the handbrake doesn't hold".

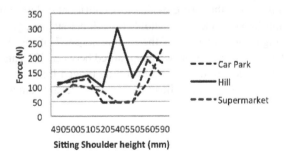

Figure 3. Force applied in response to visual cues in relation to shoulder height.

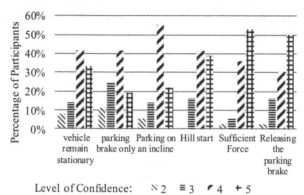

Level of Confidence: ⤫ 2   ≡ 3   ⟋ 4   ＋ 5

Figure 4.   Level of confidence in relation to parking brake task (n = 35).

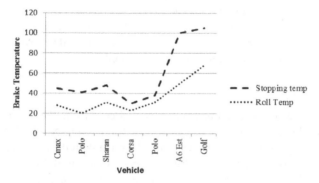

Figure 5.   Brake temperatures from stopping to rolling per vehicle observed.

In part one of the incline study, 28 (80%) of the participants rated their confidence parking on an incline to be 4 or 5 indicating they were very confident conducting this task. Thirty two (91%) also considered they were very or extremely confident that they could apply sufficient force to the parking brake to operate it. Thirteen drivers (37%) indicated they were moderately confident or less (rating of 2 or 3) that the parking brake alone would hold the vehicle stationary. The results are illustrated in Figure 4. In part 2 of the incline study, 7 of the 9 vehicles fitted with disc brakes rolled (5 of these were registered between 2010 and 2013). Brake temperature differences from stopping to rolling ranged from 7°C to 50°C with a mean difference of 20.9°C and is illustrated for the relevant vehicles in Figure 5.

Discussion

The fault tree analysis illustrates that the task of applying the parking brake to hold a vehicle stationary is multifaceted with a complexity of variables which could lead to failure. Discussed here are just 2 of these factors – physical and mechanical and the studies presented are part of a larger project.

Force applied

Early results indicate that the force which is applied to the parking brake varies across individuals and vehicles. This could concur with the findings of Chateauroux and Wang where the force applied depended on the subject group and location of the parking brake, in that a greater force was recorded when the parking brake was low (Chateauroux and Wang, 2012). Larger forces recorded when the "parking on a hill" visual cue was displayed suggest that drivers' perception of the gradient affects the degree of parking brake application. The mean force of 255 N recorded using driver's own vehicles is less than the figure used for testing purposes where the vehicle should remain stationary on a 20% incline for 5 minutes and the force should not exceed 40 daN (UNECE, 2008; Curry, Suffolk, 2011). Observation of practice demonstrated that drivers re-apply the parking brake as necessary to hold the vehicle and may or may not park in gear. The 8 drivers observed who did not park in gear, parked on the flat overnight and 2 had experienced a roll away where they had forgotten to apply the parking brake. There did not appear to be a relationship between age, gender or years driving experience but further investigation is required.

Confidence in the system

If the driver applies the parking brake sufficiently to hold the vehicle stationary at the time of stopping and leaving the vehicle, will it remain stationary if not in gear? Drivers reported a reduced level of confidence as to whether the vehicle would remain stationary on an incline and whether it would remain stationary with only the parking brake applied i.e. not in gear. This reduced confidence may not be isolated to lever operated systems and may be a factor as to whether drivers park in gear.

Mechanical compromise

Responses from drivers who had experienced a roll away and a review of media reports (Noble, Frampton, Richardson, 2013), suggests that there is a delay in the time from stopping the vehicle to it rolling. It is possible this is due to a "cooling effect". The results communicate how with a relatively small drop in disc brake temperature, friction contact can be reduced, compromising the function, and the vehicle could roll. The key factor may not be the length of time itself but the time that it takes for the brake system temperature to reduce to the critical point where the brake force is no longer sufficient to hold the vehicle stationary. Where a driver is unaware of this effect and does not park in gear or park on an incline, the risk of roll away is increased. Although an increase in the fitting of Electro-mechanical parking brake systems, negating the application of force by the driver, is projected, further research is required into the mechanical efficiency of and driver interaction with parking brake systems employing disc brakes.

Conclusion

The project continues to uncover further areas for investigation and discussion with key stake holders such as Association of Driving Instructors (ADI), Vehicle Operators and Service Agency (VOSA) and Driving Standards Agency (DSA). It may be difficult to draw conclusive significance to any one causative factor but, exploring the potential contributory factors could provide the evidence base for recommendations. Despite the constraints, exploring driver interaction with the parking brake system in "real life" has provided useful data for future work.

Acknowledgements

St Luke's Hospice, Plymouth; Eddie Curry, Ben Suffolk, MIRA; Cornwall Mobility Centre; Allan Morgan – electronics expertise.

References

Borg, G. (1998). Borg's Perceived Exertion and Pain scales. Champaign, Leeds: Human Kinetics.

Chateauroux E.W.X., (2012). *Comparison between static maximal force and handbrake pulling force.* Work 41 (2012), 1305–1310: IOS Press.

Department for Transport (DFT) & Driving Standards Agency (DSA) (2007), *The Official Highway Code. Waiting and Parking* (pp. 239, 252). Stationery Office Books.

Limpert, R., (1999). Design and Analysis of Friction Brakes. *In Brake Design and Safety, Second Edition* (pp. 37–110). Society of Automotive Engineers, USA.

McKinlay, A.J., (2007). The phenomenon of vehicle park brake rollaway. PhD Thesis, University of Leeds.

Noble, V.G., Frampton, R.J., & Richardson J.H., (2013), Ergonomics of Parking Brake Application: An Introduction. In Dorn, L., Sullman, M., (Eds.) *Driver Behaviour and Training Vol. VI* (pp. 187–206). Ashgate, UK.

Standing Committee on Road Accident Statistics (2005). *STATS19*. Retrieved from http://www.stats19.org.uk

United Nations Economic Commission for Europe (UNECE). (2008). Regulation No 13-H Revision1. Braking of Passenger Cars.

Vehicle and Operator Services Agency (VOSA) (2011). *Vehicle Recalls – Handbrake*. [Homepage of Department for Transport (DFT)], [Online]. http://www.dft.gov.uk/vosa/apps/recalls/searches/search.asp [March].

DEVELOPMENT OF A HUMAN FACTORS ASSESSMENT FRAMEWORK FOR LAND VEHICLES

Rebecca Saunders Jones, Kimberly Strickland & Will Tutton

Defence Science Technology Laboratories, UK Ministry of Defence

It was ascertained that within the MOD there was no systematic framework for conducting Human Factors (HF) assessments of military land vehicles. This paper outlines the work Dstl conducted in order to develop a Human Factors Assessment Framework (HFAF) for application to military land vehicles. As part of this study, a review was conducted to identify a range of existing HF assessment tools and methods, which might be applicable for use under the framework. In order to establish the efficacy of the methods that sit under the framework, the mounted systems HFAF needs to be applied during a range of military vehicle HF assessments, this will also enable review by industry, academia and wider MoD.

Introduction

The purpose of this study was to develop a Human Factors Assessment Framework (HFAF) for land vehicles and the sub-systems mounted to them (referred to, within the military, as 'mounted systems'). The requirement for the work came from UK Ministry of Defence (MOD) (Army Headquarters), who asked Dstl to ascertain whether a HFAF[1] previously developed for the evaluation of Dismounted Close Combat (DCC) equipment, could be expanded for use in the assessment of mounted systems. If it was found that the HFAF could be extended for use on Mounted Systems a further requirement would determine what methods should be applied at the different levels of the assessment framework.

Background

As part of the support that Dstl provides to the Ministry of Defence's (MoD) rapid procurement process for Urgent Operational Requirements (UORs), HF practitioners are regularly tasked to conduct a wide range of HF assessments on military vehicles and equipment. Whilst these have always been conducted to a high standard,

---

[1]The original HFAF (Humm et al., 2010) was developed between 2008 and 2009 and is used to assess the HF elements of Dismounted Close Combat (DCC) systems in a rigorous and systematic manner, whilst remaining flexible to the project context and the customer needs.

it was apparent that there was no standardised framework for HF practitioners to structure the assessments or to identify the most appropriate methods and tools to use, which was sufficiently flexible to accommodate the variety of vehicles to be assessed. A lack of a common assessment framework, consisting of approved methodologies has a number of implications for the design and evaluation of mounted systems making it more difficult to:

• Compare data collected for different systems;
• Control consistency of assessors;
• Quickly and systematically identify and mitigate HF issues with the system;
• Conduct repeatable HF assessments throughout the design process.

Method

The method used to develop the mounted systems HFAF was a five stage process, each of which are described in further detail below.

Stage 1: Reviewing the original DCC HFAF

The DCC HFAF provides HF practitioners with a robust and auditable means of gathering data on the people related aspects of a system or capability (Humm et al., 2010). The structure provided by the DCC HFAF enables better discrimination between proposed systems or capabilities through the application of suitable HF methods at the correct points in the project. Applying the HFAF, as part of research and acquisition programmes, improves HF SME's ability to provide advice and aid evidence based decision making by aiding the selection of designed, standardised and repeatable data collection activities.

The DCC HFAF utilises a three level approach:

1. Level 1. Initial HF assessment – this is a paper-based, specification or parameter review, visual inspection and/or functional assessment of a piece of equipment, system or idea for the DCC domain.
2. Level 2. Functional performance assessment – this assesses the function and performance for a number of users during a range of simulated DCC military tasks, using HF data collection methods with representative DCC users.
3. Level 3. Controlled environment assessment – this involved lab-based assessments which ensures a controlled environment. This is in order to provide evidence to answer specific questions or where more control is required to ensure reliable physiological, psychological or performance related data is captured.

The DCC HFAF has become a recognised and standardised framework that can be used not only by Dstl, but also by industry and academia. To date it has been applied across a range of DCC projects and it has been endorsed at MOD 1* level as the 'Gold Standard' framework for assessing the HF component of dismounted solider

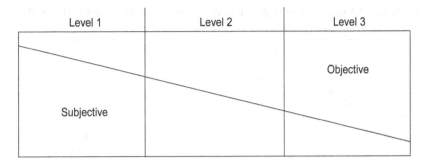

Figure 1.   How the data type differs with the HFAF Level.

systems. The DCC HFAF is promoted for use in support of current UK military operations, and for future MOD acquisition programmes (Humm et al., 2010).

At each level of the HFAF there is a shift from subjective to increasingly objective data collection. There is also a corresponding increase in the level of confidence that can be applied to the conclusions of the study. This can be simplistically presented as the reliance on subjective information at Level 1 through to a focus on objective data at Level 3 as shown in Figure 1.

Following the review of the DCC HFAF, an analysis was conducted to establish those people-related elements that need to be included in any HF assessment (in either the dismounted or mounted domain). From this simplistic analysis, it became clear that there are many similarities and overlaps between the elements that should be covered as part of DCC HF assessments and HF assessments for the mounted systems domain.

Stage 2: Development of the mounted systems HFAF (alpha)

On completion of the DCC HFAF review to identify whether the DCC HFAF could be simply applied to the mounted domain, an initial version of the mounted systems HFAF was developed. The following paragraphs outline the new framework.

The new mounted systems HFAF adopts the same three level approach and like the DCC HFAF, the three levels do not need to be conducted in a sequential order i.e. a methodology from level 3 of the HFAF could be applied alongside one from level 1. As with the DCC HFAF, as the level number increases the scientific rigour increases and the time the assessments take to complete increases. An overview of each level in the mounted systems HFAF is provided in Table 1.

Stage 3: Identification of HF assessment methods

In order to ensure that the mounted systems HFAF contained useful and understandable guidance for HF practitioners, suggested HF methods, which could be used at each of the assessment levels, were required.

Table 1.   An illustration to give a breakdown of the mounted systems HFAF Levels.

Level 1	Level 2	Level 3
An inspection and evaluation of a vehicle system or individual component parts	Deeper analysis of the vehicle system and/or component parts	In depth HF assessment of system wide capabilities of a vehicle platform
HF SME led	User involvement	Trials with many users and or more mounted systems i.e. group formations
Subjective data drawing upon HF standards	Subjective and objective data	Primarily objective data
Aims to identify HF issues early in the design process that may impact of performance, safety or comfort	Aims to assess individual aspects (cab lighting, nose levels) as well as some system wide capabilities (visibility)	Aims to collect reliable data on the system wide HF issues. May involve specialist equipment such as vibration rigs

A systematic literature review was conducted to identify a range of HF assessment tools, including commercial HF assessment tools and defence specific HF methods, which would be applicable to sit under the new mounted systems framework. The review was initiated to ensure that the mounted systems HFAF would draw on existing, and where possible standardised, methods. The literature review also included assessment methods that have either been developed within Dstl or those which have previously been used on vehicle projects by Dstl HF practitioners. The level descriptions shown in table 1 formed the criteria against which the methods were selected for each level of the framework.

The literature review identified 32 potential methods that could be included and that would sit under the three different HFAF assessment levels (1–3). As with the DCC HFAF, the mounted systems HFAF includes a compendium of standardised HF best practice methods that are recommended for use, at the time of writing.

Table 2 shows a small sample of the HF methods that were identified and where they fit within the mounted systems HFAF.

Stage 4: Review by Dstl HF SMEs and UK military

The mounted system HFAF (alpha) was produced in a user guide format, together with the compendium of methods. It was important at this stage in the study that these were reviewed by Dstl HF practitioners with vehicle ergonomics experience and military advisors. This review aimed to obtain an initial assessment of the framework's efficacy and the standardised methods that could be employed.

Table 2.   A selection of the HF methods chosen to fit under the mounted systems HFAF.

Level 1	Level 2	Level 3
HFAF Functional Assessment Method (Humm et al., 2010)	Questionnaire for User Interface Satisfaction (QUIS) (Stanton et al., 2005)	Full User Trials (Stanton et al., 2005)
Layout Analysis (Stanton et al., 2005)	HMI Heuristic Analysis (HFIDTC, 2004)	
Human-Machine Interface (HMI) Checklists (Stanton et al., 2005)	Basic Visibility Assessment (polar plot) (Smith, 2012)	Full Visibility Assessment (polar plot) (Smith, 2012)
Instantaneous Self Assessment-workload (HFIDTC, 2004)	NASA Task Load Index (HFIDTC, 2004)	Cognitive Reliability and Error Analysis (HFIDTC, 2004)
	Noise and Lighting Assessments (Humphrey, 2007, DEF STAN 00-250)	Thermal Environmental Assessments (EBRE, 1972)
Static Anthropometric Measures	Dynamic Anthropometric Assessment	Whole body Vibration Assessments (tuvps, 2012)
	Rapid Entire Body Assessment (Hignett & McAtamney, 2000)	

Stage 5: Development of the mounted systems HFAF (beta)

Following the peer reviews, a second version (beta) of the mounted systems HFAF was produced and delivered to the MoD customer together with the full compendium of methods.

Designed in a user guide format, the mounted systems HFAF includes information at each level (1, 2 and 3) on the assessments, the objectives, the scope, advantages and disadvantages of using the different levels, the expected duration, likely equipment required, recommended methods, reporting format, and the impact of the assessment.

The mounted systems HFAF now needs to undergo assessment exercises by being applied during a range of military mounted system HF assessments to source views from industry, academia and wider MOD. These exercises should be conducted by a partnership of Dstl, military personnel and industry HF practitioners to ensure that the methods within the framework are comprehensive and fit for purpose.

Following the UK military's withdrawal from Afghanistan in 2014, the mounted systems HFAF is likely to have a role in providing a framework for the assessments of military vehicles that can be used to support design modifications to make the vehicles effective for use in the core vehicle fleet which will be used in many different environments. The mounted systems HFAF will be used for military vehicle

HF assessments as the opportunities arise, thus help in improving an understanding of the framework's efficacy, as well as the validation state of methods which sit within it.

In addition to the military uses, this framework could also have utility for emergency services vehicles, such as ambulances and police vehicles and these exploitation routes should also be explored in the future.

Conclusions

Dstl have developed a Human Factors Assessment Framework (HFAF) for military land vehicles (mounted systems) that aids HF practitioners to apply appropriate HF methods and tools at different stages of the design process. The mounted systems HFAF is a development of the existing DCC HFAF, which is a framework for the way HF practitioners gather data on the HF component of DCC systems in order to identify human-related issues and risks, which can then be mitigated in capability management.

A compendium of appropriate HF methods that could be applied at each level of the framework has been generated to aid HF practitioners when using the framework. It should not be seen as a fully comprehensive list, but merely a list of suggested methods for each level. The most appropriate methods for assessing the cognitive element of human performance and for conducting thermal assessments in particular, will need further investigation.

Recommendations

Should the mounted systems framework and methods be found to be a useful framework, it will be advocated more widely to obtain endorsement from industry, academia and wider MoD. Furthermore, it is recommended that identifying additional methods, which sit within the mounted systems HFAF, should be a continual activity, particularly for cognitive assessments and for thermal environment assessments.

The mounted systems HFAF will need to continually evolve to adapt to changing defence requirements, to address more niche areas and keep pace with industrial and academic best practice. Dstl should lead the continual review of the framework; however, the practical application of the mounted systems HFAF should by undertaken by partnerships involving Dstl, MoD, Training and Development Units, industry and academia.

References

DEF STAN 00-250 *Human Factors Guidance for Designers of Systems* Part 3-Section 14, Part 14

Department of the Environment Building Research Establishment. (1972). Thermal comfort and moderate heat stress. *Proceedings of the CIB Commission W45 (Human requirements) Symposium* held at the Building Research Station. 1 (1)

HFIDTC (2004) *Human Factors Design & Evaluation Methods Review*

Hignett, S. & McAtamney, L. (2000). Rapid Entire Body Assessment (REBA), *Applied Ergonomics*, **31**(1) 201–205

http://www.tuvps.co.uk/home_psuk/industries/defence/defence_environmental_ testing/shock_and_vibration_testing, Date accessed November 2012

Humm. E et al., (2010) '*Human Factors Assessment Framework Working Paper*' DSTL/WP40683 V1.0

Humphrey, P. (2007). *Workplace Noise Assessment of HMS ML Gleaner*. Institute of Naval Medicine Report. 1 (1)

Smith, I., (2012) *How to conduct a visibility assessment*. DSTL/WP69457

Stanton. N., Salmon. P., Walker. G., Baber. C., and Jenkins. D (2005) *Human Factors Methods. A guide for engineering and design*. HFIDTC

Author index

Abdullah, S.C. 417, 425
Ahmad, I.N. 417, 425
Andresen, G. 135
Ashalatha, K.V. 236
Astwood, J. 290

Baber, C. 251, 327, 435
Baker, W. 69
Balfe, N. 77
Bao, B. 41
Begley, J. 308
Berkeley, N. 316
Berman, J. 7
Best, N. 385
Brehmen, M. 229
Bridger, R.S. 269
Bruder, R. 229
Burford, E.-M. 229
Burton, J. 35

Carey, M. 343
Carlisle, M. 183
Carr, K. 3
Cats, O. 316
Chang, C.-C. 49
Chang, W.-R. 49
Charalambous, G. 116
Ciocoiu, L. 277
Clarkson, J. 211
Clegg, C. 15
Coplestone, A. 343
Covill, D. 143
Croq, M. 124

Davies, P. 461
Day, P.N. 183
Devereux, J. 16
Diels, C. 301
Durando, D. 377

Eaves, S. 191
Ellegast, R. 229
Elton, E. 143, 151
Endo, A. 53, 477

Ferguson, G. 183
Fioratou, E. 203

Fletcher, S.R. 85, 93, 116
Frampton, R.J. 481
Frankova, K. 308
Fung, W. 385
Furukawa, T. 469

Gibb, A.G.F. 191
Gilmore, D.J. 11
Golightly, D. 335
Goto, A. 53, 101, 469
Groneberg, D. 229
Grundy, C. 159
Gyi, D. 413
Gyi, D.E. 191

Hamada, H. 101
Hamilton, W.I. 259
Henshaw, M. 277
Horberry, T. 211
Hrin, G.R. 316
Hubbard, E.-M. 277

Isobe, S. 477

Jenkins, D.P. 167
Johnson, J.P. 183
Johnson, T.L. 85, 93
Jordan, P.W. 461

Kam, A.C.S. 443
Karali, S. 413
Karim, S.A. 417, 425
Karunarathne, B. 443
Kawabata, S. 195
Kida, N. 469
Kikuchi, T. 101
Krömker, H. 367
Kuwahara, N. 53, 195,
  469, 477
Kwok, T.C. 203

Lang, A.R. 203
Large, D.R. 335
LeJeune, I. 203
Lesch, M.F. 49
Leva, M.C. 77
Li, W.-C. 377

Lin, J.-h. 241
Liu, H. 435
Lowe, E. 351

Ma'arof, M.I.N. 417, 425
Maior, H.A. 450
Male, T. 327
Malinge-Oudenot, A. 124
Mansfield, N. 413
Markucevičiūtė, I. 316
Martinsen, T.J.S. 359
Matz, S. 49
McCaig, R. 215
McDermott, H. 59
McGorry, R.W. 241
Meeks, R. 27
Mile, M. 143
Morris, R. 143, 159
Mowbray, E. 215
Munir, F. 59

Naghiyev, A. 343
Naik, R. 236
Narul, K.F.K. 251
Nasu, M. 195
Newman, P. 27
Nishimura, H. 53
Noble, V.G. 481
Nock, P. 35, 385
Novak, B.B. 173

O'Reilly, O. 316
Omar, A.R. 417, 425
Osmond, J. 308
Ota, T. 477
Øvergård, K.I. 359

Patil, V. 211
Pemberton, L. 159
Pickup, L. 351
Pike, M. 450
Pimentel, T. 316
Pisula, P. 269
Pollard, C. 390

Rashid, H. 417, 425
Richardson, J.H. 481

Roquelaure, Y. 124
Ryan, B. 343

Salunke, R. 236
Saunders Jones, R. 489
Sawkar, S. 236
Sharples, S. 203, 343, 450
Siemieniuch, C.E. 110, 283
Sinclair, M.A. 110, 283
Sinn-Behrendt, A. 229
Skelton, C. 27
Smith, S. 351
So, R.H.Y. 41, 443
Sorensen, L.J. 359
Spundflasch, S. 367
Stanton, N.A. 395
Strickland, K. 290, 489

Sumangala, P.R. 236
Susilo, Y. 316

Tainsh, M. 403
Takai, Y. 101, 469
Tang, G. 116
Tani, Y. 101
Tatlock, K. 290
Taylor, E.L. 335
Taylor, H. 259
Teng, Y.-C. 211
Ting, L.-Y. 377
Tutton, W. 290, 489

Vail, C. 259
Veland, Ø. 135

Walters, E. 385
Wang, Z. 53
Ward, J. 211
Waterson, P. 215
Webb, P. 116
Weber, B. 229
Williams, C. 461
Wilson, I. 59
Wilson, M.L. 450
Woodcock, A. 219, 308, 316

Xu, X. 241

Yamamoto, A. 195

Zare, M. 124